D0687379
WITHDRAWN

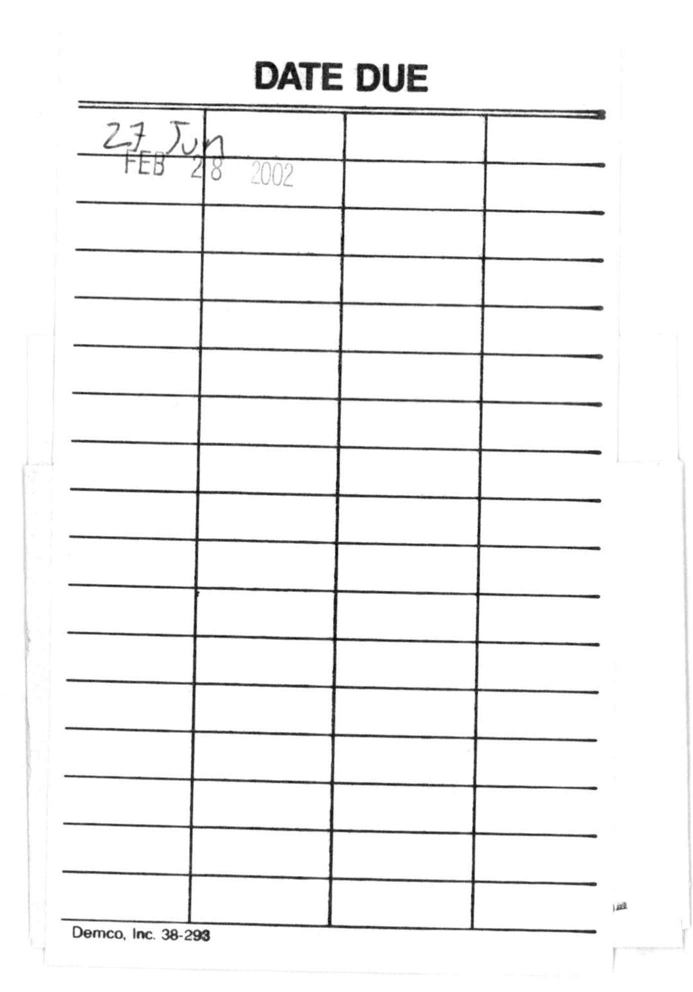
DATE DUE
27 Jun
FEB 28 2002
Demco, Inc. 38-293

DYNAMIC ANALYSIS AND FAILURE MODES OF SIMPLE STRUCTURES

DYNAMIC ANALYSIS
AND FAILURE MODES OF SIMPLE STRUCTURES

DANIEL SCHIFF
Assurance Technology Corporation
Carlisle, Massachusetts

A WILEY-INTERSCIENCE PUBLICATION
JOHN WILEY & SONS
New York / Chichester / Brisbane / Toronto / Singapore

TA 654 .S35 1990

Schiff, Daniel, 1925-

Dynamic analysis and failure modes of simple structures

Copyright © 1990 by John Wiley & Sons, Inc.

All rights reserved. Published simultaneously in Canada.

Reproduction or translation of any part of this work beyond that permitted by Section 107 or 108 of the 1976 United States Copyright Act without the permission of the copyright owner is unlawful. Requests for permission or further information should be addressed to the Permissions Department, John Wiley & Sons, Inc.

***Library of Congress Cataloging in Publication Data*:**

Schiff, Daniel, 1925-

Dynamic analysis and failure modes of simple structures/Daniel Schiff.

p. cm.

"A Wiley-Interscience publication."

Includes bibliographical references.

1. Structural dynamics. 2. Structural analysis (Engineering) 3. Structural stability—Mathematical models. I. Title.

TA654.S35 1990
624.1′71—dc20 89-38193
ISBN 0-471-63505-7 CIP

Printed in the United States of America

10 9 8 7 6 5 4 3 2 1

To my wife, Lonny

PREFACE

Mechanical analysis of structures is required to ensure that they maintain their integrity and provide predictable and acceptable performance and dynamic response throughout the specified lifetime profile of loads and acceleration environments. This analysis includes vibration- and shock-induced displacements and stresses, isolation of sensitive components, and the evaluation of elastic instability, fatigue, and fracture as potential failure modes. In aerospace applications, as in other applications, proper design of vents is required to maintain safe pressure differentials in venting air from a compartment. When the structure contains electronics, thermal analysis may be utilized to achieve adequate heat transfer and avoid temperature-related performance degradation or failure of these components. This book addresses these technical areas in a manner most useful to the engineer who is involved in a mechanical design and requires timely quantitative answers to the problems that arise.

The organization of the subject matter is in a logical sequence, introducing the concepts of loads and failure modes in the first chapter, and dealing with the natural frequency of components and structures in the next two chapters. With an estimate of natural frequency, the specified acceleration environment (vibration, Chapter 4, and shock, Chapter 5) may be utilized to obtain displacements and stresses, and isolation may be designed where required (Chapter 6). The acceleration-induced stresses may then be evaluated in terms of fatigue, fracture, elastic instability, and material strength limitations (Chapters 7 through 10). Chapters 11 and 12 cover venting and thermal analysis.

The approach taken is to describe the problem in physical terms, defining the known (given) parameters and the unknown quantities that must be evaluated. The methodology is then provided for obtaining quantitative estimates for these unknowns and relating them to specified limiting values and failure modes to obtain an understanding of the acceptability of the design. The methodology used may include equations, data tables, and figures, and can be rapidly implemented with the use of a pocket calculator. This type of analysis is particularly helpful in the early design configuration stage, in the comparison of different design approaches, and in monitoring the results of more massive and detailed computer analyses. It will provide the engineer with a better understanding of the system being designed and will pave the way for the development of an effective finite element model for computer analysis.

I am indebted to Larue Renfroe, President of Assurance Technology Corporation, for his support and encouragement throughout the writing of this book. Additional thanks go to Gail Nash for word processing; Jim Poe, Laurie Olson, and Leslie McIntosh for the illustrations; and to my wife, Lonny, for proofreading.

DANIEL SCHIFF

Carlisle, Massachusetts
January 1990

CONTENTS

1

MECHANICAL LOADS AND FAILURE MODES

1.1 FORCES

This chapter describes and characterizes the types of mechanical loads, deformations, stresses, and failure modes that will be addressed in subsequent chapters. The mechanical loads are the forces that act on a structure to produce a change in the deformation and stress in the structure. These forces are specified by four characteristics: their point of application on the structure, their direction, their magnitude, and their time dependence. Because forces have both magnitude and direction, they are defined as vector quantities, in contrast to scalar quantities such as mass, which have only magnitude and no direction.

The ways in which these four characteristics of a force affect a structure are illustrated as follows:

1. *Point of Application*. A force of constant magnitude acting in a fixed direction normal to a slender cantilever beam and applied at the free end of the beam will produce twice the stress and strain in the cantilever material at the fixed end of the beam than will be produced when the force is applied to the midpoint of the cantilever.

2. *Direction*. A force of constant magnitude applied normal to the free end of a horizontal cantilever beam stressed by its own weight will result in a greater total stress and displacement in the cantilever when the force is pointing down in the direction of the gravitational force than when it is pointing up.

3. *Magnitude*. A force applied at a fixed point on a cantilever beam and acting in a fixed direction normal to the beam will cause a stress and strain in

the cantilever material that increases directly as the magnitude of the force increases, within the elastic limits of the cantilever material.

4. *Time Dependence*. If a force of given magnitude, applied in a given direction normal to the end of a cantilever beam, is applied suddenly, the maximum transient stress and displacement in the cantilever is twice that produced by the same force applied slowly. If the same force produces stresses that are below the yield strength of the cantilever material and cannot cause failure, repeated application of this force over a period of many cycles may cause the cantilever to fail (rupture) by fatigue or by fracture.

Static (time-independent) forces acting on a structure can produce static (constant) stresses and strains in the structure unless the structure contains materials that creep. Creep is a slow and continuous increase in deformation under constant or decreasing stress, or a slow decrease in stress when the deformation is maintained constant (relaxation). Time-dependent forces can be caused by time-dependent accelerations imposed on a structure. The acceleration results in inertial forces acting throughout the structure. These inertial forces are opposite to the direction of acceleration, are applied everywhere throughout the mass of the structure, and are proportional to the product of the material mass density and the magnitude of the acceleration. These inertial forces are distributed forces, which act on every particle of mass in the structure, since, ideally, every part of the structure undergoes the same acceleration. As the acceleration increases beyond 1 g ($g =$ earth's gravity acceleration $= 386$ in./sec^2), the situation is equivalent to uniformly increasing the weight of the structure and orienting it so that "down" is opposite to the direction of acceleration. In the treatment of accelerations and inertial forces in later chapters, the propagation through a structure of stress/strain waves, due to time-dependent accelerations of a mounting surface supporting the structure, is ignored. This simplification assumes that the entire structure accelerates instantaneously. From a design viewpoint, the most important time-dependent inertial forces acting on a structure are due to vibration and shock.

Vibration (Chapter 4) is an oscillating stress and deformation of the structure defined by either a fixed frequency (sinusoidal vibration) or a time-varying set of frequencies and amplitudes conforming to an amplitude/frequency envelope (a spectrum). The latter phenomenon is called random vibration, and the spectrum is specified in terms of a power spectral density (PSD). It is called random vibration because the frequencies and amplitudes are not predictable except in terms of probabilities, as defined by the PSD. Major sources of random vibration are ground transportation and launch operations. Although the random-vibration environment due to surface transportation has the longest duration, often lasting hours, its amplitude is normally less severe than that due to launch operations. Space-system launch vehicles use rocket-engine firings that last for minutes and present the most severe random-vibration environment to which an aerospace system will be exposed. Since the type of acceleration time history due to rocket-engine firings is difficult to reproduce in the laboratory, an equivalent approach is taken. A shaker table is controlled to yield the same time-averaged acceleration, or power per unit mass (PSD), at every frequency as would result from a time average of the complete rocket-motor firing. The shaker table produces its own random-vibration time history, and over a very short time interval its average power versus frequency may be quite different from the average values for the rocket motor; but over a period of several minutes the shaker table can be controlled to yield an average power spectrum within a few decibels of the rocket-motor time-averaged spectrum.

Shock (Chapter 5) is a sudden, severe, and transient acceleration of the structure. In aerospace applications shock can be caused by pyrotechnic separation of a system from a supporting structure with the additional sudden release of stored energy in the previously strained system or supporting structure. Shock can also occur when a vehicle lands with a payload still on board. Pyrotechnic shocks last milliseconds or less and can produce accelerations of thousands of g's. The acceleration time history of a shock represents the actual acceleration versus time in the structural member that contains the pyrotechnic device or that transmits the shock to another structure. This acceleration time history is a complex, nonanalytical function of time. For purposes of analysis, specification, and testing, the shock acceleration time history is converted to a shock acceleration response spectrum. This spectrum represents, at each frequency, the magnitude of the acceleration in g units that would be experienced by a simple harmonic oscillator at that frequency when exposed to the shock. Any specified amount of damping may be incorporated into the shock acceleration response spectrum. This spectrum is much easier to use in design and analysis than the shock acceleration time history. It is also much easier to specify a laboratory shock test using a very simple half-sine or sawtooth shock acceleration time history to approximate the actual shock acceleration response spectrum than it is to approximate the actual shock acceleration time history in the laboratory.

In design and analysis it is usually assumed that the acceleration environment (limit load, vibration, or shock) may be imposed on a structure with the acceleration vector acting in either direction along any three orthogonal axes. This is because the acceleration environment is not always specified as to directionality with respect to the structure, and subsystem structures sometimes change orientation with respect to the primary structure between successive flight units. For the same reason, laboratory tests usually require random vibration along three orthogonal axes and shocks in both directions along three orthogonal axes. The test axes are usually chosen to be parallel to the subsystem housing walls in the case of rectangular housings.

1.2 STRESSES

The application of forces to a structure produces deformations and stresses. These stresses in the structural material may be sufficiently severe in magni-

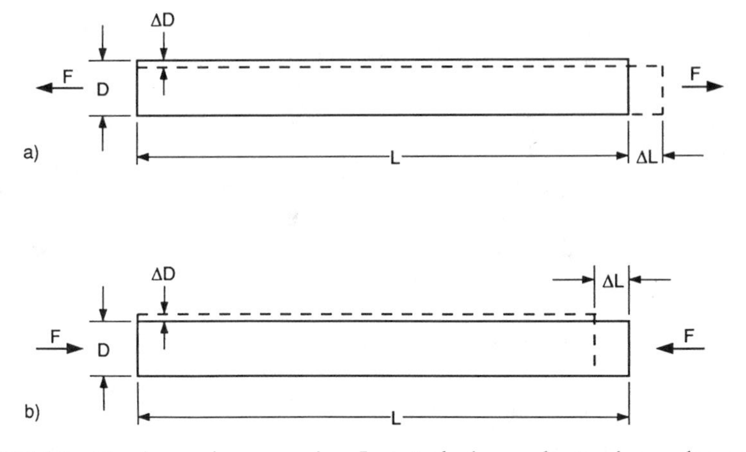

FIGURE 1.1 Tension and compression. In (*a*), the bar under tension undergoes axial elongation and lateral contraction. In (*b*), the bar under compression undergoes axial contraction and lateral expansion.

tude, duration, or repetition to cause the structure to fail. In Chapter 4 analysis techniques are provided for estimating material stresses based on the structural design and the load environments to which the structure is subjected. The stresses may be tension, compression, or shear, or a combination of them. The stresses will also reflect the time-varying behavior of the loads that cause them, sometimes resulting in alternating tensile and compressive stresses and other time-varying combinations. A brief review of some of the simple relationships among forces, elastic constants, and stresses follows.

A straight, homogeneous bar of length L in. having a uniform circular cross section of diameter D in. is subjected to a tensile force of F lbf parallel to its long axis and uniformly distributed over the bar cross section. See Fig. 1.1*a*. The tensile stress in the bar is

$$\sigma = F/A \text{ psi (pounds force per square inch)} \tag{1-1}$$

where $A = (\pi/4)D^2$ in.^{2}, the cross-sectional area of the bar. Equation (1-1) shows that the stress is proportional to the load (force) and is also a function of structural geometry, in this case the cross-sectional area A. Equation (1-1) is usually applied only when the magnitude of the stress is in the elastic range of the bar material, that is, σ does not exceed the yield stress of the bar material. In other cases, more complex geometric parameters such as moments of inertia may be involved. The axial elongation of the bar under tension is

$$\Delta L = \varepsilon L \text{ in.} \tag{1-2}$$

where ε is the strain in the bar (dimensionless) and

$$\varepsilon = \sigma/E \tag{1-3}$$

where E is the modulus of elasticity, or Young's modulus, of the bar material (psi). Equation (1-3) shows that the strain is proportional to the stress and inversely proportional to the modulus of elasticity. Equation (1-3) is usually applied only when the magnitude of σ is less than the yield stress of the bar material. In other cases other elastic moduli may be involved.

Combining the preceding relationships yields

$$\Delta L = FL/AE \text{ in.} \tag{1-4}$$

As long as the material stress is in the elastic range, the elongation of the bar is directly proportional to the length and the tensile force and inversely proportional to the cross-sectional area and modulus of elasticity. This relationship is independent of the shape of the cross section, provided the cross section is constant over the length of the bar.

When the bar under tension undergoes axial elongation, it also undergoes lateral contraction and a decrease in cross-sectional area. The linear contraction at right angles to the load, for the round bar of diameter D, is

$$\Delta D = -\nu\varepsilon D \text{ in.} \tag{1-5}$$

where

$$\nu = -(\Delta D/D)/\varepsilon = -(\Delta D/D)/(\Delta L/L) \equiv \text{Poisson's ratio} \tag{1-6}$$

Poisson's ratio is the ratio of the unit lateral contraction to the unit axial elongation. The new cross-sectional area of the loaded bar is

$$A' = (1 - \nu\varepsilon)^2 A \text{ in.}^2 \tag{1-7}$$

and its new volume is

$$V' = (1 + \varepsilon)(1 - \nu\varepsilon)^2 V \text{ in.}^3 \tag{1-8}$$

where V is the original volume of the unloaded bar. Neglecting the second and third powers of ε, the unit volume expansion of the bar is

$$(V' - V)/V = \varepsilon(1 - 2\nu) \tag{1-9}$$

Since it is unlikely that materials of interest will diminish in volume when under tension, ν must be less than $\frac{1}{2}$. Most structural metals have a value of ν

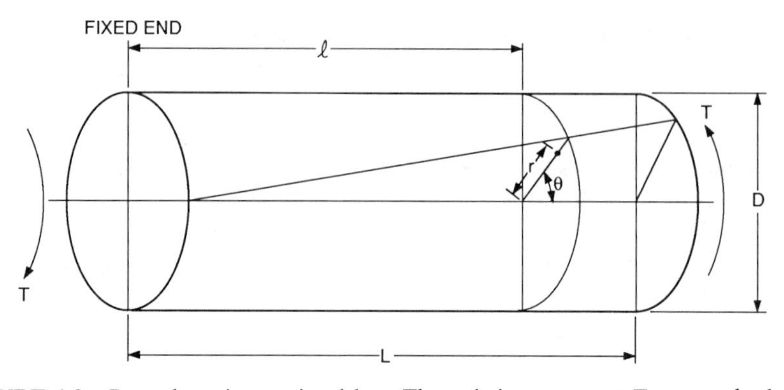

FIGURE 1.2 Pure shear in a twisted bar. The twisting moment T causes the bar to twist through an angle θ at distance l from the fixed end.

in the range 0.28–0.35, while plastics used in printed wire boards have lower values, in the neighborhood of 0.1.

When the bar is subjected to compressive forces as in Fig. 1.1*b*, it undergoes a decrease in length and an increase in lateral dimensions, causing a change in algebraic sign in Eqs. (1-2), (1-4), and (1-5) and resulting in a unit volume contraction of

$$(V' - V)/V = -\varepsilon(1 - 2\nu) \qquad (1\text{-}10)$$

For most materials of interest, the modulus of elasticity and Poisson's ratio are the same for tension and compression.

A pure shear stress may be produced by subjecting the same bar to equal and opposite twisting couples (torsion), T in.-lb, at its ends in planes normal to its longitudinal axis, as in Fig. 1.2. The bar will twist, with every section (i.e., every plane perpendicular to the axis) rotating about the axis through an angle

$$\theta = (Tl)/JG \text{ rad} \qquad (1\text{-}11)$$

where l = distance along the axis from the fixed end of the bar (in.)
J = polar moment of inertia
= 2(moment of inertia about central axis)
= $2I = 2(\pi/64)D^4$ (in.^{4})
G = shear modulus
= modulus of rigidity of the bar material
= $E/2(1 + \nu)$ psi.

The shear stress at a radius r from the axis of the bar is

$$\tau = Tr/J \text{ psi} \qquad (1\text{-}12)$$

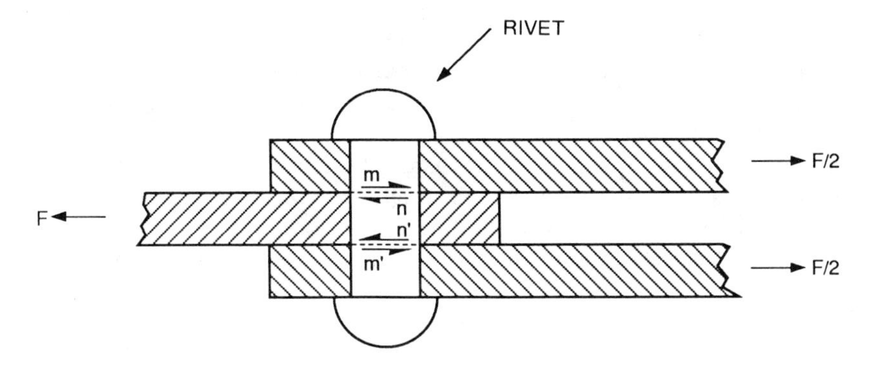

FIGURE 1.3 A common loading producing primarily shear stress. The rivet is subjected to shear in planes *mn* and *m′n′*, but other stresses are also present.

At the point (l, r) there is a shear stress in the plane of the section and perpendicular to the radius through the point. There is also an equal longitudinal shear stress in the radial plane through the same point. The shear strain is given by

$$\gamma = \tau/G \text{ rad} \tag{1-13}$$

There are other common forms of loading that produce primarily shear stress, with a small degree of bending. Figure 1.3 shows a rivet subjected to shearing forces $F/2$ at planes *mn* and *m′n′*, which ideally would result in a shear stress, at each plane, of magnitude

$$\tau = F/2A \text{ psi} \tag{1-14}$$

where A is the cross-sectional area of the rivet (in.^{2}). However, there are also stresses normal to planes *mn* and *m′n′* due to prestress and fit, and the shear stress is usually distributed nonuniformly over the section.

Tension, compression, and shear stresses can all occur in the same structural element under a simple loading condition. The beam of Fig. 1.4 is simply supported near its ends and loaded in the middle. The beam bends under the force F, causing the material in the bottom convex part of the beam to be

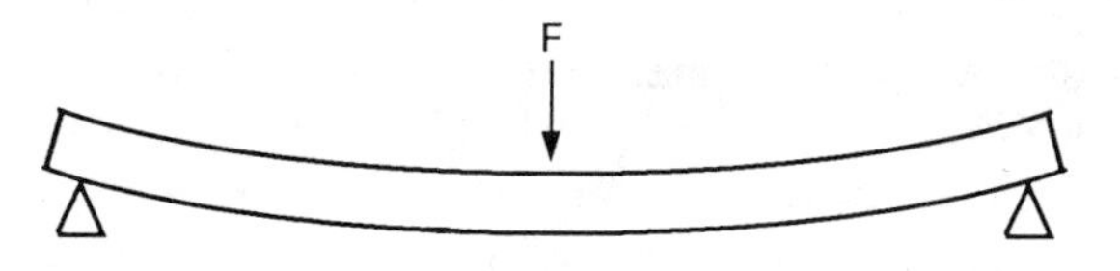

FIGURE 1.4 Simply supported, loaded beam. Tension, compression, and shear stresses all occur in the beam under this loading condition.

elongated and in tension, and causing the material in the concave upper part of the beam to be shortened and in compression. At any point in the beam there is a longitudinal shear stress on the horizontal plane and an equal vertical shear stress on the transverse plane.

In evaluating the capability of a structure to withstand loads successfully, the maximum stresses that occur in the structure are often of greatest concern. (This may not be the case when buckling is a consideration.) In the example shown in Fig. 1.4, the maximum tensile stress in any section has the magnitude

$$\max \sigma = Mc/I \text{ psi} \tag{1-15}$$

where M = bending moment at the section (in.-lb)
c = distance from the section centroid to most remote point in section in bottom convex part of the beam (in.)
I = moment of inertia of the section of the beam with respect to the section centroid (in.^4)

The maximum compressive stress in the section is given by Eq. (1-15) with c being the distance from the section centroid to the most remote point in the section in the top, concave, part of the beam. The maximum shear stress in the section usually occurs at the section centroid and is given approximately by

$$\max \tau = (dM/dx)/KA \text{ psi} \tag{1-16}$$

where dM/dx is the transverse shearing force (lbf) supported by the section; A is the section area (in.^2); x is the distance along the longitudinal axis (in.); and K is a dimensionless constant whose value depends on the section shape, with a value in the neighborhood of 0.8.

1.3 FAILURE MODES

There are several ways in which structural stresses can cause a system to malfunction or fail. These failure modes are classified in the following sections in order of decreasing stress severity.

1.3.1 Elastic Failure

The majority of the structural materials of concern in this book are ductile metals (e.g., steel and aluminum) in which the material deforms elastically (strain proportional to stress) under simple, uniaxial tension until a stress level is reached where plastic deformation occurs. This stress level, σ_y, represents the yield strength or yield point of the material. The yield strength is the stress at which the stress–strain diagram changes slope from the initial straight-line portion to a lower slope value, and the material exhibits a permanent deformation (set). The yield point is the lowest stress at which the strain increases without an increase in stress. When plastic deformation occurs, the material is considered to have undergone elastic failure. The ultimate strength of a material, σ_u, is the stress at which rupture occurs when the stress is uniaxial in tension, compression, or shear. Elastic failure also occurs in compression, and for ductile materials the stress level for plastic deformation is lower in compression than in tension. When a structure is subjected to combined stresses (biaxial or triaxial stresses), it is not so simple to predict failure. See Chapter 7.

It should be noted that elastic failure can occur locally in the material of a structural component without resulting in structural failure if the volume of affected material is too small to significantly reduce the component strength. However, local elastic failure can affect fatigue properties and impair strength and should be interpreted as a danger sign that may lead to failure in certain materials and applications. When elastic failure of the material significantly reduces the structural strength of a component, it may result in excessive strain, which may lead to malfunction or catastrophic failure of the component or of the entire structure. In any case, the structure should be designed so that the largest stress that the ductile material will see in service, including stress concentrations, is significantly less than the yield stress of the material, σ_y. This design requirement is often expressed as a factor of safety,

$$FS = \sigma_y/\sigma_D > 1 \tag{1-17}$$

where σ_D is the design value of the maximum service stress, or as a margin of safety,

$$MS = FS - 1 = \sigma_y/\sigma_D - 1 > 0 \tag{1-18}$$

Values for FS typically range from 1.2 to 3 or more depending on the degree of uncertainty in material properties, fabrication tolerances, and analytical results and on whether structural failure affects human safety or incurs significant economic loss.

1.3.2 Failure of Brittle Materials

Brittle failure occurs when a material breaks with little strain or deformation prior to the break. Brittle materials of interest in this book include cast iron and ceramics. Failure can occur through a tensile fracture in which the tensile stress reaches the ultimate strength, σ_{UT}, of the material, or in a shear fracture when the compression stress reaches the ultimate strength, σ_{UC}. Shear fracture occurs on a plane oblique to the plane of maximum compressive stress but not usually on the plane of maximum shear stress, and, consequently, this type of

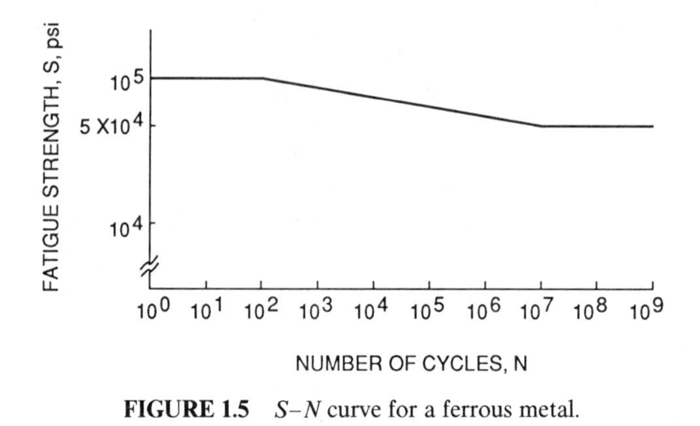

FIGURE 1.5 *S–N* curve for a ferrous metal.

fracture cannot be considered a pure shear failure. In brittle materials the compressive strength is greater than the tensile strength, contrary to the situation in ductile materials. Three theories of failure for brittle materials have been found applicable: the maximum principal stress theory, the maximum principal strain theory, and the Mohr's circle theory.

The maximum principal stress theory for brittle materials states that fracture will occur when the maximum principal tensile stress reaches σ_{UT} or the maximum principal compressive stress reaches σ_{UC}, whichever occurs first:

$$\max \sigma_T = \sigma_{UT}, \qquad \text{or} \qquad \max \sigma_C = \sigma_{UC} \tag{1-19}$$

In calculating the stress of fracture in a brittle component, it is customary to use an elastic stress formula to calculate the maximum stress, as in Eq. (1-15). The result, known as the modulus of rupture, is not a true ultimate stress, but changes value depending on the geometry, size, and manner of loading and support of the component. This variation is taken into account by means of a rupture factor, which is the ratio of the calculated modulus of rupture to the ultimate strength of the material, σ_u. The rupture factor is an important consideration in establishing the factor of safety (or margin of safety) in a specific design utilizing brittle materials.

1.3.3 Fatigue Failure

Cyclic stresses that are below the stress value for elastic failure can cause fatigue failure, or rupture, if the number of stress cycles is sufficiently large. See Chapter 7. The number of cycles N required to produce failure decreases as the maximum absolute stress value* S increases. The stress S is called the

*In fatigue analysis, stress is designated by S instead of by σ.

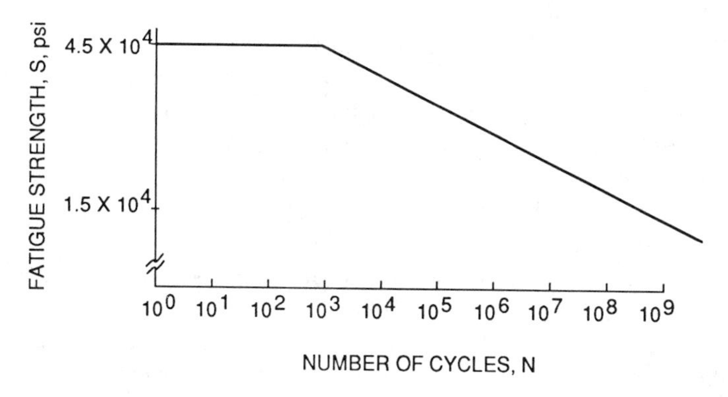

FIGURE 1.6 *S–N* curve for a nonferrous metal.

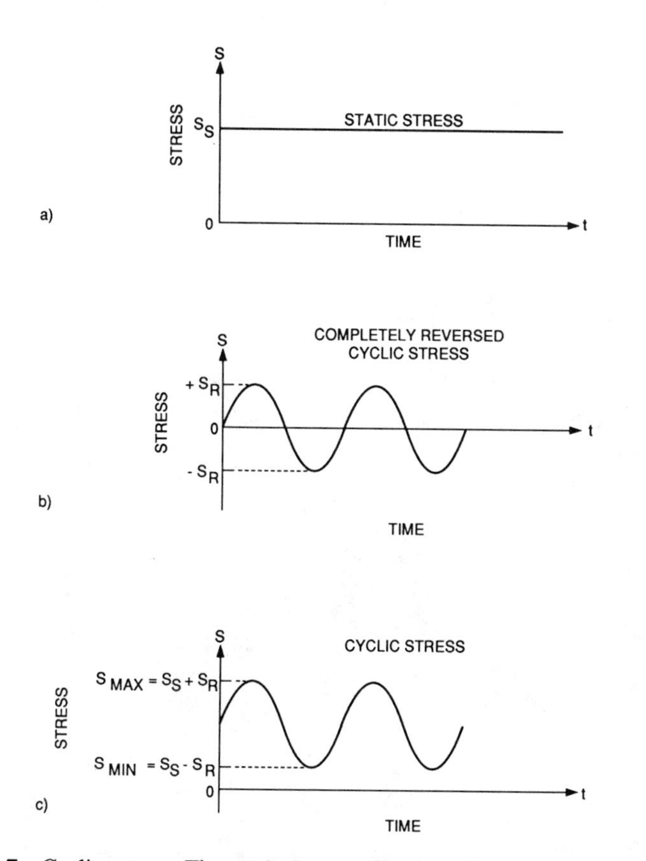

FIGURE 1.7 Cyclic stress. The typical case of a completely reversed cyclic stress superimposed on a static stress to produce a cyclic stress with maximum and minimum values S_{max} and S_{min}.

fatigue strength. Plots of S versus N, called S–N fatigue curves, or just S–N curves, are shown in Figs. 1.5 and 1.6. Fatigue testing is usually done on a rotating beam under the action of a constant bending moment that produces a complete stress reversal in the material, where the magnitudes of the maximum positive and negative stresses are equal and the algebraic mean stress is zero. For many ferrous metals and titanium alloys there is an endurance limit, which is the largest magnitude of completely reversed stress that the material will endure without failure no matter how many stress cycles are applied. The endurance limit for steel is about 50% of its ultimate tensile strength, and occurs at about 10^6–10^7 cycles. See Fig. 1.5. For most nonferrous metals, the S–N curve does not level off with increasing number of cycles, and there is no endurance limit. For aluminum, the fatigue strength at $N = 5 \times 10^8$ is about 33% of its ultimate tensile strength. See Fig. 1.6. In the high-stress, low-cycle region the S–N curve is flat with the fatigue strength equal to the ultimate

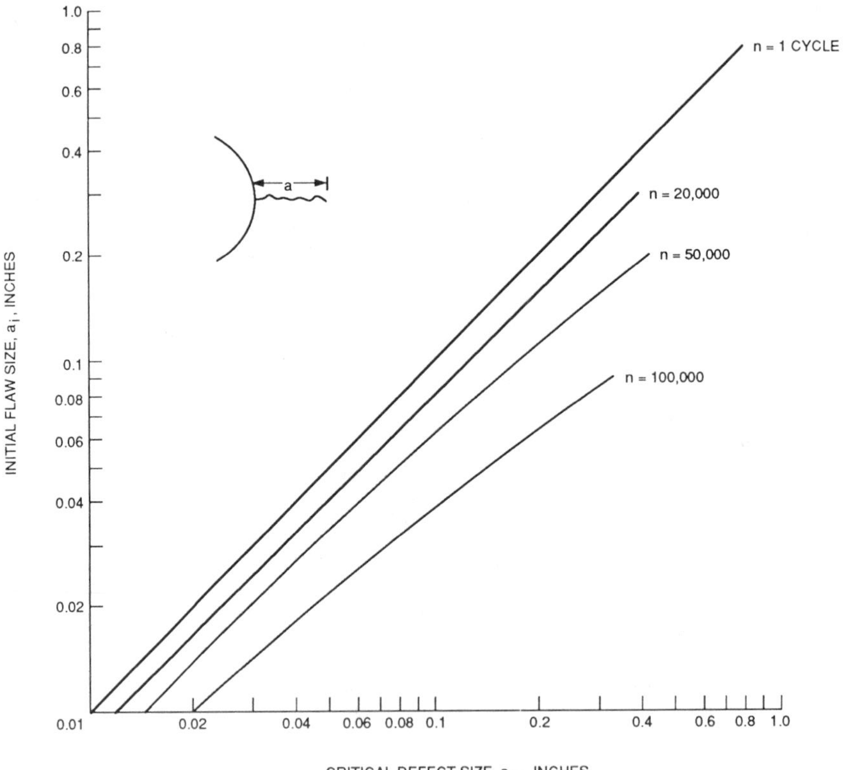

FIGURE 1.8 Number of cycles to grow defect to critical size at $\sigma = 30{,}000$ psi for low-alloy steel ($\sigma_y = 94{,}000$ psi). (*Source*: H. Liebowitz (Ed.), Fracture: An Advanced Treatise, Vol. V, New York: Academic Press, 1969.)

tensile strength up to about 100 cycles for ferrous alloys and 1,000 cycles for nonferrous alloys.

Most of the fatigue data have been obtained experimentally with complete stress reversal, while many design applications involve a static stress component S_S superimposed on a completely reversed cyclic stress of maximum magnitude S_R. This situation is illustrated in Fig. 1.7, where the maximum and minimum values of the resultant cyclic stress are given by

$$S_{max} = (S_S + S_R); \qquad S_{min} = (S_S - S_R) \tag{1-20}$$

Conversely, the values of S_S and S_R may be expressed as

$$S_S = \tfrac{1}{2}(S_{max} + S_{min}); \qquad S_R = \tfrac{1}{2}(S_{max} - S_{min}) \tag{1-21}$$

1.3.4 Brittle Fracture

Brittle fracture is the failure of a normally ductile material without any appreciable plastic deformation. Brittle fracture is caused by the application of cyclic stresses that have magnitudes well below the stress at which plastic deformation occurs, and is a result of the growth of minute flaws in the material into larger cracks to the point where the material fractures. Figure 1.8 is an example of crack growth as a function of the initial flaw size a_i and the number of stress cycles N for a local maximum cyclic stress of magnitude 30,000 psi in a low-alloy steel with σ_y = 94,000 psi. The critical defect size a_c at which fracture occurs depends on the defect geometry. Several methodologies have been developed for the prediction of crack growth rate under cyclic stress, and these are covered in Chapter 8.

1.3.5 Failure Due to Elastic Instability

When the maximum load that can be sustained by a structural component is determined by the stiffness of the component and not by the strength of the component material, an increase in the load will cause failure by elastic instability, or buckling. The Euler column is the best-known example of buckling. When a straight, slender column is loaded under axial compression only, it will deform by becoming shorter as in Fig. 1.9*a*. If at the same time it is loaded at its midpoint by a transverse force F_T, it will deflect as in Fig. 1.9*b* and return to its original straight position after removal of the transverse force, as in Fig. 1.9*c*, provided the axial load is less than the critical load F_c. When the axial load equals the critical load F_c, it will hold the column in the deflected position after the transverse force is removed, as in Fig. 1.9*f*. When this condition of balance occurs at material stresses less than the elastic limit, it is called elastic instability or buckling. An increase of the axial load beyond

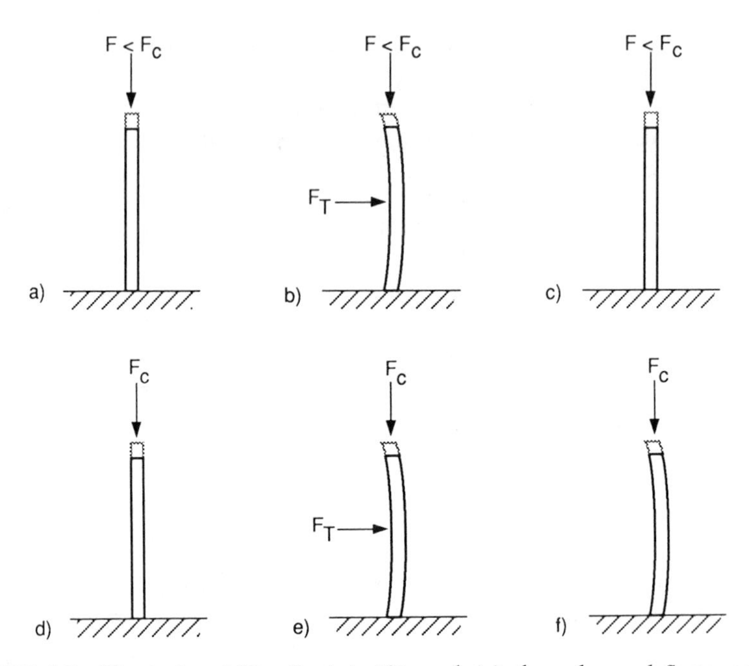

FIGURE 1.9 Elastic instability. In (*a*), (*b*), and (*c*) the column deflects under a transverse load and straightens out upon removal of the transverse load, for axial loads less than critical. For a critical axial load the deflection remains after removal of the transverse load, as shown in (*f*).

the critical value will cause the column to collapse. This common example occurs whenever the bending or twisting moment of an applied load is proportional to the deformation produced by the load: a thin plate under edge compression or edge shear; a deep, thin cantilever beam under an end load applied parallel to its depth dimension; a thin cylinder under external pressure. In some configurations the component will exhibit increasing resistance as buckling progresses and not fail under the load that initiates buckling. This type of postbuckling behavior is used in lightweight shell designs, which utilize structural components to their known strength limits by safely allowing buckling of the component without failure of the total structure. Common examples are the skin of an aerospace structure and the web of a deep beam.

The critical load for the slender column is determined by the material modulus of elasticity, the moment of inertia of the column cross section about the central axis perpendicular to the plane of buckling, the length of the column, and the type of support of each end of the column: free, guided, pinned, hinged, or fixed. For other shapes of structural components, the component geometry and dimensions, modulus of elasticity, Poisson's ratio, types and locations of end/edge supports and intermediate supports, and location and direction of the applied load determine the critical load. In many aerospace applications the maximum material stress at which buckling occurs is substantially less than the yield strength of the material. Chapter 9 contains formulas for calculating critical loads for some cases of interest.

1.3.6 Excessive Deflection

The deformation produced by the stress in a structure may result in the malfunction or failure of a system even when the stress levels are otherwise acceptable. This can happen when the displacement, or deflection, of a structural element exceeds the allowable clearance designed into the structure. For this reason the estimate of deflections under various stress and loading conditions is important. In Chapter 4 methods are provided for estimating deflections in different types of structural elements under a variety of loading conditions. However, it should be remembered that deflections that exceed design clearances can lead to malfunction or failure at stress levels which are completely acceptable in terms of structural integrity. For example, deflection of a printed wire board (PWB) in excess of design clearances can cause components on the board to contact adjacent PWBs or structural elements and result in an electrical short circuit or impact damage to a sensitive electronic part.

2

NATURAL FREQUENCY OF SIMPLE COMPONENTS

This chapter presents formulas and data for estimating natural frequencies of simple components. The natural frequency F of a component or a structure establishes the loads and deformations due to the acceleration environments of vibration and shock. The natural frequency (or frequency of free vibration) is the structural parameter that determines a specific value of acceleration and deformation when a broad-frequency spectrum of acceleration is imposed. The power spectral density (PSD) of the random-vibration environment and the acceleration response spectrum of the shock environment will have specific values at the structure's natural frequency, and these values will determine the deformations and stresses that occur in the structure. Generally, there will be several natural frequencies of interest for a structure. The lowest frequency is referred to as the fundamental natural frequency, F_1, and higher frequencies are labeled $F_2, F_3, \ldots,$ in order of increasing magnitude.

The words "natural frequency" in the present context are associated with the free vibration of the structure, that is, the small, oscillatory deformations of parts of the structure from the position of stable equilibrium in which all parts of the structure are at rest with respect to coordinate axes embedded in the structure. The frequency is the number of times per second that the oscillatory deformation of a part or parts of the structure occurs, and is expressed in units of cycles per second, or hertz (Hz). Any structure composed of elastic materials will vibrate naturally when external forces cause a relative motion between structural parts or a deformation from the equilibrium configuration and the external forces are then removed. The word natural refers to the case when the external forces are removed after the initial application and the resultant structural vibrations are governed only by the geometry and materials of the structure, the manner in which it is supported and loaded, and the gravitational or acceleration field in which the structure is immersed. In contrast, forced vibration of the structure occurs when the external force is applied in a periodic manner to cause structural deformations at a frequency that may be unrelated to the natural frequencies of the structure.

2.1 VIBRATIONAL ENERGY EXCHANGE AND MODE SHAPE

The natural frequency reflects the periodic exchange of kinetic and potential energies within the structure. The sum of the potential and kinetic energies will remain constant if dissipative processes such as internal friction are ignored. The potential energy is directly related to the elastic strain or deformation of the structure from the equilibrium position, and the kinetic energy is proportional to the square of the speed with which the structural mass is deforming. Figure 2.1 illustrates this periodic energy exchange for the example of a cantilever beam. In Fig. 2.1*d* and 2.1*f*, when the cantilever beam passes through the equilibrium position in its vibratory motion, the deformation and potential energy are zero, and all the energy is in the form of kinetic energy, with the speed of beam motion at a maximum. When the beam is at maximum deformation, Figs. 2.1*c* and 2.1*e*, all the energy is in the form of potential energy, and since the kinetic energy is zero, the speed of the beam motion is

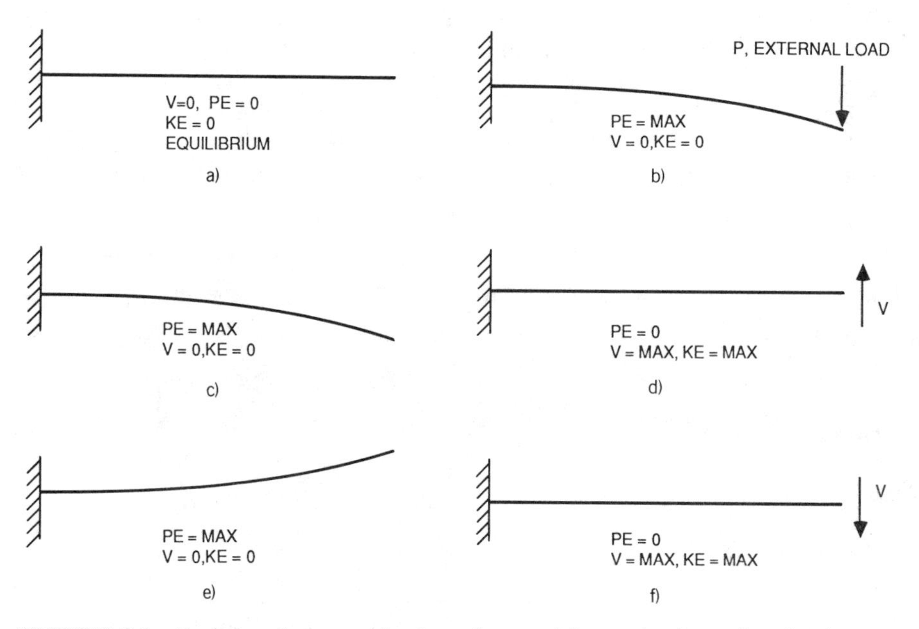

FIGURE 2.1 Exchange between kinetic and potential energies in a vibrating beam. V = velocity; PE = potential energy; KE = kinetic energy.

also momentarily zero as the motion reverses direction. At intermediate positions in the vibratory cycle the beam has a mixture of potential and kinetic energies, with both deformation and speed being less than their maximum values.

As the beam continues to vibrate, it will lose energy by internal friction (converted to heat) and air resistance (work done on the air) if the beam is not in a space environment, and its total energy will decrease, resulting in a decrease in maximum deformation and maximum speed. This energy loss will continue until the beam remains in its equilibrium position with both the kinetic energy and potential energy equal to zero. In the linear elastic region, with the original deformation proportional to the load (Fig. 2.1*b*), the natural frequency is constant and independent of the amplitude of vibration. This type of vibrational energy exchange, illustrated for the cantilever beam, applies to all structural vibrations.

The shape that the structural deformation takes during vibration is known as the vibration-mode shape. There is a specific natural frequency associated with each vibration mode, and the number of different vibration modes is equal to the number of degrees of freedom of the structure. The number of degrees of freedom is the number of spatial coordinate values required to specify the deformation of the structure from equilibrium. For structural shapes such as beams and plates there are an infinite number of degrees of freedom and vibration modes. This is illustrated in Figs. 2.2, 2.3, and 2.4 in which the first three flexural (bending) modes of vibration are shown for a uniform slender straight beam with various end conditions. The vibration mode with the lowest frequency is called the fundamental natural frequency (or sometimes the resonant natural frequency) of the structure. The fundamental frequency usually has the simplest mode shape, the largest amplitudes of structural distortion and stress, and the fewest nodes. Exceptions occur in curved shells, Section 2.8, where the lowest-frequency mode may be associated with more complex node shapes. A node is an unsupported location in the

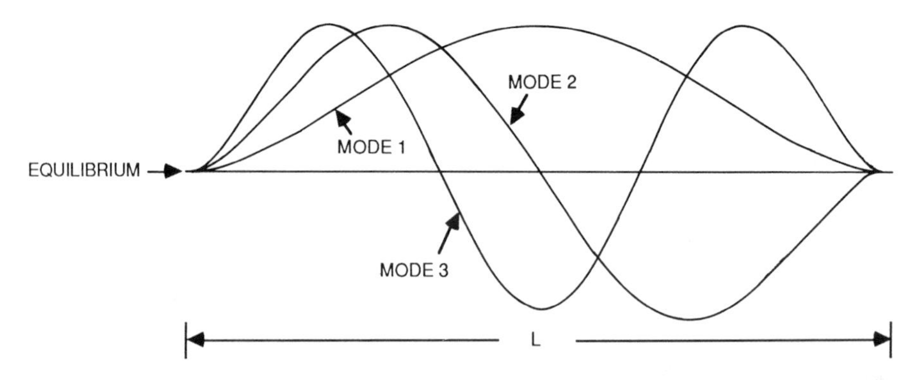

FIGURE 2.2 First three mode shapes of a slender beam clamped at both ends.

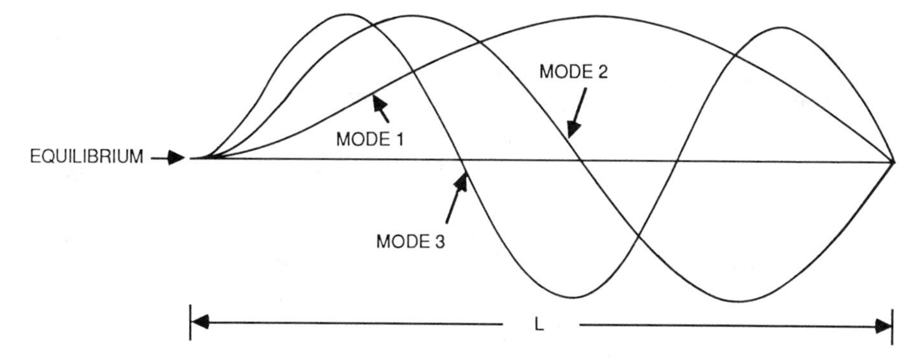

FIGURE 2.3 First three mode shapes of a slender beam clamped at the left end and pinned at the right end.

structure that maintains its equilibrium position while the rest of the structure undergoes vibratory deformation. Nodes generally occur in structures that are undergoing bending or torsion. In Figs. 2.2–2.4 the fundamental mode has no nodes along its unsupported length. The ends of the beam are supported and are not nodes; the second mode has one node; and the third mode has two nodes. The number of nodes can increase indefinitely in this case, corresponding to an unlimited number of degrees of freedom. This unlimited number of degrees of freedom corresponds to the fact that there are an infinite number of points along the length of the beam and an infinite number of coordinate values may be used to specify an arbitrary displacement of each of these points from their equilibrium position. The vibration of the beam can include any number of vibration modes, each of which may have any amplitude. The values of the vibration-mode amplitudes and the phase relations between vibration modes is determined by the means used initially to deform the beam or set the beam in motion by the use of external forces, which are subsequently removed.

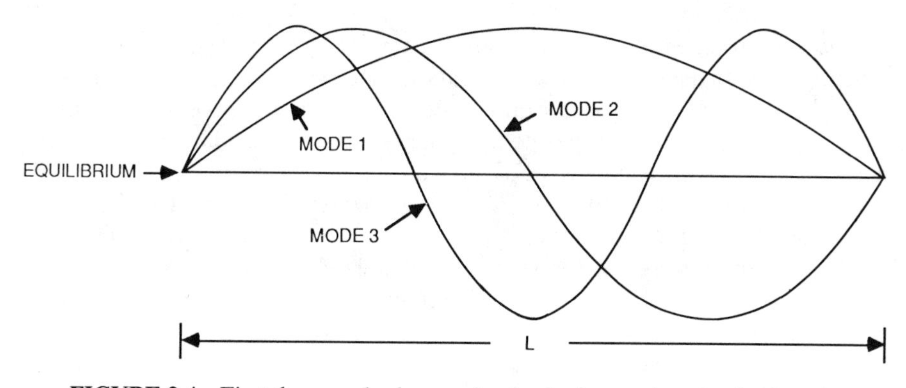

FIGURE 2.4 First three mode shapes of a slender beam pinned at both ends.

A major concern in this book is the design of structures that will safely tolerate the maximum and repeated stresses which are imposed by service in the specified environment. Consequently, the fundamental natural frequency, which has the greatest displacements and stresses and is usually the most easily excited, is of primary concern, and higher frequencies corresponding to higher vibration modes are of lesser interest. In the material presented in subsequent sections of this chapter the emphasis will be on the fundamental natural frequency, although data relating to higher modes are often included. Mode shapes are not dealt with here since maximum material stress can usually be estimated from the fundamental natural frequency and acceleration environment in the cases of interest. The fundamental modes of interest here usually have simple shapes, with the maximum displacement usually occurring at the geometric midpoint between support points or edges. For more complete data on higher modes and mode shapes, the references should be examined.

2.2 RELATION BETWEEN NATURAL FREQUENCY AND STRESS IN ACCELERATION ENVIRONMENTS

In a random-vibration environment or a shock environment the acceleration experienced by a structure is determined by the natural frequency of the structure and the magnitude of the environmental acceleration at that frequency. Vibration and shock environments consist of accelerations covering a broad band of frequencies, referred to as an acceleration spectrum. In a vibration environment, the time-averaged value of acceleration will depend on the frequency and is expressed as a PSD in units of g^2/Hz, where g is the gravitational acceleration of 32.2 ft/sec^2 or 386 in./sec^2 at the earth's surface (Chapter 4). In a shock environment, the value of acceleration as a function of frequency is found from the shock acceleration response spectrum (Chapter 5), in units of g. In each case, given the acceleration environment, it is the natural frequency of the structure that will determine what magnitude of acceleration will be experienced.

If the environmental acceleration must pass through other structures before reaching the structure under investigation, the acceleration spectrum will usually be altered, often significantly. Because of the intervening structure, acceleration magnitudes at specific frequencies may be much larger (amplified) or much smaller (attenuated) when arriving at the structure being analyzed than they were in the original environment (Chapters 4 and 5). The deformations that a structure undergoes are proportional to the accelerations imposed on it,[1] and the stresses in the structure are related to the deformations (or strains) through the elastic moduli. Therefore, in the linear range of material elasticity, the stresses in the structure are directly related to the accelerations imposed on the structure, which in turn are determined by the natural frequencies of the structures as well as the acceleration environment and other intervening structures.

2.3 DEPENDENCE OF NATURAL FREQUENCY ON MATERIAL, GEOMETRY, SUPPORT, LOADING, AND MODE OF VIBRATION

The natural frequency of a structure will depend on its material properties, its geometry, the way in which it is supported and loaded, and its mode of vibration. See Fig. 2.5. Isotropic material properties may include modulus of elasticity E, Poisson's ratio ν, and mass density μ. For orthotropic materials and structures, additional elastic constants are required (Section 2.7). Geometry may include dimensions, wall thickness, and shape. The structure may have various boundaries free, clamped, pinned, or sliding. Loading can be in flexure (bending), tension, compression, shear, or torsion. Natural frequencies may be divided into three categories: those associated with extensional deformation, those associated with torsional deformation, and those associated with flexural (bending) deformation. Extension of a structure refers to stretching or compression, in contrast to bending. In extensional vibrations, a straight beam is stretched or compressed along its longitudinal axis, and a flat plate is stretched or compressed in the plane of the plate. Torsional deformations occur, for example, when a beam twists around its longitudinal axis. Inextensional, or flexural vibrations, refer to deformations (bending) normal to the undeformed beam axes or plate surface. In pure flexural deformations, no axial loads are supported by the beam and the in-plane load on plates is zero. Structures may be subjected to a combination of extensional and flexural deformations that will result in different natural frequencies than occur with pure flexural or pure extensional deformations. The flexural modes are of most interest since their natural frequencies are much lower than those of the extensional and torsional modes, and, consequently, their resulting stresses are usually greater.

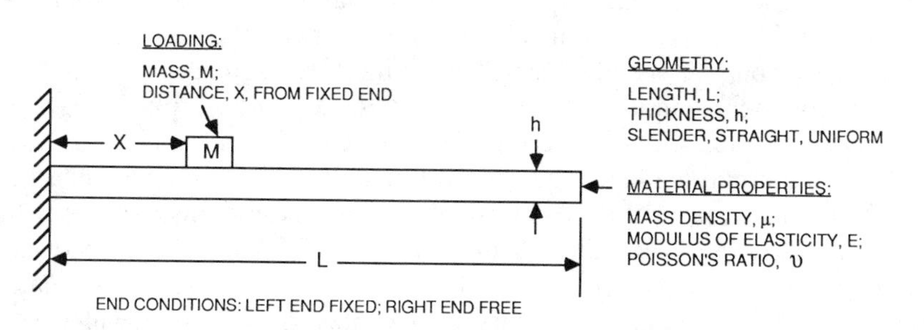

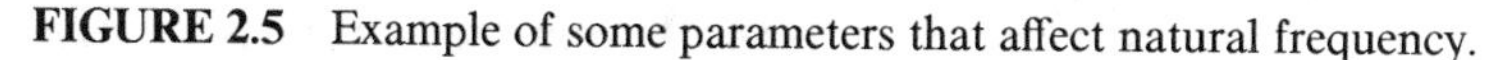

FIGURE 2.5 Example of some parameters that affect natural frequency.

The flexural natural frequencies for many of the structures of interest in this book have a common form. For slender beams, arcs, and rings, the flexural natural frequency has the form

$$F = (\lambda/2\pi L^2)(EIg/w)^{1/2} \tag{2-1}$$

where λ is a constant determined by the type of structure support and mode number, L is the length of the beam or the radius of the arc or ring, E is the modulus of elasticity, I is area moment of inertia about the neutral axis or a principal axis, w is the weight per unit length of structure, and g is the gravitational acceleration. For plates, the flexural natural frequency has the form

$$F = (\lambda/2\pi L^2)\left[Eh^3 g/\rho(1 - \nu^2)\right]^{1/2} \tag{2-2}$$

where λ, g, and E are as previously defined, L is a dimension of the plate, h is the plate thickness, ρ is the weight per unit area of the plate, and ν is Poisson's ratio. For curved shells the dependence of F on the geometry is more complex. In many references both λ and λ^2 are used in the frequency equations (2-1) and (2-2), the choice depending on the specific structure under consideration. Because mode shapes are not covered in this book, it is practical to simplify matters and use only λ throughout. This practice will reduce the chance of making errors when using data from the tables. Note in Eqs. (2-1) and (2-2) the natural frequency decreases with increasing structural dimension and increasing weight per unit length or area; and the frequency increases with increasing modulus of elasticity and moment of inertia or shell thickness. The terms EI/L and $Eh^3/12(1 - \nu^2)$ are sometimes referred to as the stiffness of the beam or plate. A larger stiffness leads to a higher natural frequency.

A structure may be mechanically connected to the rest of the system in several ways. For example, a straight beam may have the same or different boundary conditions on each end. These boundary conditions can include clamped (no translation or rotation of the end allowed), pinned (end rotation allowed, but not translation), sliding (end translation allowed, but not rotation), and free (both translation and rotation allowed). The beam may also have "point" supports (rotation allowed at point, but not translation) along its span. These boundary conditions will affect the natural frequency of the beam. Similar comments apply to the boundary conditions at plate edges and internal point supports within the plate perimeter.

A structure may be loaded in several different ways. For example, a beam may be unloaded (except for its own mass under accelerations), or it may bear external transverse loads, shear loads, longitudinal loads of tension or compression, or torsional (twisting) loads. These loads will affect the natural frequency of the beam. A straight beam in flexural vibration will have its natural frequency increased by an axial tensile load, decreased by an axial compressive load, and decreased by a transverse load.

2.4 STRAIGHT BEAMS

Formulas and tables are available for the natural frequencies of straight beams, both uniform and tapered, single span and multispan, with various types of loadings and boundary conditions, in flexural, transverse shear, longitudinal, and torsional vibration modes, and combinations of these modes. From this data base are selected the cases most often encountered in design work. In all cases presented it is assumed that the beam consists of a single span of homogeneous isotropic, linear elastic materials.

2.4.1 Slender, Uniform Beams

Slender beams are those in which deformation perpendicular to the beam axis is due primarily to bending (flexing), and shear deformation perpendicular to the beam axis can be neglected. For the fundamental mode ($i = 1$) of transverse flexural vibration, a metal beam is considered slender if the ratio R of the span of the beam to its depth meets the following conditions[2]:

$$R = L/h \geq 8 \text{ for compact cross sections} \qquad \geq 15 \text{ for relatively thin webs} \tag{2-3}$$

where L is the span of the beam and h is the depth or transverse dimension in the direction of flexural vibration. Transverse shear deformation substantially lowers the natural frequency of the beam below the pure flexural value when the quantity R/i approaches unity, where i is the flexural vibration mode number.[3]

Transverse Flexural Vibration

In Fig. 2.6, the slender beam geometry and coordinate system are illustrated for flexural deformation in the Y direction, transverse to the beam's longitudinal axis. The X coordinate corresponds with the longitudinal axis of the undeformed beam, passing through the centroid of the beam cross section. The centroid is the geometric center of the cross-sectional area: the sum over the

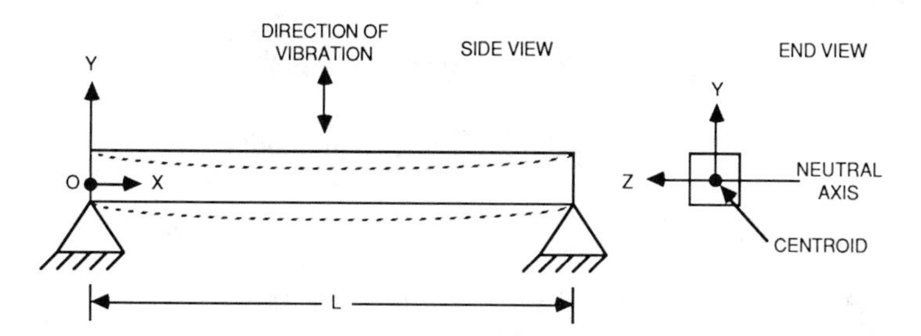

FIGURE 2.6 Coordinate axes for flexural deformation of the beam in the Y direction.

area of all elements of area multiplied by the distance from any axis through the centroid is zero. For example, for beams that have cross sections that are circular, elliptical, square, rectangular, or regular polygons (equal sides and angles), the centroid is at the center of the figure. For other beam cross sections, the location of the centroid with respect to any pair of orthogonal (perpendicular) axes U and V is given by

$$U_c = \frac{1}{A}\int_A U\,dA; \qquad V_c = \frac{1}{A}\int_A V\,dA \tag{2-4}$$

where A is the beam cross-sectional area and the integral is over the area A.

The neutral axis, Fig. 2.6, is the axis in the plane of the beam cross section, parallel to the Z axis, that has zero stress due to the bending and, consequently, undergoes no lengthening or shortening. If there are no axial loads on the beam and the beam is in pure flexure, the neutral axis passes through the centroid of the beam cross section. The neutral surface of the beam is the longitudinal surface perpendicular to the X–Y plane and containing all the neutral axes; it is a surface of zero stress and zero deformation.

Three of the boundary conditions for the flexure beam ends that are commonly encountered are:

Free Boundary

The beam end may move in the Y direction and may be at an angle to the X axis. The mathematical constraints at $x = 0$ or L are

$$\frac{\partial^2 y}{\partial x^2} = 0 \qquad \text{(zero end moment)}$$

$$\frac{\partial^3 y}{\partial x^3} = 0 \qquad \text{(zero shear load at end)}$$

Pinned Boundary

Also called simply supported boundary, the beam end may not move in the Y direction but may be at an angle to the X axis. The mathematical constraints at $x = 0$ or L are

$$y = 0 \qquad \text{(zero end displacement)}$$

$$\frac{\partial^2 y}{\partial x^2} = 0 \qquad \text{(zero end moment)}$$

Clamped Boundary

Also called fixed boundary, the beam end may not move in the Y direction and must maintain a zero angle with the X axis. The mathematical constraints at $x = 0$ or L are

$$y = 0 \qquad \text{(zero end displacement)}$$

$$\frac{\partial y}{\partial x} = 0 \qquad \text{(zero end angle with } X \text{ axis)}$$

Three boundary conditions are used in Table 2-1, where four combinations of end supports are presented for slender flexural beams. The values of area moment of inertia I, required in Table 2-1, are provided in Table 2-2. Table 2-1 provides the formula for calculating the uniformly loaded beam natural flexural frequency[4] as a function of the vibration mode number i, where $i = 1$ is the fundamental (or first-harmonic) mode, $i = 2$ is the second-harmonic mode, and so forth. The λ_i values in the table are dimensionless, numerical constants that incorporate the modal dependence of the frequency. Note that the λ_i values, and, consequently, the frequencies, increase as i increases, and that for a given i value the λ_i values (and frequencies) increase in going from the cantilever (Case 1) to the clamped–clamped supports (Case 4). This latter effect corresponds to a stiffening of the beam as the end supports become more constraining, leading to the higher frequencies produced by increasing stiffness.

Sandwich Beams

A slender beam may consist of a single material with a solid or hollow cross section of any shape that satisfies Eq. (2-3), or it may be a sandwich beam consisting of two or more materials bonded together at longitudinal surfaces and maintaining a uniform cross section along the span. See Fig. 2.7 for an example of a symmetric sandwich beam. The natural frequency of a sandwich beam can be approximated using the formulas developed for homogeneous beams by defining the equivalent stiffness and weight per unit length of the symmetric sandwich beam as[6]

$$EI = \frac{2b}{3} \sum_{i=1}^{n} E_i \left(h_i^3 - h_{i-1}^3 \right) \text{ lbf} \cdot \text{in.}^2$$

$$w = 2b \sum_{i=1}^{n} \mu_i \left(h_i - h_{i-1} \right) \text{ lbf/in.} \tag{2-5}$$

where E_i = modulus of elasticity of material i (lbf/in.^{2})

h_{i-1} = distance from the neutral surface to the inner surface of layer i (in.) with $h_0 = 0$

h_i = distance from the neutral surface to the outer surface of layer i (in.)

μ_i = weight density of material i (lbf/in.^{3})

TABLE 2-1 Slender, Uniformly Loaded Beams in Transverse Flexural Vibration

$$F_i = (\lambda_i/2\pi L^2)(EIg/w)^{1/2} \text{ Hz}$$

where i = mode number = 1, 2, 3, 4, 5, ...
λ_i = dimensionless constant provided in the table
L = beam span (in.)
E = modulus of elasticity of beam material (lbf/in.^{2})
I = area moment of inertia of beam about neutral axis (in.^{4}); see Table 2-2
w = weight per unit length of beam (lbf/in.)
g = gravitational acceleration; on earth = 386 in./sec^2

Method of Support	λ_i
1. Clamped–free (cantilever) 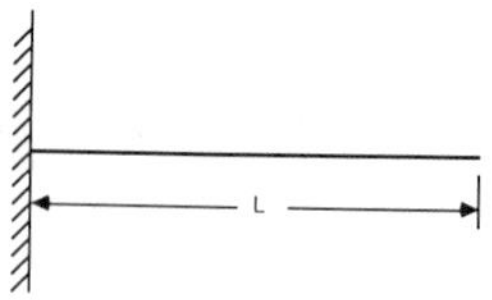	$\lambda_1 = 3.5160$ $\lambda_2 = 22.035$ $\lambda_3 = 61.698$ $\lambda_4 = 120.90$ $\lambda_5 = 199.86$
2. Pinned–pinned (simply supported) 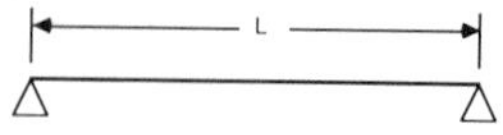	$\lambda_1 = \pi^2$ $\lambda_2 = 4\pi^2$ $\lambda_3 = 9\pi^2$ $\lambda_4 = 16\pi^2$ $\lambda_5 = 25\pi^2$
3. Clamped–pinned 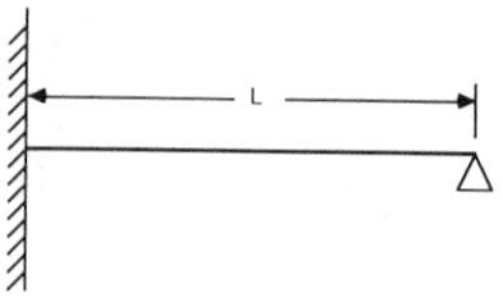	$\lambda_1 = 15.418$ $\lambda_2 = 49.965$ $\lambda_3 = 104.25$ $\lambda_4 = 178.27$ $\lambda_5 = 272.03$
4. Clamped–clamped 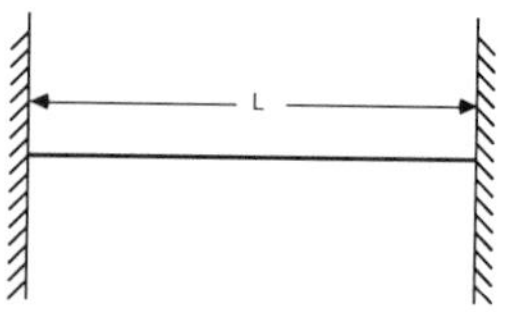	$\lambda_1 = 22.373$ $\lambda_2 = 61.673$ $\lambda_3 = 120.90$ $\lambda_4 = 199.86$ $\lambda_5 = 298.56$

TABLE 2-2 Values of Area Moments of Inertia and Torsional Constants[a]

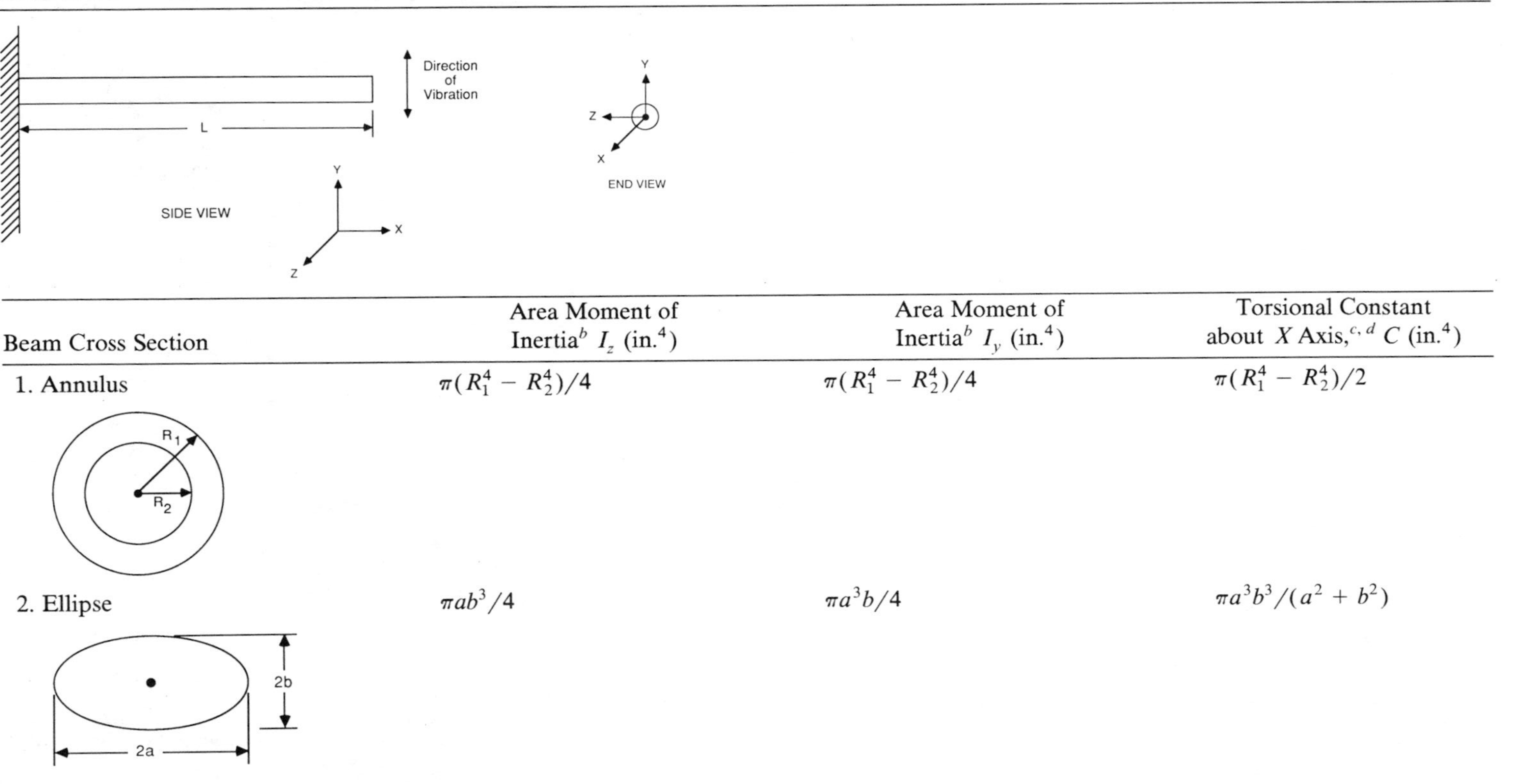

Beam Cross Section	Area Moment of Inertia[b] I_z (in.^{4})	Area Moment of Inertia[b] I_y (in.^{4})	Torsional Constant about X Axis,[c, d] C (in.^{4})
1. Annulus	$\pi(R_1^4 - R_2^4)/4$	$\pi(R_1^4 - R_2^4)/4$	$\pi(R_1^4 - R_2^4)/2$
2. Ellipse	$\pi ab^3/4$	$\pi a^3b/4$	$\pi a^3b^3/(a^2 + b^2)$

TABLE 2-2 *(Continued)*

Beam Cross Section	Area Moment of Inertia[b] I_z (in.^{4})	Area Moment of Inertia[b] I_y (in.^{4})	Torsional Constant about X Axis,[c,d] C (in.^{4})
3. Rectangle 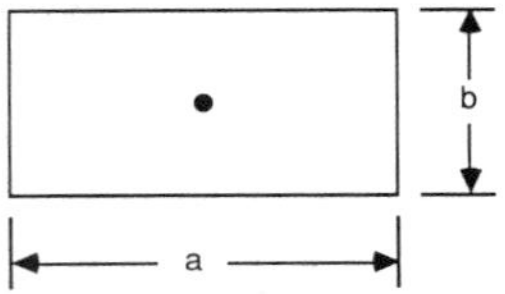	$ab^3/12$	$a^3b/12$	$ca^3b^3/(a^2+b^2)$ a/b — c 1 — 0.281 2 — 0.286 4 — 0.299 8 — 0.312 ∞ — 0.333
4. Triangle 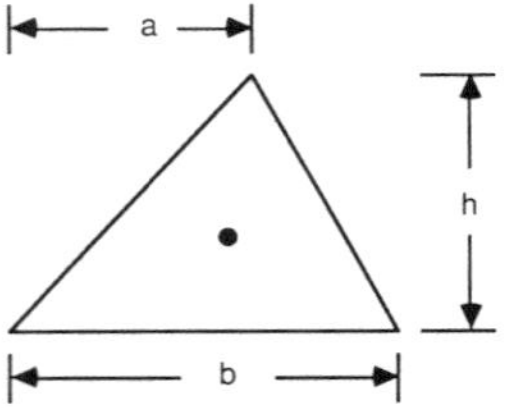	$bh^3/36$	$bh(b^2 - ab + a^2)/36$	—
5. Equilateral triangle 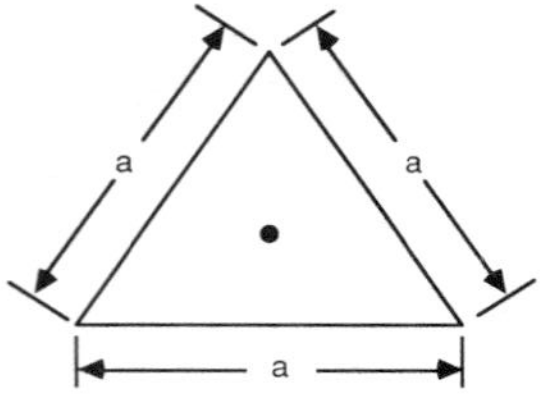	$3^{1/2}a^4/96$	$3^{1/2}a^4/96$	$3^{1/2}a^4/80$

6. Hexagon 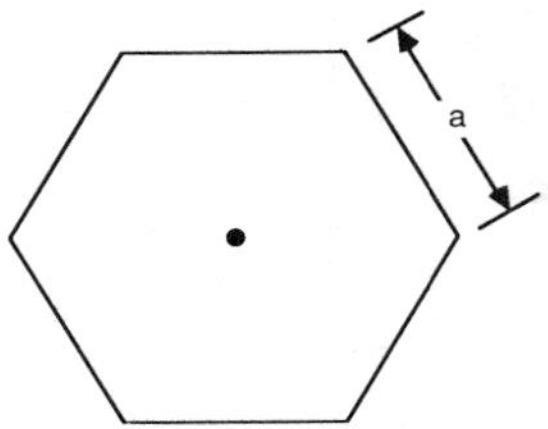	$0.541a^4$	$0.541a^4$	$1.03a^4$
7. Hollow rectangle with thin walls 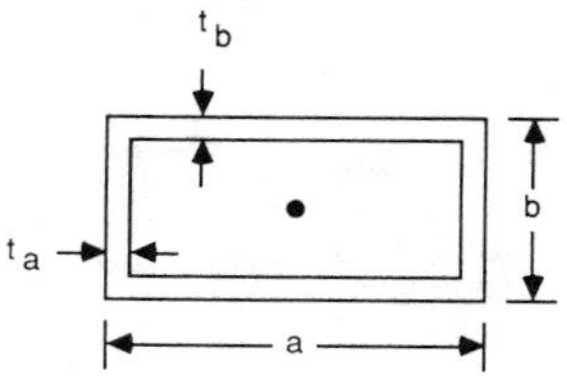	$[ab^3 - (a - 2t_a)(b - 2t_b)^3]/12$	$[a^3b - (a - 2t_a)^3(b - 2t_b)]/12$	$\dfrac{2t_a t_b(a - 2t_a)^2(b - 2t_b)^2}{at_a + bt_b - t_a^2 - t_b^2}$
8. Channel 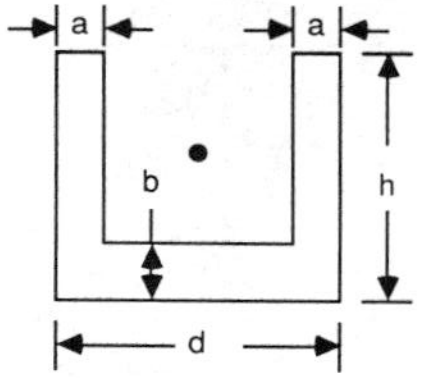	$(2/3)ah^3 + (1/3)(d - 2a)b^3 - \dfrac{(2ah^2 + db^2 - 2ab^2)^2}{4(2ah + db - 2ab)}$	$[d^3h - (h - b)(d - 2a)^3]/12$	$2c(h/a)h^3a^3/(h^2 + a^2) + c(d/b)d^3b^3/(d^2 + b^2)$ [See Case 3 for values of $c(h/a)$ and $c(d/b)$]

TABLE 2-2 *(Continued)*

Beam Cross Section	Area Moment of Inertia[b] I_z (in.^{4})	Area Moment of Inertia[b] I_y (in.^{4})	Torsional Constant about X Axis,[c,d] C (in.^{4})
9. I section 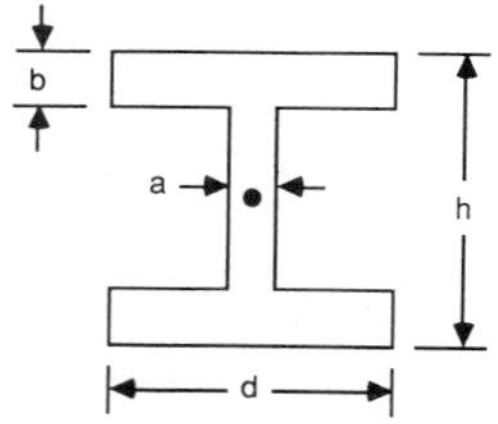	$[dh^3 - (d-a)(h-2b)^3]/12$	$[2bd^3 + (h-2b)a^3]/12$	$2c(d/b)d^3b^3/(d^2+b^2) + c(h/a)h^3a^3/(h^2+a^2)$ [See Case 3 for values of $c(d/b)$ and $c(h/a)$]
10. T section 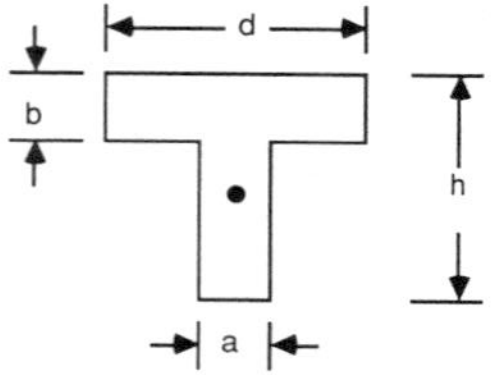	$(1/3)[dh^3 - (d-a)(h-b)^3] - \dfrac{(db+ah-ab)}{3} \times \left[h - \dfrac{ah^2+db^2-ab^2}{2(db+ah-ab)}\right]^2$	$[d^3b + (h-b)a^3]/12$	$c(d/b)d^3b^3/(d^2+b^2) + c(h/a)h^3a^3/(h^2+a^2)$ [See Case 3 for values of $c(d/b)$ and $c(h/a)$]
11. Z section 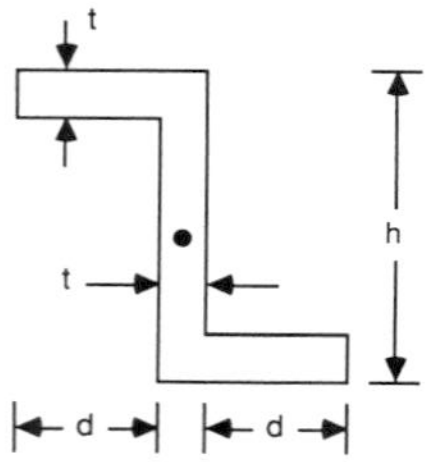	$[(d+t)h^3 - d(h-2t)^3]/12$	$[h(2d+t)^3 - 2d^3(h-t)]/12$	$2c(d/t)d^3t^3/(d^2+t^2) + c(h/t)h^3t^3/(h^2+t^2)$ [See Case 3 for values of $c(d/t)$ and $c(h/t)$]

12. L section	$[ab_1^3 + (a_1 - a)b^3]/3 - \dfrac{(a_1^2 b + a^2 b_1 - a^2 b)^2}{-4(a_1 b + ab_1 - ab)}$	$[ba_1^3 + (b_1 - b)a^3]/3 - \dfrac{(ab_1^2 + a_1 b^2 - ab^2)^2}{4(a_1 b + ab_1 - ab)}$	$c(a_1/a)a_1^3 a^3/(a_1^2 + a^2) + c(b_1/b)b_1^3 b^3/(b_1^2 + b^2)$ [See Case 3 for values of $c(a_1/a)$ and $c(b_1/b)$]

[a]The X axis is the longitudinal axis of the beam and passes through its centroid. The Y axis is vertical and perpendicular to the beam, passes through its centroid, and is in the direction of vibration. The Z axis is horizontal and perpendicular to the beam and passes through its centroid.

[b]I_Y and I_Z are the area moments of inertia of the beam cross section, in units of in.^{4}, about the Y and Z axes, respectively. I_X is the polar area moment of inertia of the beam cross section, in units of in.^{4}, about the X axis, and is the sum of the area moments of inertia: $I_X = I_Y + I_Z$. For hollow beam cross sections where a_1 and b_1 are the exterior dimensions of the cross section and a_2 and b_2 are the dimensions of the internal cavity in the cross section, $I = I(a_1, b_1) - I(a_2, b_2)$ where I may be either an area or polar moment of inertia.

[c]Torsional constants C of the beam cross section are about the X axis.[5] For a beam composed of thin walls of uniform thickness: $C = 4A^2t/S$ in.^{4} for closed (hollow) beam cross sections; $C = St^3/3$ in.^{4} for open beam cross sections where A = area enclosed by mid-wall perimeter (in.^{2}), t = uniform wall thickness (in.), and S = length of mid-wall perimeter (in.).

[d]For an open beam cross section composed of rectangles with large aspect ratios, $a/b \gg 1$,

$$C = \sum_i c_i(a_i/b_i) \cdot (a_i^3 b_i^3)/(a_i^2 + b_i^2)$$

where the values of $c(a, b)$ are provided in Case 3 of the table.

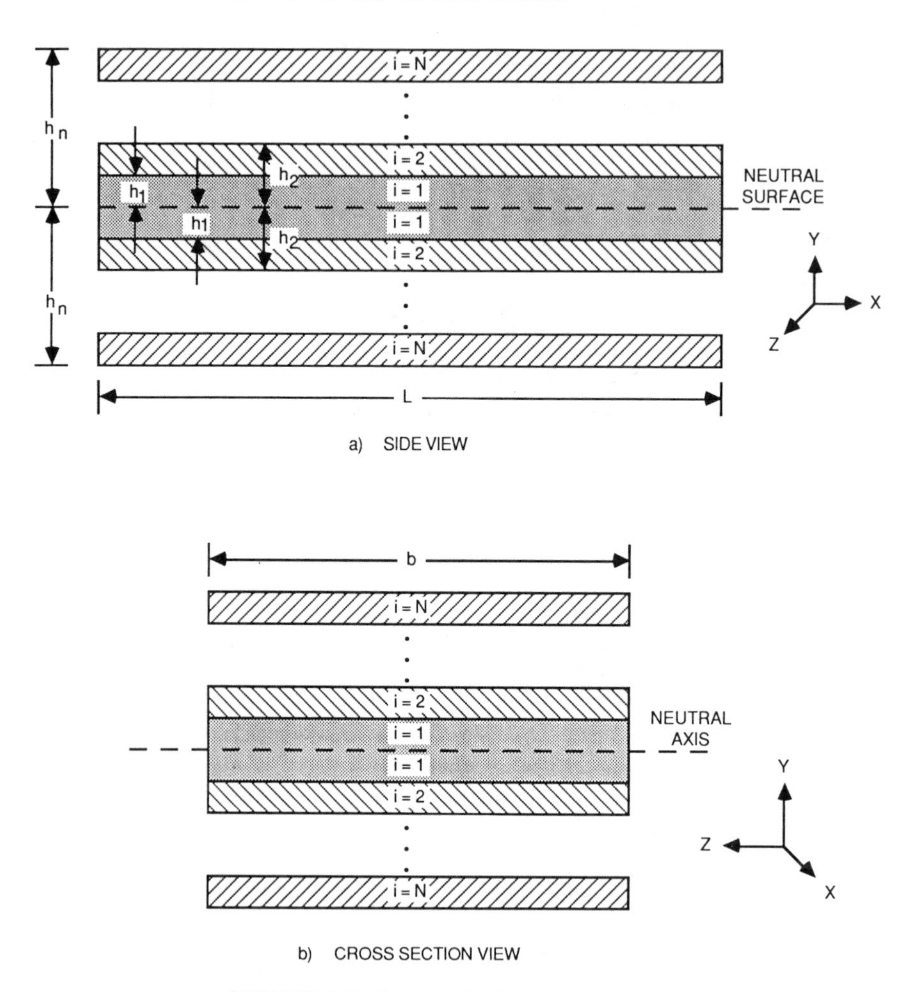

FIGURE 2.7 Symmetrical sandwich beam.

Axial Loads

Slender flexural beams are sometimes subjected to axial loads in addition to the transverse loads that cause bending. The load may be either in tension, as shown in Fig. 2.8, or in compression. For an axial load P, which is constant along the length of the beam, the natural frequencies for the clamped–pinned beam and the clamped–clamped beam, Cases 3 and 4 of Table 2-1, are given by

$$F_i(P) = F_i(P = 0)\left[1 + (P/P_{\mathrm{b}})(\lambda_1/\lambda_i)\right]^{1/2} \text{ Hz} \tag{2-6}$$

FIGURE 2.8 Slender beam under a tensile axial load.

where $F_i(P = 0)$ = frequency found from Table 2-1 (Hz)
P = axial load, positive for a tensile load, negative for a compressive load (lbf)
P_b = critical buckling load (lbf)

λ_1 and λ_i are taken from Table 2-1.

For the clamped–pinned beam, Case 3 of Table 2-1, the critical buckling load is[7]

$$P_b = 2.05\pi^2 EI/L^2 \text{ lbf} \tag{2-7}$$

For the clamped–clamped beam, Case 4 of Table 2-1, the critical buckling load is

$$P_b = 4\pi^2 EI/L^2 \text{ lbf} \tag{2-8}$$

Note that tensile loads will increase the natural frequencies and compressive loads will decrease them.

Concentrated Weights

The slender flexural beam may be loaded transversely with concentrated weights in addition to its uniformly distributed beam weight. See Fig. 2.9. The addition of such loads always decreases the natural frequency. Three cases are presented with formulas for finding the fundamental natural frequency F_1.[8-10] The notation is the same as in Table 2-1, with the addition of w_C = concentrated weight (lbf) and w_B = total weight of beam (lbf).

1. *Cantilever with Concentrated Weight at Free End* (Fig. 2.9*a*).

$$F_1 = \frac{1}{2\pi}\left[\frac{3EIg}{L^3(w_c + 0.24w_B)}\right]^{1/2} \text{ Hz} \tag{2-9}$$

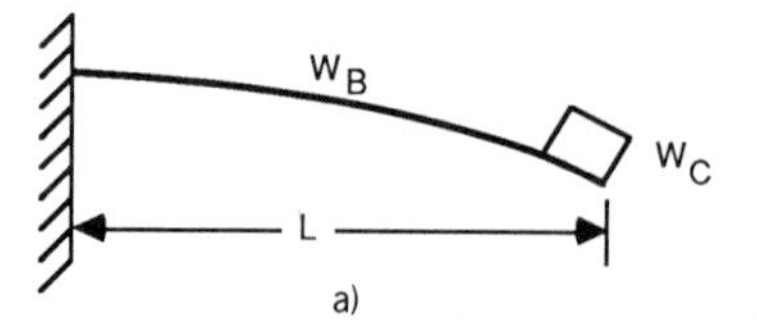

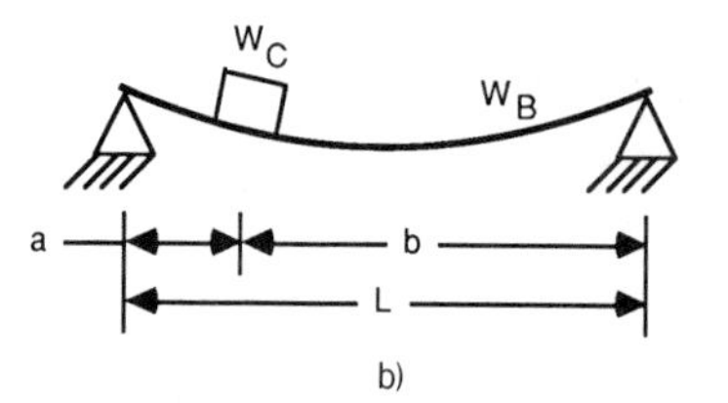

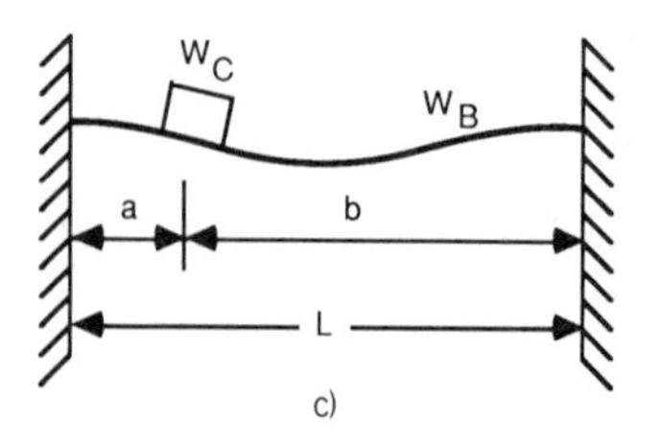

FIGURE 2.9 Slender beams with concentrated weight and uniformly distributed beam weight.

2. *Pinned–Pinned Beam* (Fig. 2.9*b*).

$$F_1 = \frac{1}{2\pi}\left\{\frac{3EIg(a+b)}{a^2b^2[w_C + (\alpha+\beta)w_B]}\right\}^{1/2} \text{ Hz}$$

$$\alpha = \frac{a}{a+b}\left[\frac{(2b+a)^2}{12b^2} + \frac{a^2}{28b^2} - \frac{a(2b+a)}{10b^2}\right] \tag{2-10}$$

$$\beta = \frac{b}{a+b}\left[\frac{(2a+b)^2}{12a^2} + \frac{b^2}{28a^2} - \frac{b(2a+b)}{10a^2}\right]$$

where a and b are the distances from the concentrated weight to the ends of the beam (in.). In the case of the concentrated weight at the center of the beam, $a = b$ and Eq. (2-10) reduces to

$$F_1 = \frac{2}{\pi}\left[\frac{3EIg}{L^3(w_C + 0.486w_B)}\right]^{1/2} \text{ Hz} \tag{2-11}$$

3. Clamped–Clamped Beam (Fig. 2.9*c*).

$$F_1 = \frac{4}{\pi}\left\{\frac{3EIg}{L^3[w_C + (\alpha + \beta)w_B]}\right\}^{1/2} \text{ Hz}$$

$$\alpha = \frac{a}{a+b}\left[\frac{(3a+b)^2}{28b^2} + \frac{9(a+b)^2}{20b^2} - \frac{(a+b)(3a+b)}{4b^2}\right] \tag{2-12}$$

$$\beta = \frac{b}{a+b}\left[\frac{(3b+a)^2}{28a^2} + \frac{9(a+b)^2}{20a^2} - \frac{(a+b)(3b+a)}{4a^2}\right]$$

where a and b are defined in the preceding case. In the case of the concentrated weight at the center of the beam, $a = b$ and Eq. (2-12) reduces to

$$F_1 = \frac{4}{\pi}\left[\frac{3EIg}{L^2(w_C + 0.371W_B)}\right]^{1/2} \text{ Hz} \tag{2-13}$$

Torsional Vibration

Torsional vibrations involve the twisting of a beam about its longitudinal axis. See Fig. 2.10. Frequency formulas for thin beams in torsional vibration apply when[11]

$$Lt/iD^2 > 10 \tag{2-14}$$

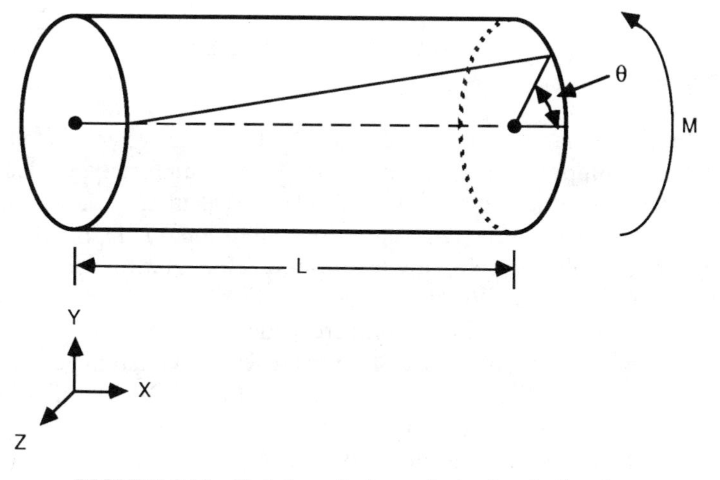

FIGURE 2.10 Twisting of a beam in torsional vibration.

where L = span length (in.)
t = a typical minimum wall thickness of the beam cross section (in.)
D = diameter or characteristic dimension of the beam cross section (in.)
i = vibrational mode number

Three of the boundary conditions for the torsion beam ends that are commonly encountered are:

Free Boundary

The end of the beam is without stress. The mathematical constraint at $x = 0$ or L is

$$\frac{\partial \theta}{\partial x} = 0 \qquad \text{(zero end torque)}$$

Inertial Mass Boundary

The end of the beam has a torque about the beam's longitudinal axis. This torque is directly proportional to the angular acceleration of the inertial mass at the end of the beam. The mathematical constraint at $x = 0$ or L is

$$\frac{\partial \theta}{\partial x} = -\frac{J}{GC} \frac{\partial^2 \theta}{\partial t^2}$$

Fixed Boundary

The end of the beam cannot twist about the beam's longitudinal axis. The mathematical constraint at $x = 0$ or L is

$$\theta = 0 \qquad \text{(zero end rotation)}$$

These three boundary conditions are used in Table 2-3, where three combinations of end supports are presented for slender, uniformly loaded torsion beams.[11–13] This table provides the formula for calculating the beam natural torsional frequency as a function of the vibration mode number, i, and provides the λ_i values required by the formula. Note that for cases 1 and 3 the λ_i values, and, consequently, the frequencies, increase as i increases, and that for a given i value the λ_i values (and frequencies) increase in going from the fixed-free end conditions (Case 1) to the more constrained and stiffer fixed–fixed end conditions (Case 3). The use of Table 2-3 to calculate torsional vibration frequencies requires the determination of the constants C and I_p, which are dependent on the beam cross-section geometry. For Case 2, it is also

TABLE 2-3 Slender, Uniformly Loaded Beams in Torsional Vibration

$$F_i = (\lambda_i/2\pi L)(CGg/\mu I_p)^{1/2} \text{ Hz}$$

where i = mode number 1, 2, 3, ...
λ_i = dimensionless constant provided in the table
L = beam span (in.)
C = torsional constant of beam cross section (in.^{4}). See Table 2-2
G = shear modulus of beam material, (lbf/in.^{2})
g = gravitational acceleration on earth = 386 in./sec^2
μ = weight density of beam material (lbf/in.^{3})
I_p = polar area moment of inertia of beam cross section about the beam axis of torsion (in.^{4}); see Eq. (2-16) and Table 2-2
J = polar mass moment of inertia about axis of torsion (lbm · in.^{2}); see Eq. (2-17) and Table 2-4
θ = angle of twist about beam longitudinal axis (rad)

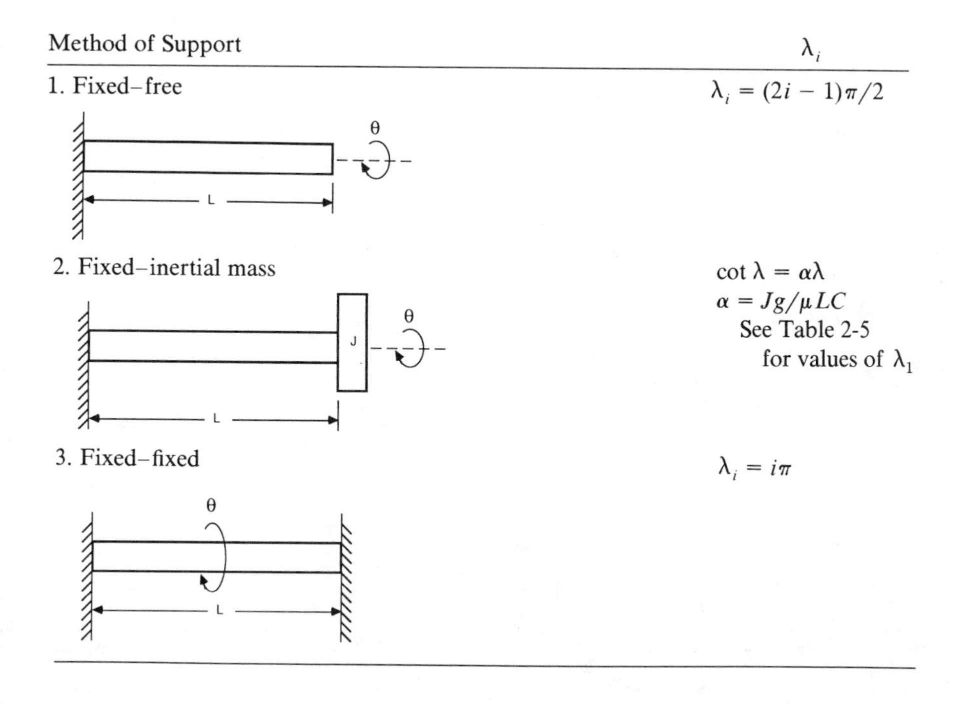

Method of Support	λ_i
1. Fixed–free	$\lambda_i = (2i - 1)\pi/2$
2. Fixed–inertial mass	$\cot\lambda = \alpha\lambda$ $\alpha = Jg/\mu LC$ See Table 2-5 for values of λ_1
3. Fixed–fixed	$\lambda_i = i\pi$

necessary to determine the constant J (Table 2-4), which is dependent on the end inertial mass geometry. Table 2-5 provides values of λ_1 for the transcendental equation in Case 2. The three constants C, I_p, and J are defined in the following paragraphs.

The torsional constant C is the moment required to produce a torsional deformation (twist) of one radian on a unit (1-in.) length of beam, divided by

TABLE 2-4 Values of *J* for Inertial End Masses on Beams in Torsional Vibration[a]

Inertial End Mass Shape	Mass Moment of Inertia about Axis of Rotation, J (lbm · in.^{2})[b]
1. Rectangular prism 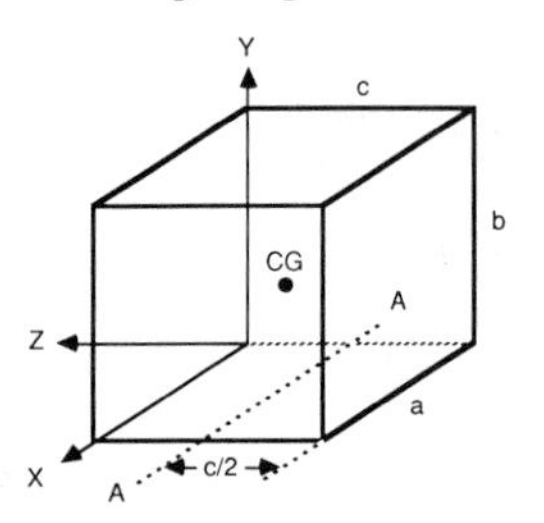	Rotation about X axis: $J_x = (M/3)(b^2 + c^2)$ Rotation about axis through CG parallel to X axis: $J_{xc} = (M/12)(b^2 + c^2)$ Rotation about axis A-A: $J_{AA} = (M/12)(4b^2 + c^2)$
2. Torus  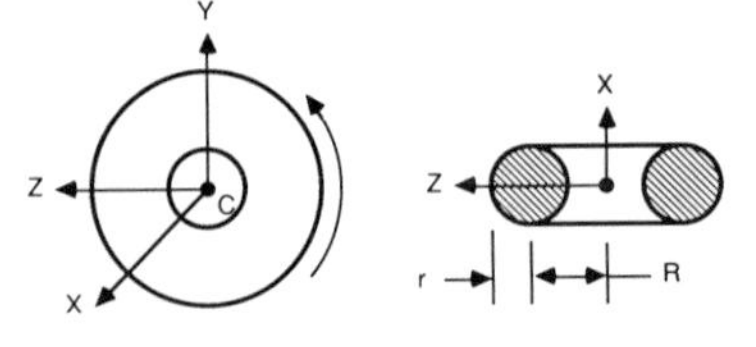	$(1/2)\pi^2 \mu R r^2 (4R^2 + 3r^2)$
3. Cone 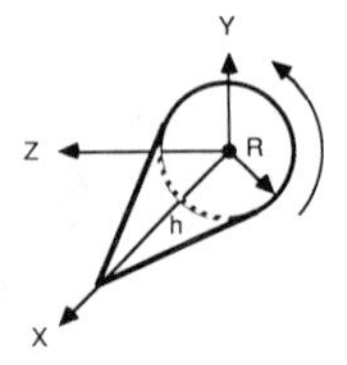	$(\pi/10)\mu R^4 h$
4. Sphere 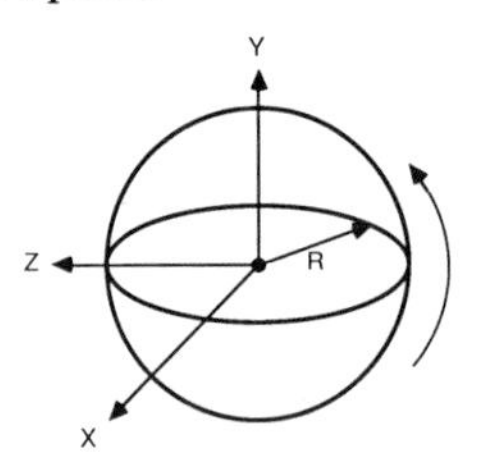	$(8/15)\ \pi\mu R^5$
5. Hollow sphere 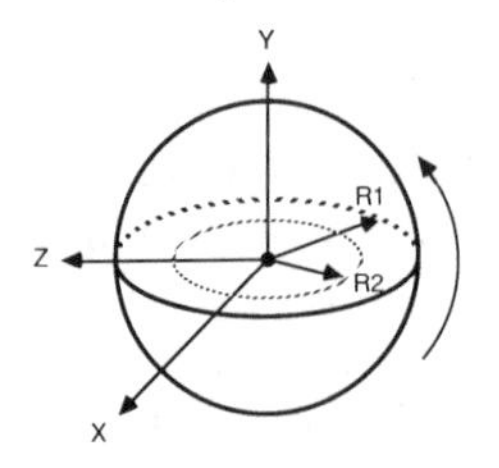	$(8/15)\pi\mu(R_1^5 - R_2^5)$

TABLE 2-4 *(Continued)*

Inertial End Mass Shape	Mass Moment of Inertia about Axis of Rotation, J (lbm · in.^{2})[b]
6. Ellipsoid	Rotation about axis $i = x, y, z$: $J_x = (4/15)\pi\mu abc(b^2 + c^2)$ $J_y = (4/15)\pi\mu abc(a^2 + c^2)$ $J_z = (4/15)\pi\mu abc(a^2 + b^2)$

[a] For shapes with constant cross section along the axis of rotation; see Eq. (2-18) and Table 2-2.
[b] $\mu = \rho/g$ = mass density of inertial end mass material (lbm/in.^{3})
ρ = weight density of inertial end mass material (lbf/in.^{3})
g = gravitational acceleration on earth (386 in./sec^2)

the shear modulus:

$$C = M/(Gd\theta/dx) \text{ in.}^4 \tag{2-15}$$

where M = torsional moment (lbf · in.)
G = shear modulus of beam material (lbf/in.^{2})
$d\theta/dx$ = twist (rad/in.)

See Fig. 2.10 and Table 2-2. For beams with circular cross sections, $C = I_p$. For beams with noncircular cross sections, $C < I_p$.

The polar area moment of inertia about the beam axis of torsion, I_p, is equal to the sum of the area moments of inertia about the principal axes of the beam cross section:

$$I_p = I_y + I_z \text{ in.}^4 \tag{2-16}$$

where it is assumed that the beam axis of torsion passes through the centroid of the beam cross section. See Fig. 2.11 and Table 2-2.

The mass moment of inertia J of the end inertial mass about its centroid (Case 2 of Table 2-3) is the moment required to produce a torsional accelera-

TABLE 2-5 Values for λ_1 in the Transcendental Equation[a] $\cot \lambda_1 = \alpha\lambda_1$

α	λ_1 (rad)
0.0	$\pi/2$
0.1	1.43
0.2	1.31
0.3	1.22
0.4	1.14
0.5	1.08
0.6	1.02
0.7	0.97
0.8	0.93
0.9	0.89
1.0	0.86
1.2	0.80
1.4	0.76
1.6	0.72
1.8	0.68
2.0	0.65
2.3	0.62
2.6	0.58
3.0	0.55
3.5	0.51
4.0	0.48
5.0	0.43
6.0	0.40
8.0	0.35
10.0	0.31
12.0	0.28
15.0	0.26
20.0	0.22
30.0	0.18
40.0	0.16
60.0	0.13
80.0	0.11
100.0	0.100
150.0	0.082
200.0	0.071
300.0	0.058
400.0	0.050
600.0	0.041
800.0	0.035
1,000.0	0.032
∞	0

[a] See Case 2 in Table 2-2 and Cases 2 and 3 in Table 2-6. The tabulated values of λ_1 vs α are solutions to the equation:

$$\cot \lambda_1 = \alpha\lambda_1$$

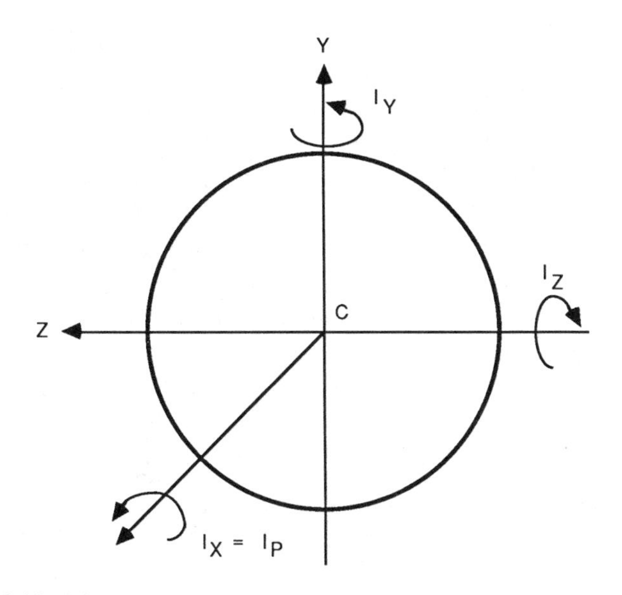

FIGURE 2.11 The polar moment of inertia I_p, about a beam's longitudinal axis.

tion of 1 rad/sec^2:

$$J = M/\alpha \text{ lbm} \cdot \text{in.}^2 \tag{2-17}$$

where M is defined following Eq. (2-15), and α is the torsional angular acceleration in rad/sec^2. Note that a value of J is only required for Case 2 in Table 2-3, and that this value needs to be determined for the inertial mass at the free end of the beam, which has a geometry independent of the beam geometry.

Values of C and I_p are provided in Table 2-2 for commonly encountered beam cross sections. For end inertial masses that have a centroid (and an axis of rotation) coincident with the beam axis of rotation, a uniform cross section along the axis of rotation, and a length (thickness) of h in. along this same axis, the J value may be found from Table 2-2 as follows:

$$J = (\rho h/g) I_p \text{ lbm} \cdot \text{in.}^2 \tag{2-18}$$

where ρ = weight density of inertial end mass material (lbf/in.^{3})
g = gravitational acceleration on earth = 376 in./sec^2
$\rho/g = \mu$ = mass density of inertial mass material (lbm/in.^{3})
h = thickness, along axis of rotation, of end mass cross section (in.)
I_p = polar area moment of inertia of the end mass about the axis of rotation (in.^{4})

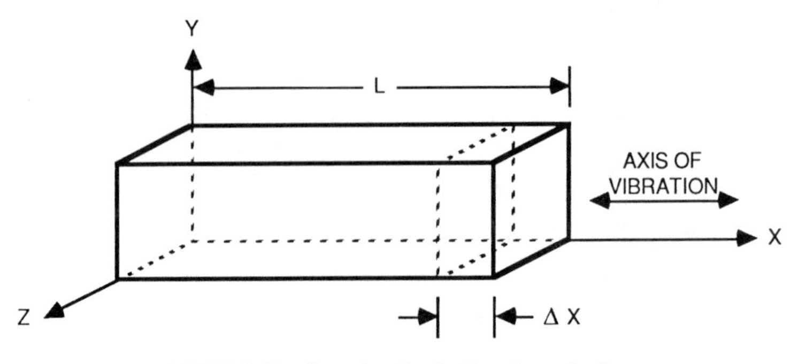

FIGURE 2.12 Longitudinal vibration of a beam.

For end inertial masses that have a centroid (and an axis of rotation) coincident with the beam axis of rotation but do not have a uniform cross section along the axis of rotation, J values are provided in Table 2-4 for some commonly used shapes.

Longitudinal Vibration

Longitudinal vibrations involve the extension and compression of the beam along its longitudinal axis, as shown in Fig. 2.12. Four of the boundary conditions for the beams ends that are commonly encountered are:

Free Boundary

The end of the beam is without longitudinal stress. The mathematical constraint at $x = 0$ or L is

$$\frac{\partial X}{\partial x} = 0$$

where $X(x)$ is the deformation of the beam parallel to the longitudinal (x) axis.

Mass Boundary

The end of the beam has a stress and deformation directly proportional to the longitudinal acceleration of the inertial mass at the end of the beam. The mathematical constraint at $x = 0$ or L is

$$\frac{\partial X}{\partial x} = -\left(\frac{M}{AE}\right)\frac{\partial^2 X}{\partial t^2}$$

where A is the beam cross section area (in.^{2}); E is the beam modulus of elasticity (lbf/in.^{2}); and M is the inertial mass at end of beam (lbm).

Spring Boundary

The end of the beam has a stress and deformation directly proportional to the product of the spring constant k and the end deformation X. The mathematical constraint at $x = 0$ or L is

$$\frac{\partial X}{\partial x} = \frac{kX}{AE}$$

Fixed Boundary

The end of the beam has zero displacement. The mathematical constraint at $x = 0$ or L is

$$X(x) = 0$$

These four boundary conditions are used in Table 2-6, where five combinations of end support are presented for slender, uniform beams.[14] This table provides the formula for calculating the beam natural longitudinal frequency as a function of the vibration mode number i and provides the λ_i values required by the formula. Note that for Cases 1 and 5 the λ_i values, and, consequently, the frequencies, increase as i increases, and that for a given i value the λ_i values (and frequencies) increase in going from the fixed-free end conditions (Case 1) to the more constrained and stiffer fixed–fixed end conditions (Case 5). Table 2-5 provides values of λ_i for the transcendental equation in Cases 2 and 3, and Table 2-7 provides values of λ_i for the transcendental equation in Case 4.

2.4.2 Deep, Uniform Beams

Deep beams are those that do not meet the conditions of Eq. (2-3): the ratio of span length L to the depth H in the direction of vibrational deformation is not very much greater than 1, or may even be less than 1. Shear deformation becomes important in deep beams and may be more dominant than flexural deformation. Figure 2.13 shows the first mode of a clamped–free deep beam in flexure and in shear. The shear deformation is due to a transverse shear stress σ_{xy}, which has a nonuniform value over the transverse cross section of the beam. The maximum value of σ_{xy} generally occurs at the centroid of the beam cross section, and the ratio of the average shear strain over the beam cross section to the maximum shear strain at the centroid is designated K:

$$\sigma_{xy}(\text{centroid}) = G\frac{\partial y}{\partial x} = \text{maximum shear stress (psi)} \tag{2-19}$$

$$\sigma_{xy}(\text{average}) = KG\frac{\partial y}{\partial x} = \text{average shear stress over section (psi)} \tag{2-20}$$

TABLE 2-6 Slender, Uniformly Loaded Beams in Longitudinal Vibration

$$F_i = (\lambda_i/2\pi L)(E/\mu)^{1/2} \text{ Hz}$$

where i = mode number $1, 2, 3, \ldots,$
λ_i = dimensionless constant provided in the table
L = beam span (in.)
A = beam cross sectional area (in.^{2})
E = modulus of elasticity of beam material (lbf/in.^{2})
k = spring constant (lbf/in.)
μ = mass density of beam material (lbm/in.^{3})
M = mass at end of beam (lbm)

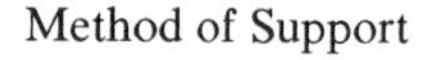

Method of Support	λ_i
1. Fixed–free 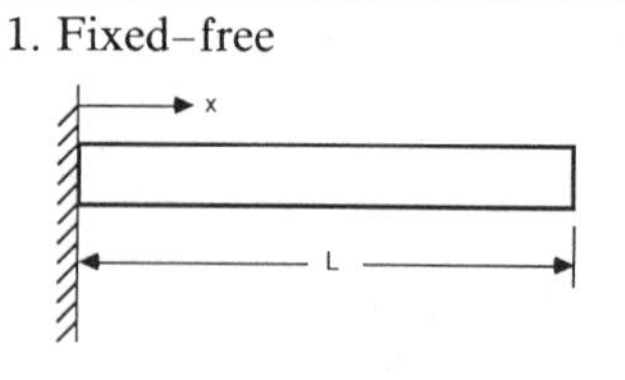	$(2i - 1)\pi/2$
2. Fixed–mass 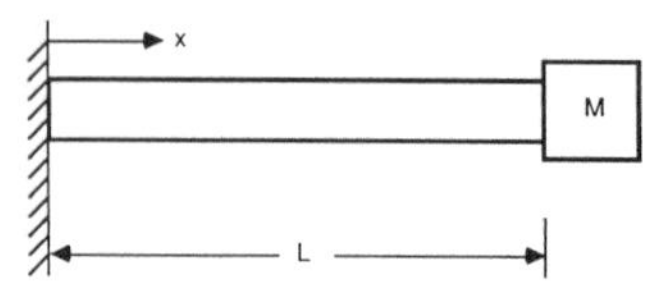	$\cot \lambda = \alpha\lambda$ $\alpha = M/\mu AL$ See Table 2-5 for values of λ_1
3. Spring–free 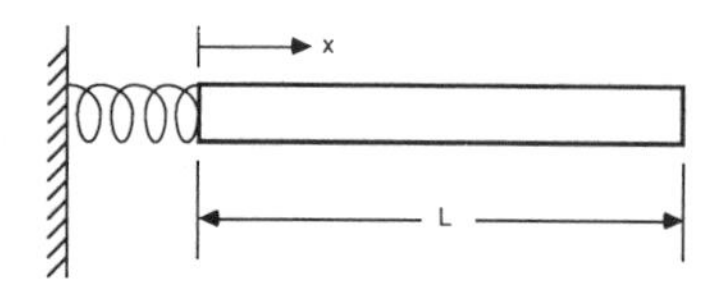	$\cot \lambda = \alpha\lambda$ $\alpha = AE/kL$ See Table 2-5 for values of λ_1
4. Fixed–spring 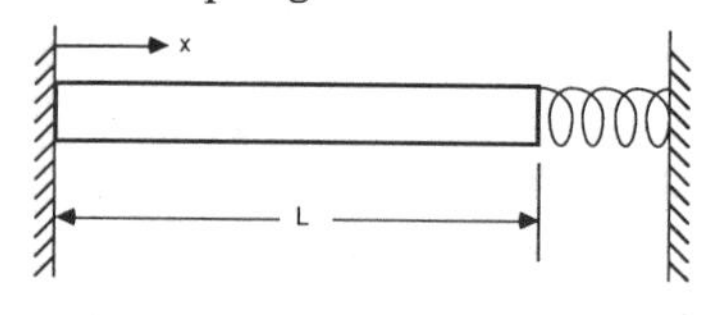	$\tan \lambda = -\alpha\lambda$ $\alpha = AE/kL$ See Table 2-7 for values of λ_1
5. Fixed–fixed 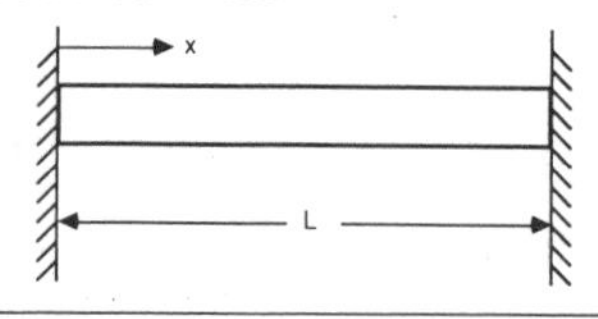	$i\pi$

TABLE 2-7 Values of λ in the Transcendental Equation[a] tax $\lambda_1 = -\alpha\lambda_1$

α	λ_1 (rad)
0.00	π
0.01	3.11
0.02	3.08
0.04	3.02
0.06	2.97
0.08	2.91
0.10	2.86
0.15	2.75
0.20	2.65
0.25	2.57
0.30	2.50
0.40	2.38
0.50	2.29
0.60	2.22
0.80	2.11
1.00	2.03
1.20	1.97
1.50	1.91
2.00	1.84
3.00	1.76
4.00	1.72
6.00	1.67
8.00	1.65
10.00	1.63
15.00	1.61
20.00	1.60
30.00	1.59
70.00	1.58
∞	$\pi/2$

[a]See Case 4 in Table 2-6. The tabulated values of λ_1 vs α are solutions to the equation: $\tan \lambda_1 = -\alpha\lambda_1$

Table 2-8 provides values of K for some commonly used beam cross sections.[15] Three of the boundary conditions for the beams ends are:

Free Boundary

There is no shear stress at the beam end. The mathematical constraint at $x = 0$ or L is

$$\frac{\partial y}{\partial x} = 0$$

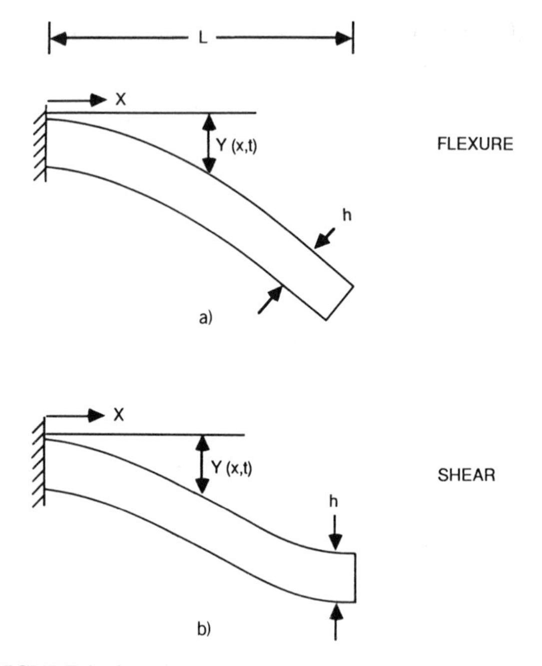

FIGURE 2.13 Flexure versus shear in a cantilever beam.

Mass Boundary

The end of the beam has a shear stress directly proportional to the transverse acceleration of the inertial mass at the end of the beam. The mathematical constraint at $x = 0$ or L is

$$\frac{\partial y}{\partial x} = -\frac{\mathrm{M}}{KAG}\frac{\partial^2 y}{\partial t^2}$$

Fixed Boundary

The end of the beam has zero displacement. The mathematical constraint at $x = 0$ or L is $y = 0$.

These three boundary conditions are used in Table 2-9, where three combinations of support are presented for deep beams in pure shear.[16] This table provides the formula for calculating the beam natural frequency as a function of the vibration mode number i and provides the λ_i values required by the formula. Flexural deformation is neglected in Table 2-9. Note that for Cases 1 and 3 the λ_i values, and, consequently, the frequencies, increase as i increases, and that for a given i value the λ_i values (and frequencies) increase in going

TABLE 2-8 Values of Shear Coefficients K^a

Beam Cross Section	Shear Coefficient K
1. Annulus 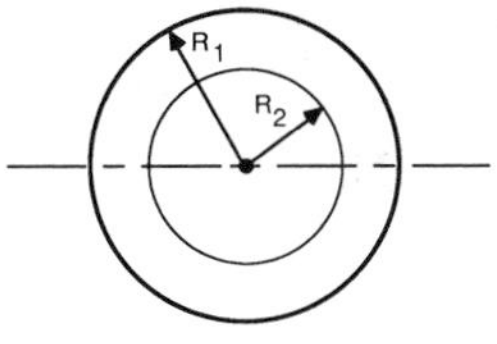	$\dfrac{6(1+\nu)(1+m^2)^2}{(7+6\nu)(1+m^2)^2+(20+12\nu)m^2}$ where $m = R_2/R_1$ For solid circle, $m = 0$, $K = \dfrac{6(1+\nu)}{7+6\nu}$ For thin annulus, $m = 1$, $K = \dfrac{2(1+\nu)}{4+3\nu}$
2. Ellipse 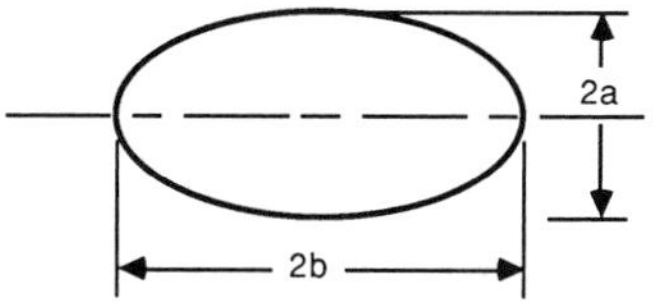	$\dfrac{12(1+\nu)a^2(3a^2+b^2)}{(40+37\nu)a^4+(16+10\nu)a^2b^2+\nu b^4}$
3. Rectangle 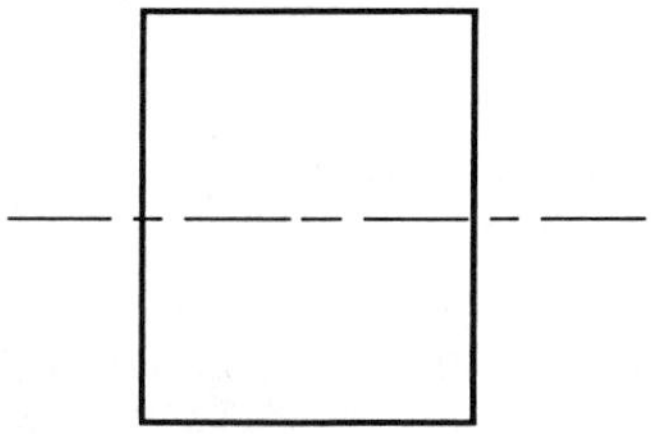	$\dfrac{10(1+\nu)}{12+11\nu}$
4. Thin-walled square 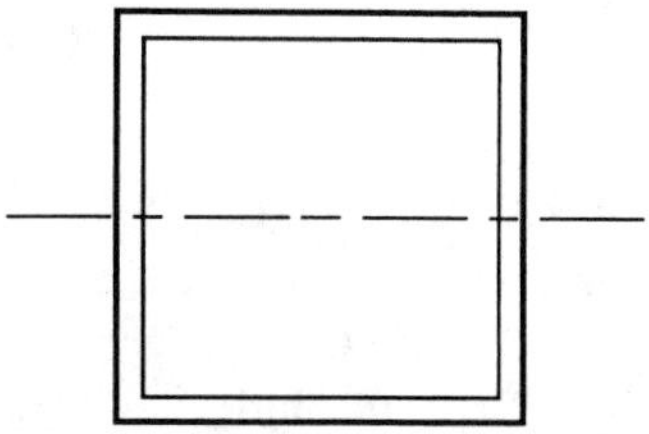	$\dfrac{20(1+\nu)}{48+39\nu}$
5. Thin-walled T section 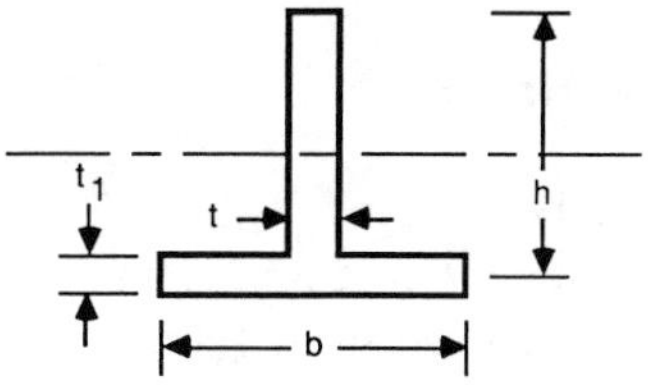	$\dfrac{10(1+\nu)(1+4m)^2}{F(m,n)}$ where $F(m,n) = 12+96m+276m^2+192m^3 + \nu(11+88m+248m^2 + 216m^3) + 30n^2(m+m^2) + 10\nu n^2(4m+5m^2+m^3)$ $m = bt_1/ht$ $n = b/h$

TABLE 2-8 *(Continued)*

Beam Cross Section	Shear Coefficient K
6. Thin-walled I section 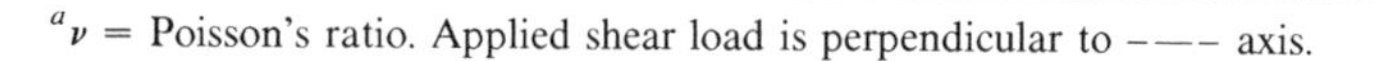	$\dfrac{10(1+\nu)(1+3m)^2}{F(m,n)}$ where $F(m,n) = 12 + 72m + 150m^2 + 90m^3 + \nu(11 + 66m + 135m^2 + 90m^3) + 30n^2(m + m^2) + 5\nu n^2(8m + 9m^2)$ $m = 2bt_1/ht$ $n = b/h$

$^a\nu$ = Poisson's ratio. Applied shear load is perpendicular to ---- axis.

from the fixed–free end conditions (Case 1) to the more constrained and stiffer fixed–fixed end conditions (Case 3). Table 2-5 provides values of λ_i for the transcendental equation in Case 2.

In most deep beams both flexure and shear are important, and the beam deformation can only be estimated accurately by adding together the flexural and shear deformations. Even in slender beams, shear deformation may become significant in the higher modes of vibration. An approximation to the ratios of flexural deformation to shear deformation for beams with simple closed cross sections is

$$\text{(shear deformation)}/\text{(flexural deformation)} = ih/L \tag{2-21}$$

where i = mode number
h = beam dimension in direction of transverse vibration (in.)
L = beam span (in.)

When the ratio in eq. (2-21) is in the neighborhood of unity, indicating comparable shear and flexural deformations, the natural frequency of the beam may be estimated by the use of Dunkerley's equation[17]:

$$F = \left(F_F^{-2} + F_S^{-2}\right)^{-1/2} \text{ Hz} \tag{2-22}$$

where F = fundamental natural frequency of the beam (Hz)
F_F = fundamental natural frequency assuming pure flexure (Hz)
F_S = fundamental natural frequency assuming pure shear (Hz).

Rotary inertia, the inertia associated with the local rotation of the beam cross section about the (----) axis shown in Table 2-8, has been ignored in

TABLE 2-9 Deep Beams in Pure Transverse Shear Vibrations

$$F_i = (\lambda_i/2\pi L)(KG/\mu)^{1/2} \text{ Hz}$$

where i = mode number 1, 2, 3 . . .
λ_i = dimensionless constant provided in the table
L = beam span (in.)
A = beam cross sectional area (in.^{2})
G = shear modulus (lbf/in.^{2})
K = dimensionless shear coefficient, given in Table 2.8
μ = mass density of beam material (lbm/in.^{3})
M = mass at end of beam (lbm)

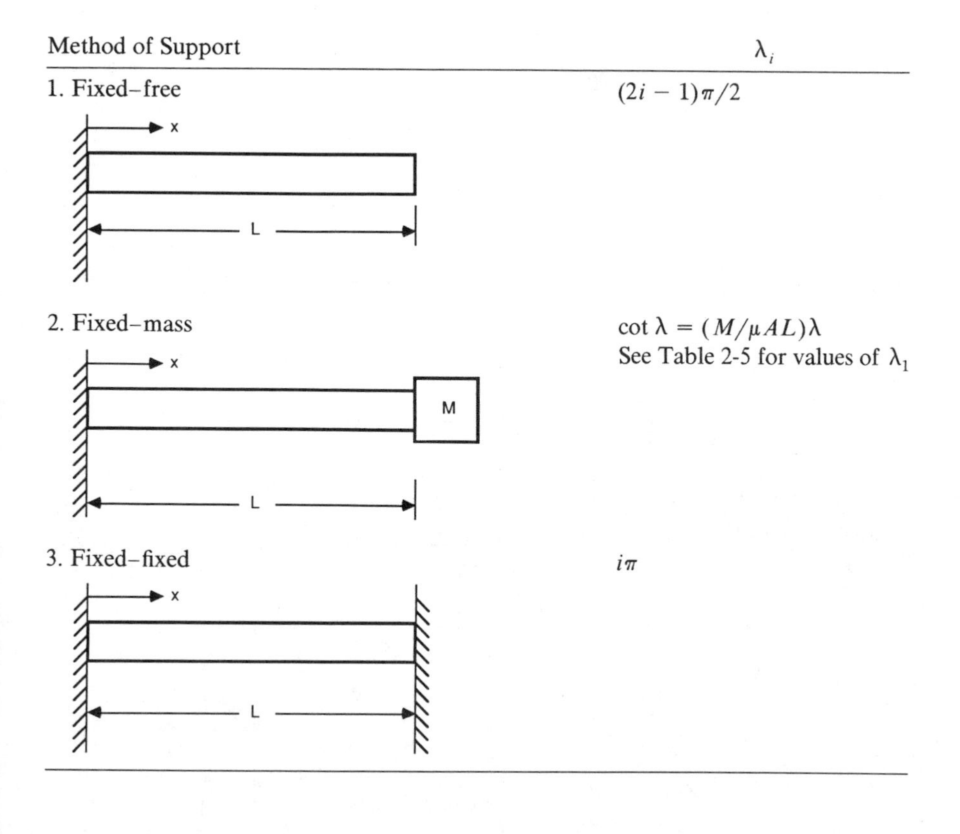

Method of Support	λ_i
1. Fixed–free	$(2i-1)\pi/2$
2. Fixed–mass	$\cot\lambda = (M/\mu AL)\lambda$ See Table 2-5 for values of λ_1
3. Fixed–fixed	$i\pi$

Eqs. (2-21) and (2-22), since it is not amenable to analytical solutions. Its effect, however, will be to reduce the fundamental natural frequency of the deep beam.

2.4.3 Slender, Tapered Beams

Table 2-10 presents values of λ for the calculation of the fundamental natural frequency of tapered beams with a variety of support conditions. The fre-

TABLE 2-10 Slender, Tapered Beams in Flexure[a]

Fundamental natural frequency:

$$F = (\lambda h_0 / 2\pi L^2)(Eg/\rho)^{1/2} \text{ Hz}$$

where λ = dimensionless constant provided in the table
h_0 = depth of beam at thick end (in.)
h_1 = depth of beam at thin end (in.)
b_0 = width of beam at thick end (in.)
b_1 = width of beam at thin end (in.)
$\alpha = h_0/h_1$
$\beta = b_0/b_1$
L = length of beam (in.)
E = modulus of elasticity of beam material (lbf/in.^{2})
ρ = weight density of beam material (lbf/in.^{2})
g = gravitational acceleration on earth (386 in./sec^2)
F = free end
S = simply supported end
C = clamped end

Elevation

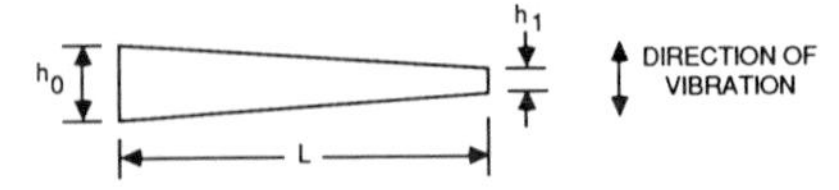

Plan

b0
b1
L
VIBRATION NORMAL TO PAPER

	End Conditions			λ				
Case Number	Thick End	Thin End		Taper Parameter Values				
—	—	—	—	$\alpha = 2$	$\alpha = 3$	$\alpha = 4$	$\alpha = 10$	
1	C	C	$\beta = 1$	4.713	4.041	3.670	2.853	
2	C	S	$\beta = 1$	3.551	3.182	2.972	2.502	
3	C	F	$\beta = 1$	1.104	1.162	1.206	1.337	
4	S	C	$\beta = 1$	2.937	2.361	2.046	1.382	
5	S	S	$\beta = 1$	2.057	1.732	1.544	1.122	
6	S	F	$\beta = 1$	3.608	3.310	3.183	2.993	
7	F	C	$\beta = 1$	0.473	0.304	0.224	0.0818	
8	F	S	$\beta = 1$	2.973	2.425	2.127	1.487	
—	—	—	—	$\beta = 1$	$\beta = 2$	$\beta = 3$	$\beta = 4$	$\beta = 5$
9	C	S	$\alpha = 1$	4.451	4.632	4.705	4.743	4.765
10	C	F	$\alpha = 1$	1.015	1.246	1.387	1.485	1.558
11	C	F	$\alpha = 2$	1.104	1.335	1.477	1.576	1.650
12	C	F	$\alpha = 3$	1.162	1.395	1.537	1.637	1.710
13	C	F	$\alpha = 4$	1.205	1.439	1.582	1.681	1.754
14	C	F	$\alpha = 5$	1.239	1.473	1.616	1.715	1.788

[a]Adapted from Refs. 18, 19, and 20.

quency formula is

$$F = \left(\lambda h_0/2\pi L^2\right)\left(Eg/\rho\right)^{1/2} \text{ Hz} \tag{2-23}$$

where the symbols are defined in Table 2-10. Cases 1–8 are for a beam of constant width that is linearly tapered in depth in the direction of vibration. Cases 9–14 are for a beam linearly tapered in both width and depth. The data in Table 2.10 were adopted from Refs. 18, 19, and 20. Reference 21 presents frequency data for beams that are nonlinearly tapered in both width and depth. References 22–24 present additional information on tapered beams.

2.5 RINGS AND ARCS

The fundamental natural frequencies of slender, complete rings, with $2i$ nodes may be calculated from Cases 1–3 of Table 2.11.[25,26] See Fig. 2.14. The fundamental natural frequencies of slender, circular arcs with end supports only may be calculated from Cases 4–6 of Table 2-11.[27–29] See Fig. 2.15. The fundamental natural frequency of in-plane flexure for a thick, circular ring with $2i$ equally spaced point supports that allow rotation but not translation, taking into account the effects of shear deformation and rotary inertia, is given by[30]

$$F = \left\{\left(1 + 4i^2\beta\right) + \left(I/AR^2\right)\left[\left(4i^2 - 1\right)^2/\left(4i^2 + 1\right)\right]\left[1/\left(1 + 4i^2\beta\right)\right]\right\}^{-1/2} \cdot F(\text{Case 1, Table 2.11}) \text{ Hz} \tag{2-24}$$

where $2i$ = number of equally spaced point supports, $i = 1, 2, 3, \ldots$
$\beta = I_y E/GKAR^2$
A = area of the thick ring cross section (in.^{2})
K = shear coefficient of the thick ring cross section, Table 2-8

I_y, E, G, and R are defined in Table 2-10, and the frequency at the right-hand end of Eq. (2-24) is found from Case 1 in Table 2-11. Out-of-plane vibrations of thick rings are treated in Refs. 31 and 32.

The fundamental natural frequency of in-plane flexure for a thick, circular arc with clamped-end supports only, taking into account the effects of shear deformation and rotary inertia, is given by[28,33,34]

$$F = \frac{s/\pi r}{\left[\left(s/\pi r\right)^2 + \gamma^2\Omega + \gamma^2/\left(1 + \Omega\pi^2 r^2/s^2\right)\right]^{1/2}} \cdot F\ (\text{Case 4, Table 2.11}) \text{ Hz} \tag{2-25}$$

TABLE 2-11 Slender Rings and Arcs

$2i$ = number of nodes
λ, σ = dimensionless constants provided in the table
E = modulus of elasticity (lbf/in.^{2})
G = shear modulus (lbf/in.^{2})
C = torsion constant (in.^{4}); Table 2-2
I = area moment of inertia about indicated axis (in.^{4}); Table 2-2
R = radius of ring or arc to centroidal axis (in.)
m = mass per unit length of ring or arc (lbm/in.)
ν = Poisson's ratio
α_0 = angle subtended by arc (rad)
a = semimajor axis of ellipse
b = semiminor axis of ellipse

Description	Fundamental Natural Frequency (Hz)	Values of Constants
1. Complete circular ring, in-plane flexure; see Fig. 2.14	$\frac{i(i^2-1)}{2\pi R^2(i^2+1)^{1/2}}\left(\frac{EI_y}{m}\right)^{1/2}$ $i = 1, 2, 3, \ldots$	
2. Complete circular ring, out-of-plane flexure	$\frac{i(i^2-1)}{2\pi R^2}\left[\frac{EI}{m(i^2 + EI_z/GC)}\right]^{1/2}$ $i = 2, 3, 4, \ldots$	
3. Complete elliptical ring, in-plane flexure two point supports only	$\frac{\lambda}{2\pi b^2}\left(\frac{EI_y}{m}\right)^{1/2}$ Point supports at interceptions of ellipse and one axis	see below

(a/b)	λ	(a/b)	λ
1.0	2.683	1.7	1.362
1.1	2.427	2.0	1.057
1.2	2.193	2.5	0.7273
1.4	1.801	3.0	0.5287

4. Circular, arc, clamped–clamped ends, in-plane flexure; see Fig. 2.15*a*	$\frac{\lambda}{2\pi(R\alpha_0)^2}\left[\frac{1 - 2\sigma^2(1 - 2/\sigma\lambda)(\alpha_0/\lambda)^2 + (\alpha_0/\lambda)^4}{1 + 5\sigma^2(1 - 2/\sigma\lambda)\cdot(\alpha_0/\lambda)^2}\right]^{1/2} \times \left(\frac{EI_y}{m}\right)^{1/2}$	$\lambda = 61.673$ $\sigma = 1.00078$
5. Circular arc, clamped–clamped ends, out-of-plane flexure; see Fig. 2.15*b*	$\frac{\pi}{2(R\alpha_0)^2}\left[\frac{3.586(\alpha_0/\pi)^2 + 1.246GC\beta/EI_z}{(\alpha_0/\pi)^2 + 1.246GC/EI_z}\right]^{1/2} \times \left(\frac{EI_z}{m}\right)^{1/2}$ $0 < EI_z/GC < 2$	$\beta = (\alpha_0/\pi)^4 - 2.492(\alpha_0/\pi)^2 + 5.139$
6. Circular arc, clamped–pinned ends, out-of-plane flexure; see Fig. 2.15*c*	$\frac{\pi}{2(R\alpha_0)^2}\left[\frac{1.080(\alpha_0/\pi)^2 + 1.166GC\beta/EI_z}{(\alpha_0/\pi)^2 + 1.166GC/IE_z}\right]^{1/2} \times \left(\frac{EI_z}{m}\right)^{1/2}$ $0 < EI_z/GC < 2$	$\beta = (\alpha_0/\pi)^4 - 2.332(\alpha_0/\pi)^2 + 2.440$

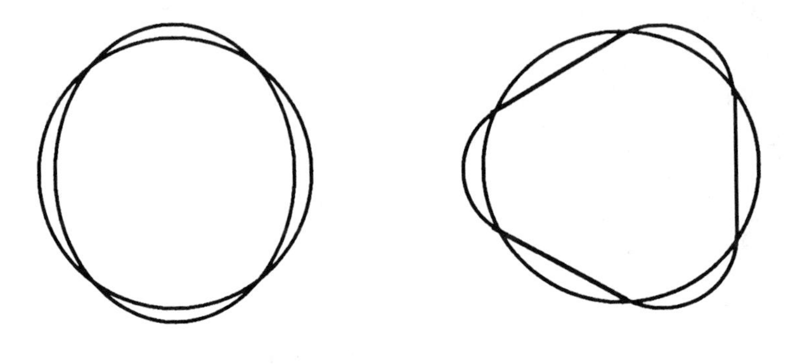

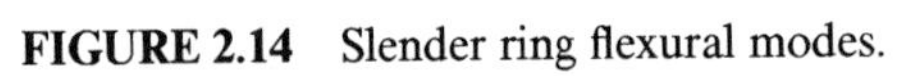

FIGURE 2.14 Slender ring flexural modes.

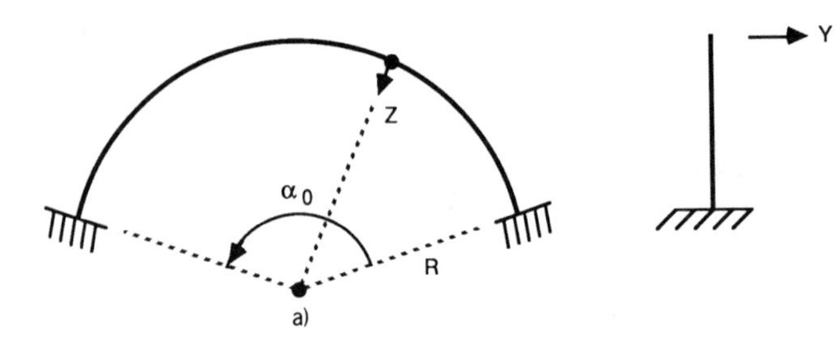

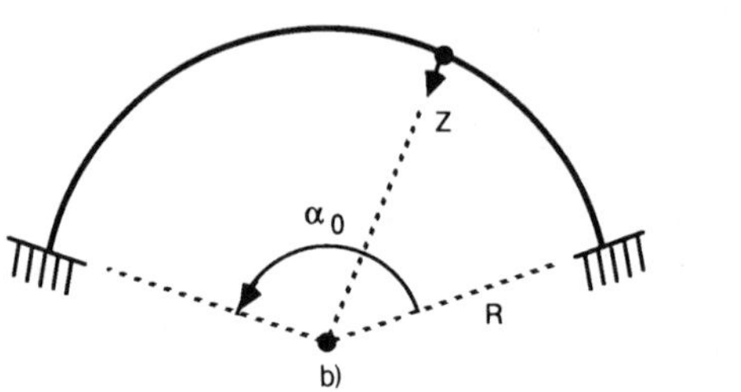

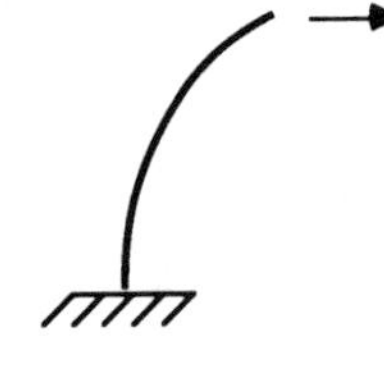

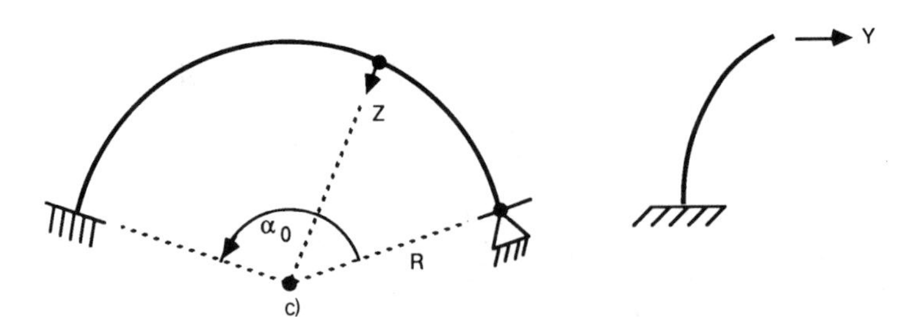

FIGURE 2.15 Slender arc coordinates and flexural modes.

where $\Omega = E/KG$ and

$$\gamma = \frac{1.5622}{1 - 0.025330\alpha_0^2} \cdot \left[\frac{(1 + 0.075991\alpha_0^2)(1 - 0.024215\alpha_0^2 + 0.00026291\alpha_0^4)}{1 + 0.060536\alpha_0^2} \right]^{1/2} \tag{2-26}$$

K = shear coefficient of the thick arc cross section, Table 2-8
S = arc length of centroidal axis = $R\alpha_0$ (in.)
r = radius of gyration of the arc cross section = $(I_y/A)^{1/2}$ (in.)
A = area of the arc cross section (in.^{2})

I_y, E, G, R, and α_0 are defined in Table 2-11, and the frequency at the right-hand end of Eq. (2-25) is found from Case 4 of Table 2.11. Note that

$$\gamma = 1.5950 \text{ for } \alpha_0 = 1 \text{ rad} = 57.3°$$
$$\gamma = 1.6965 \text{ for } \alpha_0 = 2 \text{ rad} = 114.6°$$
$$\gamma = 1.9335 \text{ for } \alpha_0 = \pi \text{ rad} = 180.0°$$

2.6 FLAT PLATES

From the large body of work on the natural vibration of flat plates have been selected those cases most often encountered in mechanical design work. Only the fundamental natural frequency of flexural vibrations is addressed, since this is the most significant mode involved in evaluating maximum stresses and fatigue life. The plates are assumed to be homogeneous and isotropic with their thickness less than one-tenth the smallest lateral dimension, with no in-plane loads. With the exception of the tapered plates in Section 2.6.1, the plates have a constant thickness.

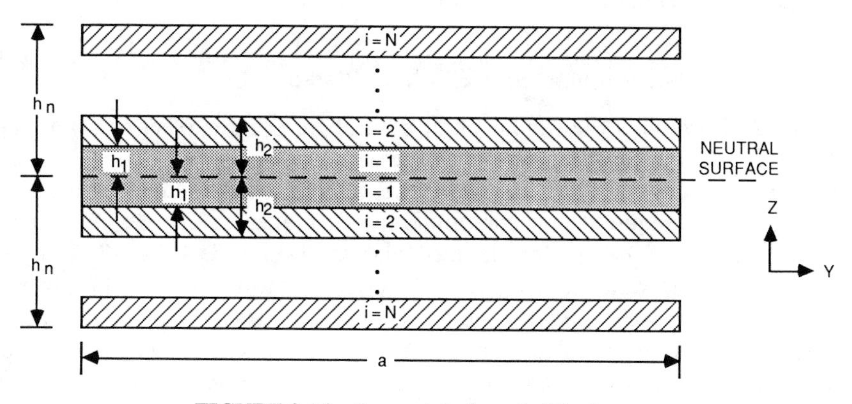

FIGURE 2.16 Symmetrical sandwich plate.

The fundamental natural frequencies of slender sandwich plates (Fig. 2.16) can be estimated, using the formulas developed in the following sections, by defining the sandwich plate equivalent stiffness and weight per unit area as[35]

$$Eh^3/12 = \tfrac{2}{3} \sum_{i=1}^{n} E_i\left(h_i^3 - h_{i-1}^3\right) \text{ lbf} \cdot \text{in.}$$
$$\rho = 2 \sum_{i=1}^{n} \rho_i\left(h_i - h_{i-1}\right) \text{ lbf/in.}^2 \tag{2-27}$$

where E_i = the modulus of elasticity of sandwich material i (lbf/in.^{2})
h_{i-1} = the distance from the neutral surface to the inner surface of layer i (in.), with $h_0 = 0$
h_i = the distance from the neutral surface to the outer surface of layer i (in.)
ρ_i = the weight density of sandwich material i (lbf/in.^{3})

2.6.1 Rectangular Plates

Table 2-12 presents the fundamental natural frequencies of rectangular plates with a variety of support conditions,[36-43] including some cases with tapered thicknesses[44-46] and some with concentric openings.[47,48] The frequency formula is the same for all, with the table providing the values of the parameter λ required for computation of frequency. The values of λ vary with the plate aspect ratio, taper, and relative size of opening. Where the value of λ is also dependent on the value of ν (Poisson's ratio), the value of ν used in the calculation is provided in the table. For the first 20 cases listed in Table 2-12, the value of λ is a function of ν when there are one or more free edges, and in these cases ν only varies by a few percent over the range of interest for metals, $0.25 \le \nu \le 0.35$. However, the materials used in printed wire boards (PWBs) generally have lower values of ν, for example, $\nu = 0.12$ for many fiberglass-reinforced epoxy boards. Several cases are presented in Table 2-12 (Cases 37–45) and subsequent tables with values of $\nu = 0.12$ specifically for use with PWBs. In general, λ decreases with increasing ν.

2.6.2 Circular, Annular, and Elliptical Plates

Table 2-13 presents the fundamental natural frequencies of circular, annular, and elliptical plates with a variety of support conditions.[37,49-56] The frequency formula is the same for all, with the table providing the values of the parameter λ required for computation of frequency. λ is a function of Poisson's ratio ν, except for clamped plates, Cases 2, 5, 8, 24–27, and 29 of Table 2-13. The value of λ varies by less than 1% to as much as 17% over the range of ν from 0.0 to 0.5 for the fundamental modes of these plates, and the value of ν used in the calculation of λ is provided in each case. See Case 28 of Table 2-13 for an example of the dependence of λ on ν. The point-supported

TABLE 2-12 Rectangular Plates

$$F = (\lambda/2\pi a^2)[Eh^3g/12\rho(1-\nu^2)]^{1/2} \text{ Hz}$$

where λ = dimensionless constant provided in the table
a = plate length (in.)
b = plate width (in.)
h = plate thickness (in.)
α = taper coefficient for tapered plates: h = thickness at thin edge; $h(1+\alpha)$ = thickness at thick edge; dimensionless
d = side of concentric square opening, or diameter of concentric circular opening (in.)
E = plate material modulus of elasticity (lbf/in.^{2})
ν = plate material Poisson's ratio, dimensionless
ρ = plate weight per unit area (lbf/in.^{2})
g = gravitational acceleration on earth = 386 in./sec^2
F = free edge
S = simply supported edge
C = clamped edge
• = point support

Description			λ		
Edge supported. Start at left edge and read clockwise:					

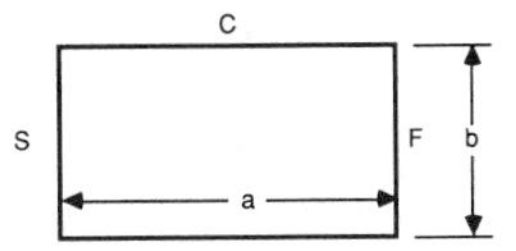

Reads: *SCFF*	$a/b = 0.4$	$a/b = 2/3$	$a/b = 1.0$	$a/b = 1.5$	$a/b = 2.5$
1. *SFFF* ($\nu = 0.3$)	2.692	4.481	6.648	9.850	14.94
2. *CFFF* ($\nu = 0.3$)	3.511	3.502	3.492	3.477	3.456
3. *SFFS* ($\nu = 0.3$)	1.320	2.234	3.369	5.026	8.251
4. *SFSF* ($\nu = 0.3$)	9.760	9.698	9.631	9.558	9.484
5. *CFSF* ($\nu = 0.3$)	15.38	15.34	15.29	15.22	15.13
6. *CFFS* ($\nu = 0.3$)	3.854	4.425	5.364	6.931	10.10
7. *CFCF* ($\nu = 0.3$)	22.35	22.31	22.27	22.21	22.13
8. *CFFC* ($\nu = 0.3$)	3.986	4.985	6.942	11.22	24.91
9. *SFSS* ($\nu = 0.3$)	10.13	10.67	11.68	13.71	18.80
10. *SFSC* ($\nu = 0.3$)	10.19	10.98	12.69	16.82	30.63
11. *CFCS* ($\nu = 0.3$)	22.54	22.86	23.46	24.78	28.56
12. *CFSS* ($\nu = 0.3$)	15.65	16.07	16.87	18.54	23.07
13. *CFSC* ($\nu = 0.3$)	15.70	16.29	17.62	21.04	33.58
14. *CFCC* ($\nu = 0.3$)	22.58	23.02	24.02	26.73	37.66
15. *SSSS* (I)[a]	11.45	14.26	19.74	32.08	71.56
16. *SSSC* (I)	11.75	15.58	23.65	42.53	103.9
17. *CSSC* (I)	16.85	19.95	27.06	44.89	105.3
18. *SCSC* (I)	12.13	17.37	28.95	56.35	145.5
19. *CSCC* (I)	23.44	25.86	31.83	48.17	107.1
20. *CCCC* (I)	23.65	27.01	35.99	60.77	147.8

TABLE 2-12 *(Continued)*

Description	λ

Square plates with concentric, free-edged openings

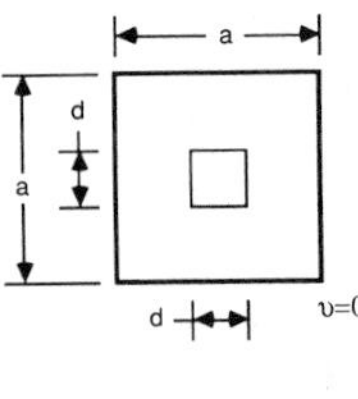

Ratio of d/a	0	1/6	1/3	1/2
21. Outside edges S	19.63	19.48	21.45	26.05
22. Outside edges C	34.85	35.80	43.25	62.40

Ratio of d/a	0	0.05	0.10	0.15	0.20	0.25	0.30
23. Outside edges S	19.9	19.75	19.5	19.4	19.3	19.35	19.5
24. Outside edges C	36.0	35.5	35.1	35.2	35.7	36.7	—

Plates linearly tapered along a direction

25.

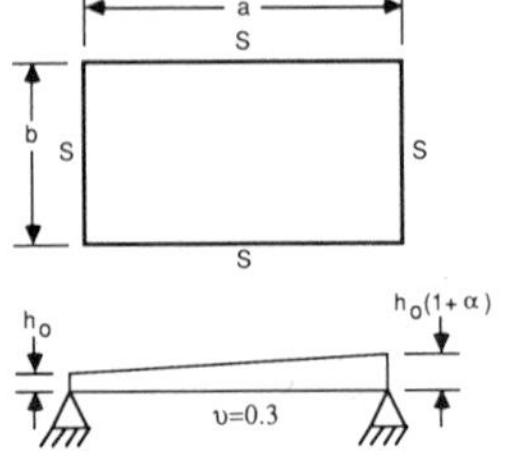

a/b	α = 0.0	0.2	0.4	0.6	0.8	1.0
0.25	—	11.51	12.48	13.42	14.34	—
0.50	—	13.55	14.73	15.87	16.99	—
1.0	19.74	21.69	23.61	25.50	27.36	29.21
1.5	32.08	35.22	38.30	41.30	44.25	—

26.

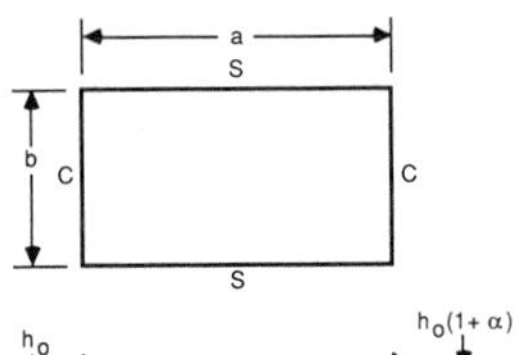

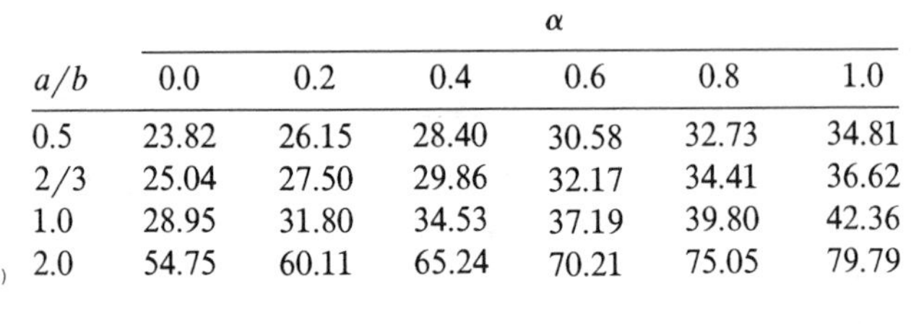

a/b	α = 0.0	0.2	0.4	0.6	0.8	1.0
0.5	23.82	26.15	28.40	30.58	32.73	34.81
2/3	25.04	27.50	29.86	32.17	34.41	36.62
1.0	28.95	31.80	34.53	37.19	39.80	42.36
2.0	54.75	60.11	65.24	70.21	75.05	79.79

27.

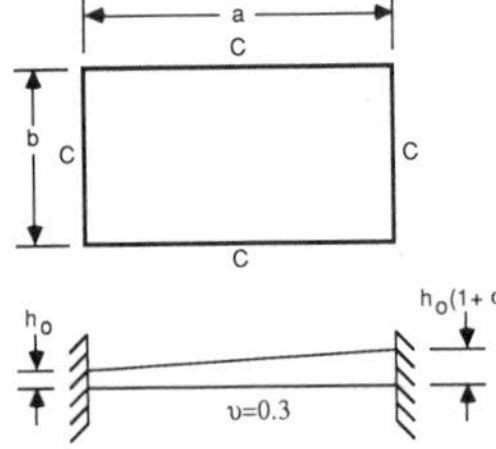

a/b	α = 0.0	0.2	0.4	0.6	0.8	1.0
0.5	24.59	27.00	29.32	31.58	33.90	35.96
2/3	27.02	29.67	32.22	34.71	37.13	39.52
1.0	36.00	39.52	42.93	46.24	49.47	52.64
2.0	98.33	107.8	116.6	124.9	132.9	140.5

TABLE 2-12 *(Continued)*

Description	λ
Point supported 28.	Ratio of a/b: 1.0, 1.5, 2.0, 2.5 λ: 7.12, 8.92, 9.29, 9.39 $\nu = 0.3$
29.	Ratio of b/a: 0.0, 0.1, 0.2, 0.3, 0.4, 0.5 λ: 7.12, 12.89, 19.69, 19.31, 13.35, 11.34 $\nu = 0.3$ $b/a = 0.5$ is a single point support at the plate center
30. N equally spaced point supports on each edge	N: 2, 3, 5, 7, 9, ∞ λ: 7.12, 18.20, 19.64, 19.71, 19.73, 19.74 $N = \infty$ is simply supported edges
31.	$\lambda = 13.5$ $\nu = 0.3$
32. N-bay plate linear array	N: 1, 2, 3, 4, 5 λ: 7.12, 16.27, 24.41, 33.02, 41.41 $\lambda \cong 8.12N,\ N > 5$ $\nu = 0.3$

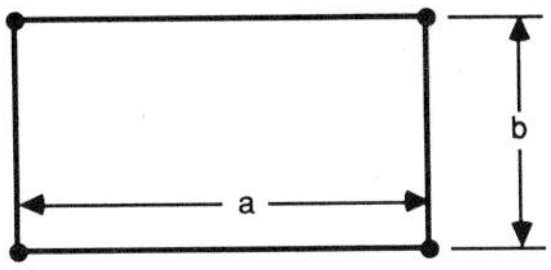

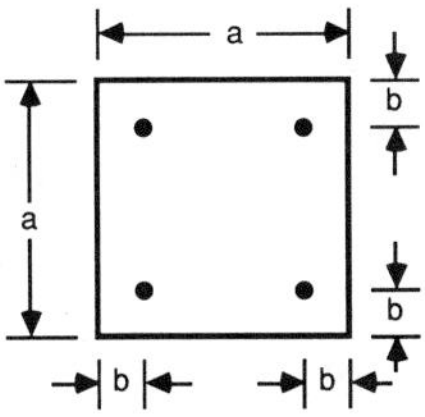

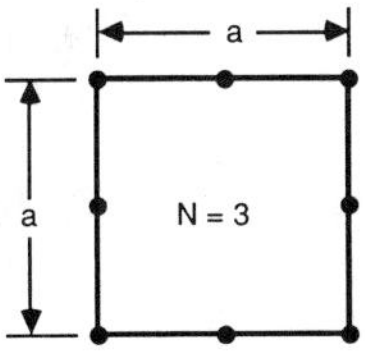

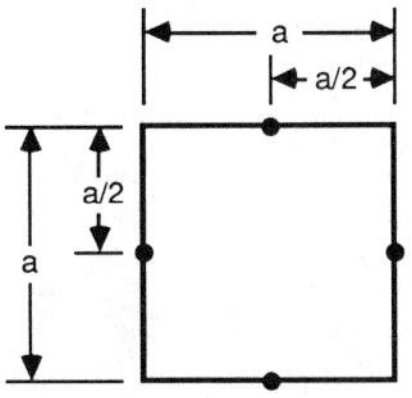

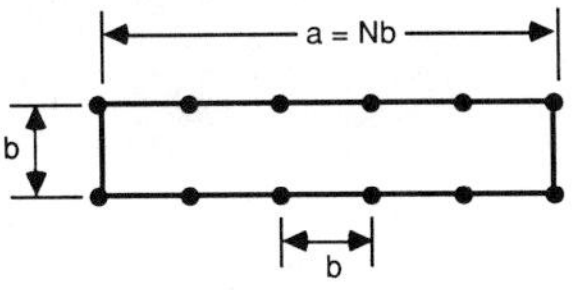

TABLE 2-12 *(Continued)*

Description	λ			
33. N^2-bay plate, square array	N	1	2	3
	λ	7.12	35.10	79.89
		$\lambda \cong N(8.12N + 1),\ N > 3$		
		$\nu = 0.3$		
34.	Ratio of a/b	1.0	1.5	2.0
	λ	52.6	73.1	91.1
35.				$\lambda = 9.00$ $\nu = 0.3$
36.				$\lambda = 13.7$ $\nu = 0.3$
37.				$\lambda = 9.22$ $\nu = 0.12$

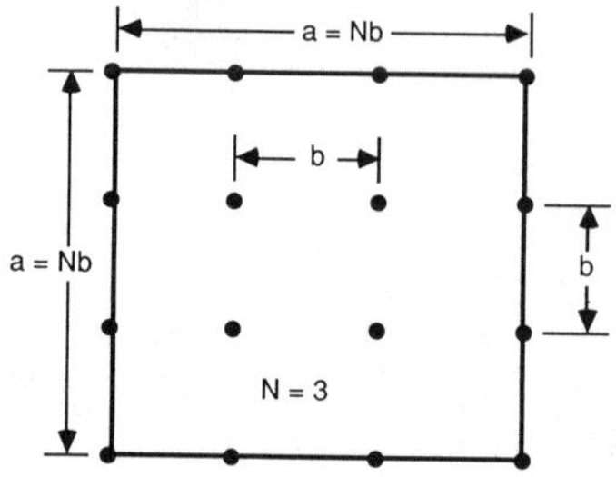

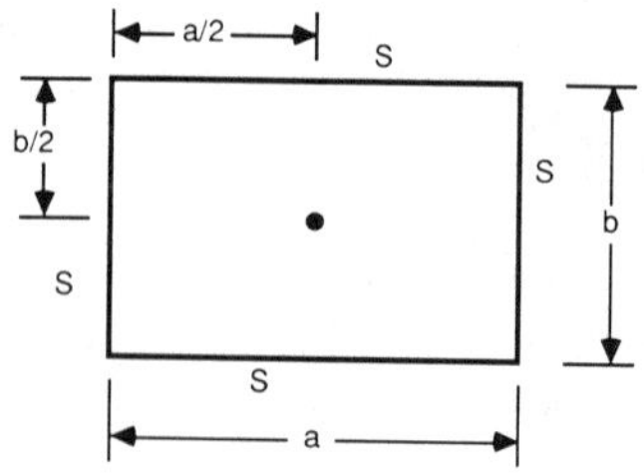

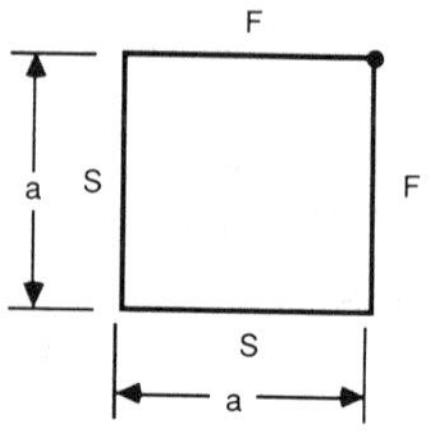

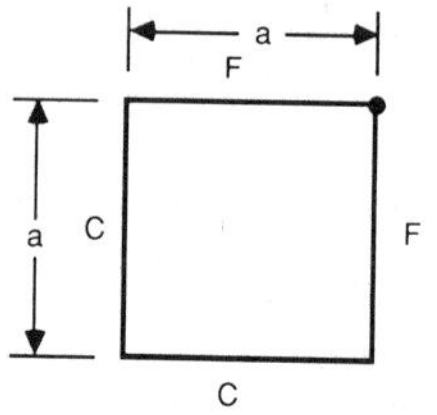

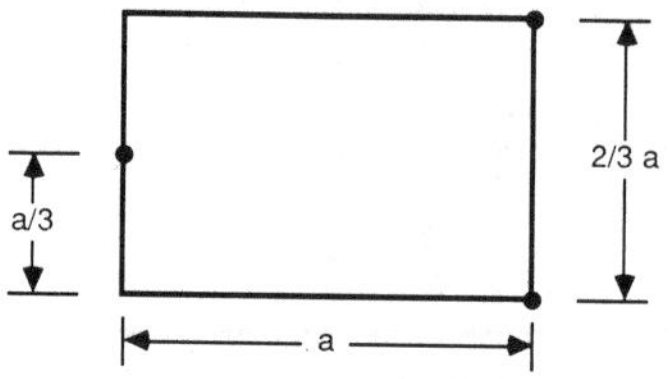

TABLE 2-12 *(Continued)*

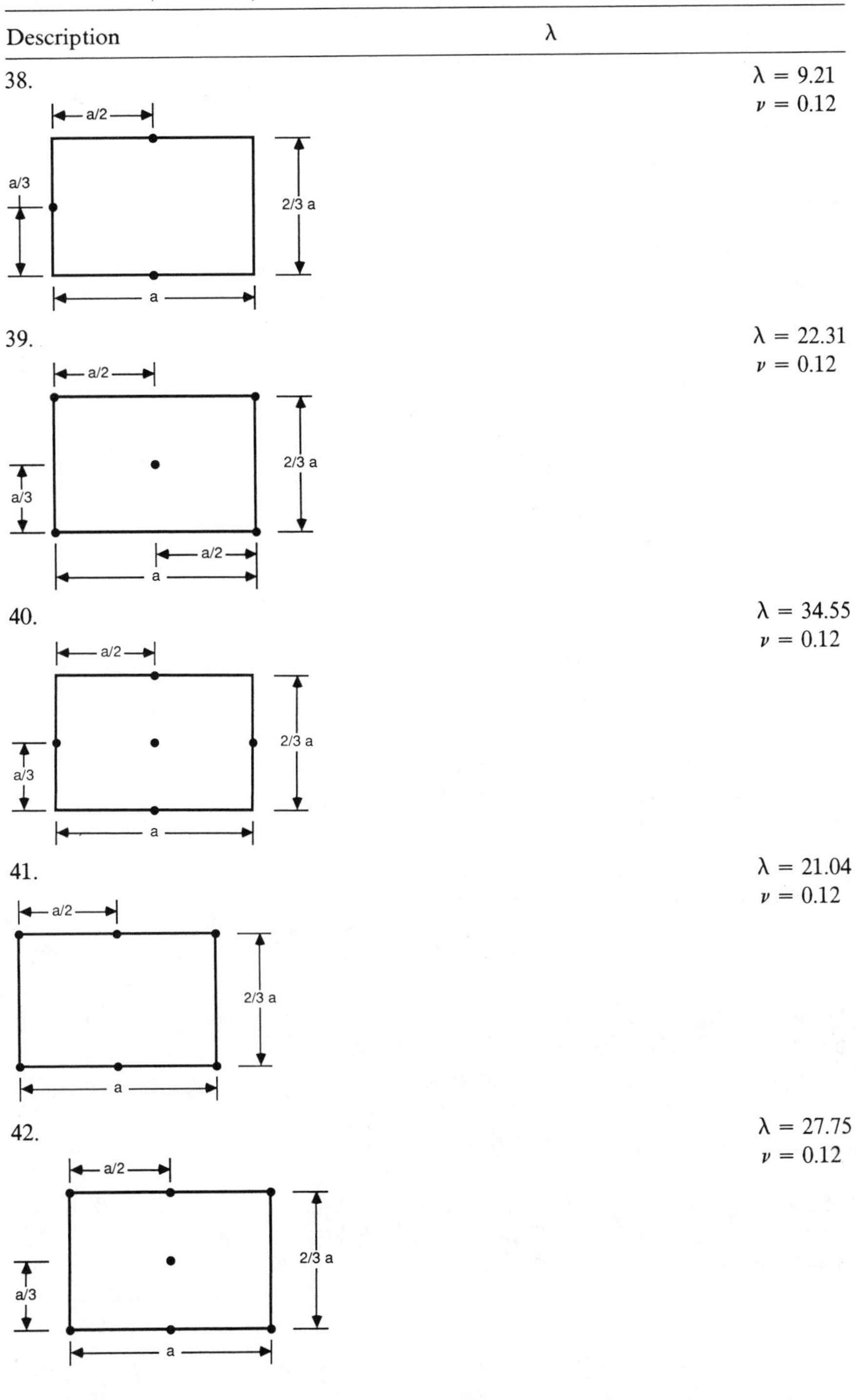

Description	λ
38.	$\lambda = 9.21$ $\nu = 0.12$
39.	$\lambda = 22.31$ $\nu = 0.12$
40.	$\lambda = 34.55$ $\nu = 0.12$
41.	$\lambda = 21.04$ $\nu = 0.12$
42.	$\lambda = 27.75$ $\nu = 0.12$

TABLE 2-12 *(Continued)*

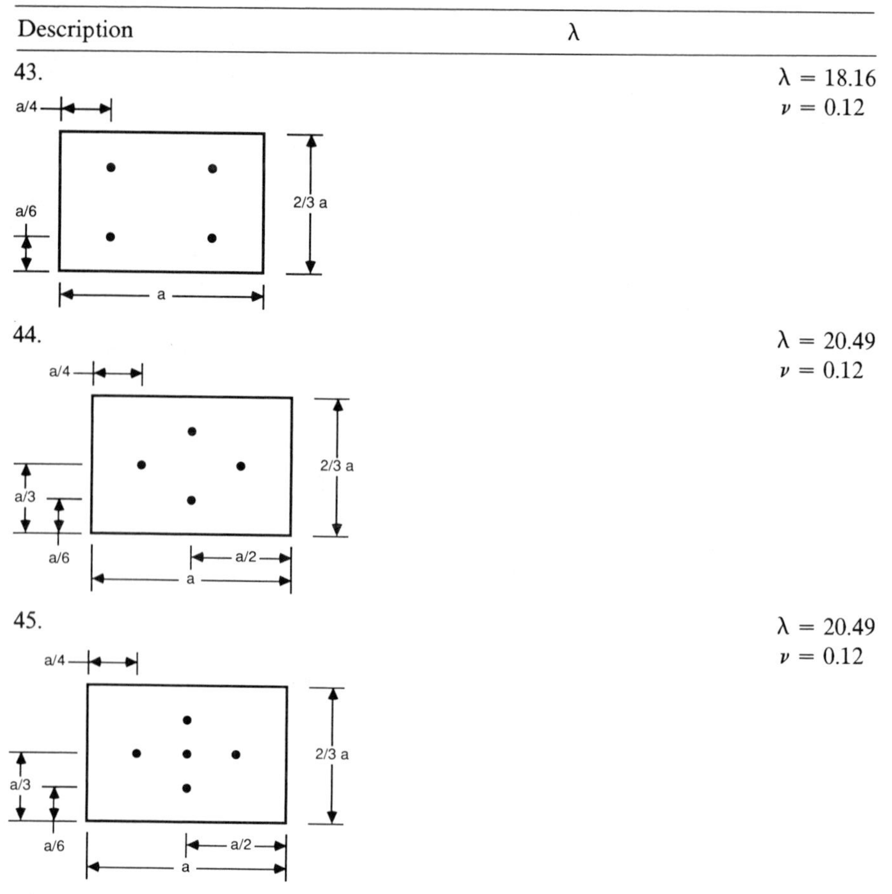

Description	λ
43.	$\lambda = 18.16$ $\nu = 0.12$
44.	$\lambda = 20.49$ $\nu = 0.12$
45.	$\lambda = 20.49$ $\nu = 0.12$

[a](I) indicates that the value of λ is independent of the value of ν.

plates in Table 2-13 (Cases 9–19) are for $\nu = 0.12$, corresponding to a typical fiberglass-reinforced epoxy printed wire board.

2.6.3 Other Plate Shapes

Table 2-14 presents the fundamental natural frequencies of plates with triangular,[37,57–60] trapezoidal,[58,59] many sided,[37,61,62] and arbitrarily shaped[63] edges. The frequency formula is the same for all, with the table providing the values of the parameter λ required for computation of frequency. λ is independent of ν for Cases 1–6, 18–21, and 23. For Cases 7–17 and 24–34, $\nu = 0.12$, the value used for fiberglass-reinforced epoxy printed wire boards.

TABLE 2-13 Circular, Annular, and Elliptical Plates

Fundamental natural frequency only:

$$F = (\lambda/2\pi a^2)[Eh^3 g/12\rho(1 - \nu^2)]^{1/2} \text{ Hz}$$

where λ = dimensionless constant provided in the table
a = plate radius, or semiminor axis (in.)
b = radius of inside opening, or semimajor axis (in.)
h = plate thickness (in.)
E = plate material modulus of elasticity (lbf/in.^{2})
ν = plate material Poisson's ratio, dimensionless
ρ = plate weight per unit area (lbf/in.^{2})
g = gravitational acceleration on earth = 386 in./sec^2
M = weight (lbf)
F = free edge
S = simply supported edge
C = clamped edge or clamped center
• = point support

Description	λ
1. Circle, S edge 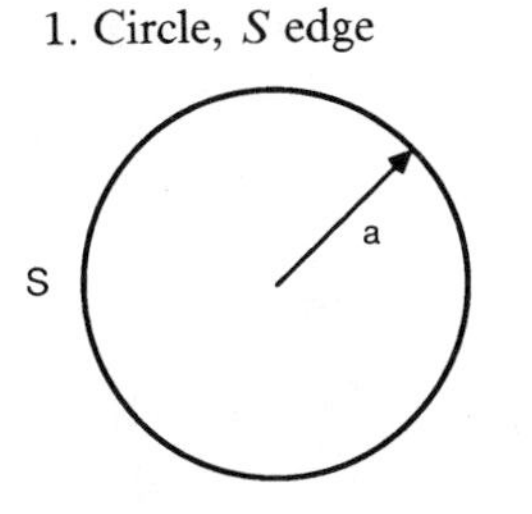	$\lambda = 4.977$ for $\nu = 0.3$ $\lambda = 4.797$ for $\nu = 0.12$
2. Circle, C edge 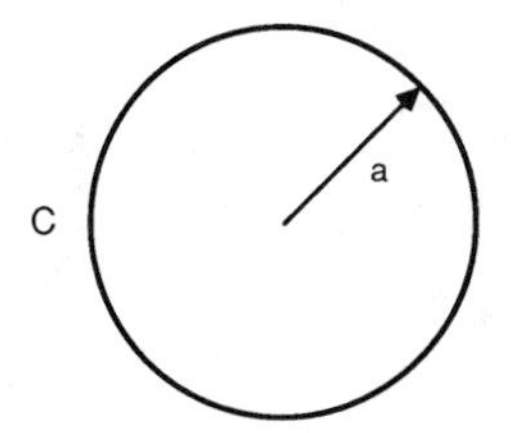	$\lambda = 10.22$ Independent of ν
3. Circle, F edge, C center 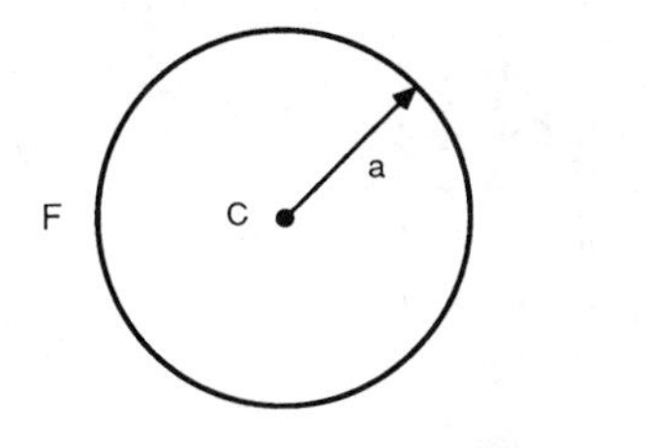	$\lambda = 3.752$ $\nu = 1/3$

TABLE 2-13 *(Continued)*

Description	λ
4. Circle, *S* edge, *C* center	$\lambda = 14.8$ $\nu = 0.3$
5. Circle, *C* edge, *C* center	$\lambda = 22.7$ Independent of ν
6. Circle, *F* edge, *S* at radius b	See below; $\nu = 1/3$
7. Circle, *S* edge, weight M at center	$\lambda \cong 4(\rho\pi a^2/M)^{1/2}/(3 + 4\nu + \nu^2)^{1/2}$ for $M > \rho\pi a^2$
8. Circle, *C* edge weight M at center	See below; $\lambda \cong 4(\rho\pi a^2/M)^{1/2}$ for $M > \rho\pi a^2$ Independent of ν

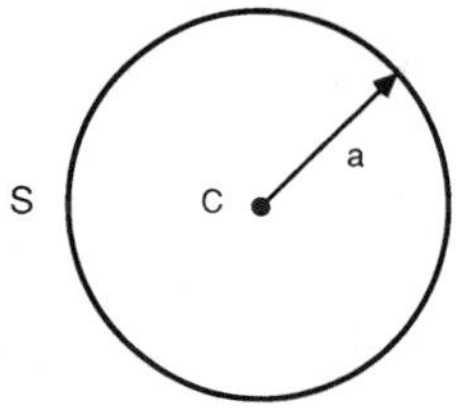

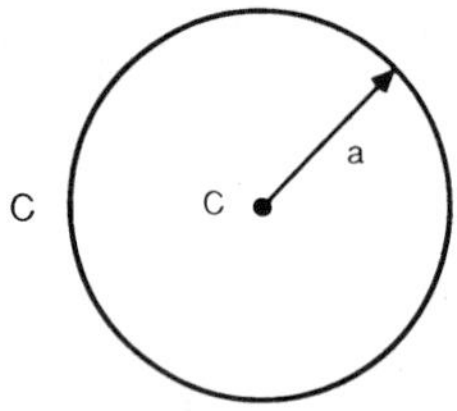

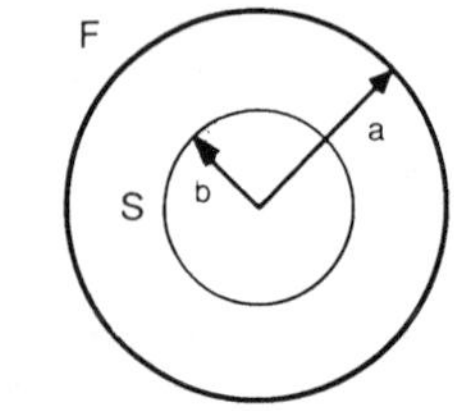

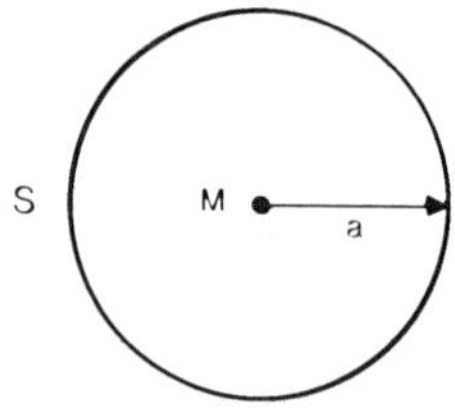

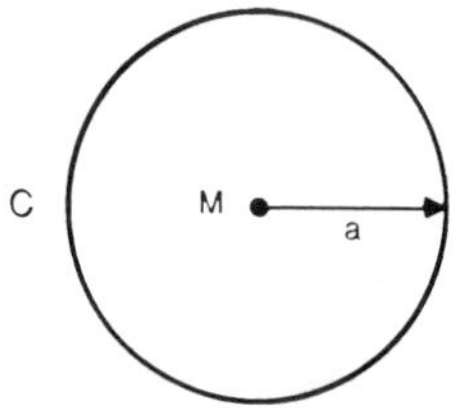

Case 6:

Ratio of b/a	0	0.2	0.4	0.6	0.8	1.0
λ	3.75	4.5	6.7	8.8	7.5	5.0

$\nu = 1/3$

Case 8:

$M/\rho\pi a^2$	0.0	0.05	0.1	0.2	0.4	0.6	1.0	1.4
λ	10.2	9.0	8.1	6.9	5.4	4.75	3.8	3.3

$\lambda \cong 4(\rho\pi a^2/M)^{1/2}$ for $M > \rho\pi a^2$
Independent of ν

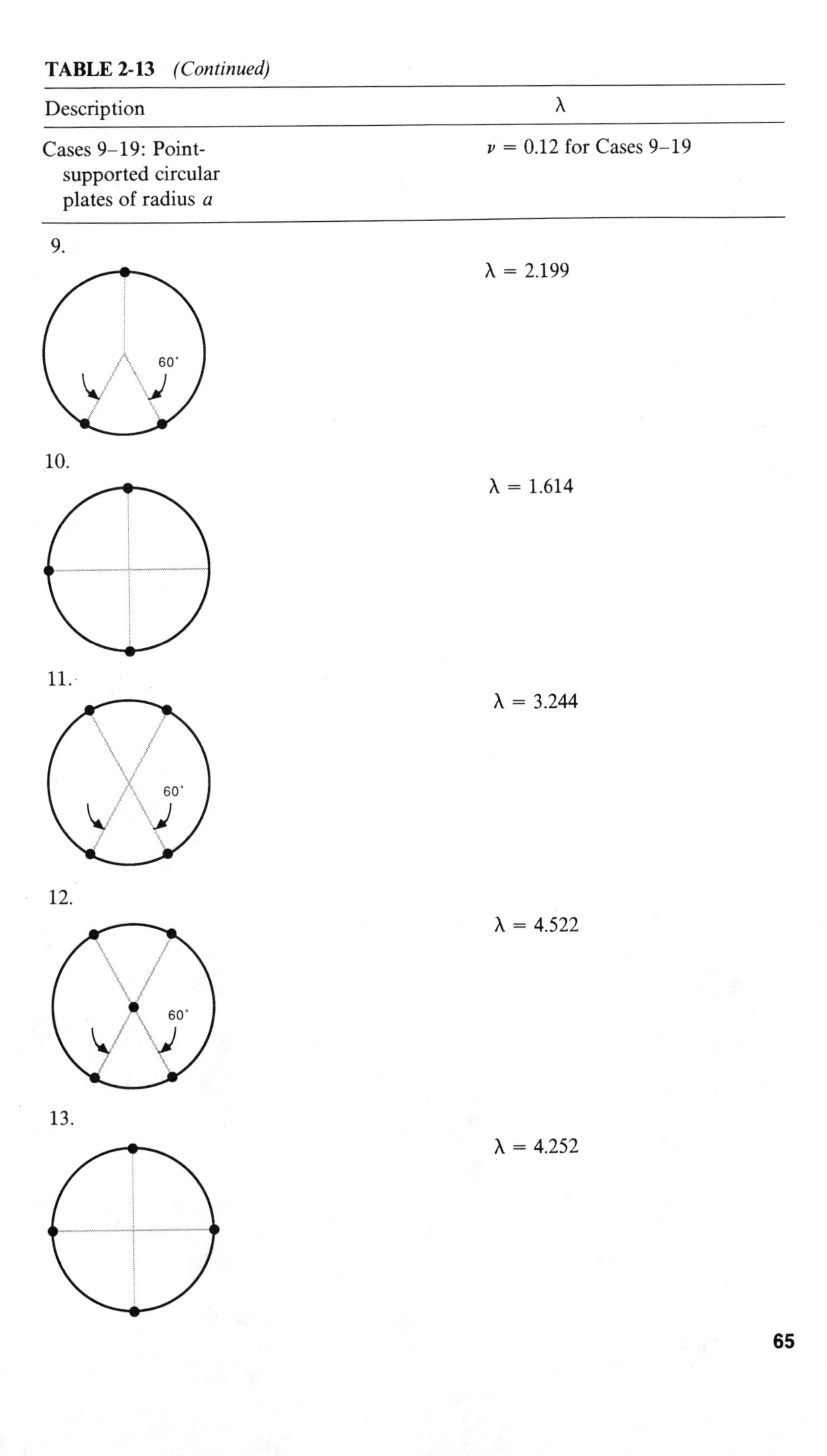

TABLE 2-13 *(Continued)*

Description	λ
Cases 9–19: Point-supported circular plates of radius a	$\nu = 0.12$ for Cases 9–19
9.	$\lambda = 2.199$
10.	$\lambda = 1.614$
11.	$\lambda = 3.244$
12.	$\lambda = 4.522$
13.	$\lambda = 4.252$

TABLE 2-13 *(Continued)*

Description	λ
14. 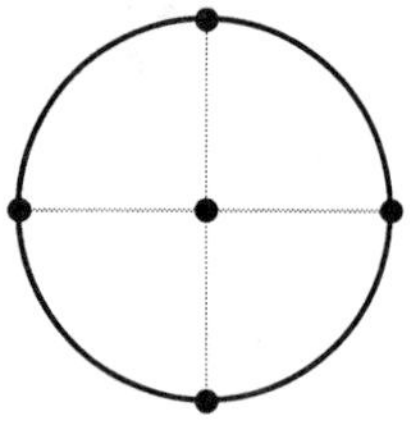	$\lambda = 5.312$
15. 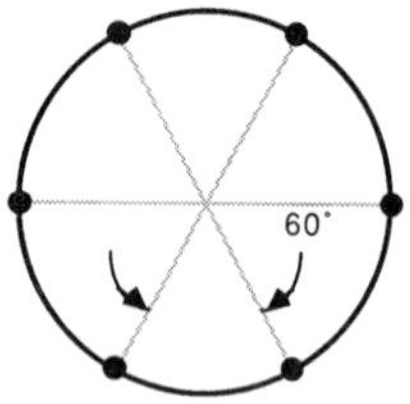	$\lambda = 4.682$
16. 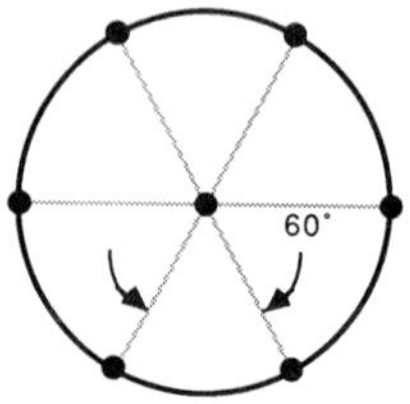	$\lambda = 11.96$
17. 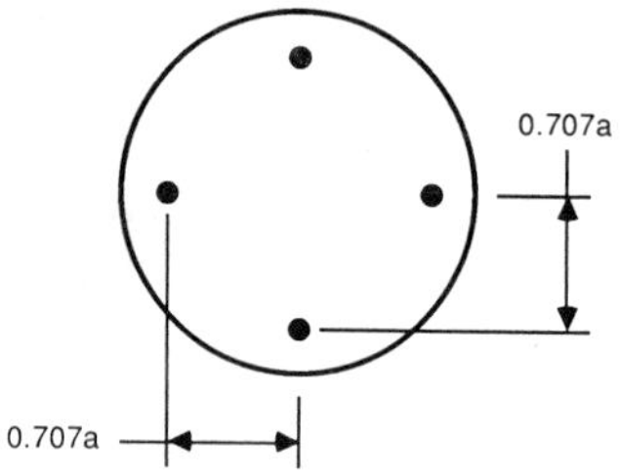	$\lambda = 5.700$
18. 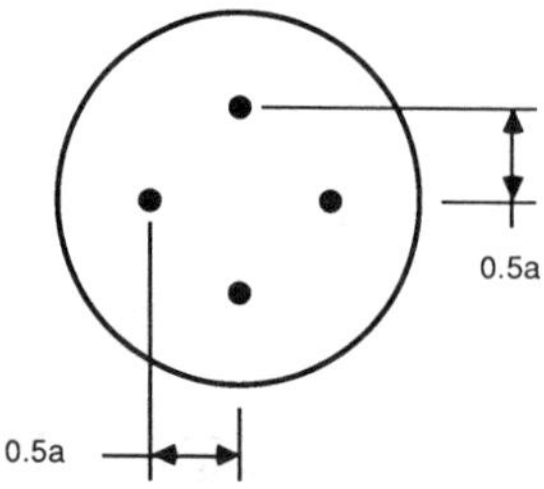	$\lambda = 5.312$

TABLE 2-13 *(Continued)*

Description	λ
19.	$\lambda = 5.312$

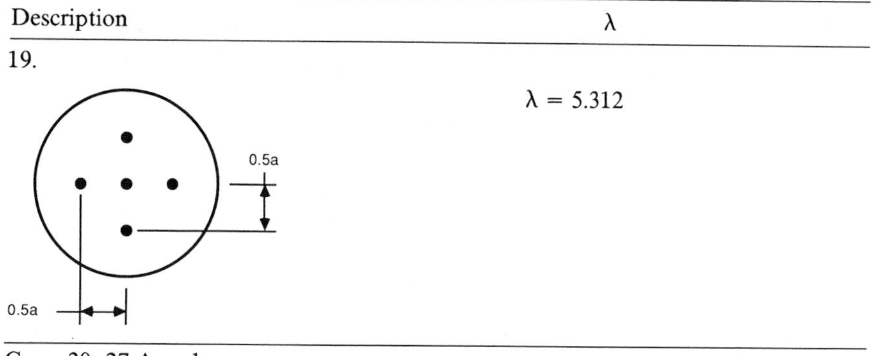

Cases 20–27 Annular

Case No.	ν	Inside Edge	Outside Edge	Ratio of b/a 0.1	0.3	0.5	0.7
20	0.3	*S*	*F*	3.45	3.42	4.11	6.18
21	0.3	*F*	*S*	4.86	4.66	5.07	6.94
22	0.3	*C*	*F*	4.23	6.66	13.0	37.0
23	0.3	*F*	*C*	10.2	11.4	17.7	43.1
24	I[a]	*S*	*S*	14.5	21.1	40.0	110
25	I	*C*	*S*	17.8	29.9	59.8	168
26	I	*S*	*C*	22.6	33.7	63.9	175
27	I	*C*	*C*	27.3	45.2	89.2	248

Cases 28 and 29 Elliptical

b/a	28. *S* on edge $\nu = 0$	$\nu = 0.25$	$\nu = 0.50$	29. *C* on edge Independent of ν
1.0	4.447	4.865	5.219	10.22
1.1	4.078	4.454	4.772	9.350
1.2	3.823	4.157	4.442	8.726
1.4	3.512	3.773	3.990	—
1.5	—	—	—	7.567
1.7	3.286	3.463	3.617	—
2.0	3.172	3.292	3.399	6.937
2.5	3.061	3.128	3.189	—
3.0	2.987	3.027	3.066	6.521
5.0	2.833	2.846	2.858	6.354
10.0	2.747	2.750	2.754	—
20.0	2.724	2.725	2.726	—

[a](I) indicates that the value of λ is independent of the value of ν.

TABLE 2-14 Other Plate Shapes

$$F = (\lambda/2\pi a^2)[Eh^3 g/12\rho(1-\nu^2)]^{1/2}\ \text{Hz}$$

where λ = dimensionless constant provided in the table
a, b = lengths (in.)
h = plate thickness (in.)
E = plate material modulus of elasticity (lbf/in.^{2})
ν = plate material Poisson's ratio
ρ = plate weight per unit area (lbf/in.^{2})
g = gravitational acceleration on earth = 386 in./sec^2
F = free edge
S = simply supported edge
C = clamped edge
• = point support

Description	λ						
Cases 1–4 Isosceles triangle							
			1. Edges a are S; Edge b is S				
	a/b	0.5	2/3	1.0	1.5		Independent of ν
	λ	24.69	30.98	45.85	73.66		
			2. Edges a are S; edge b is C				
	a/b	0.25	0.5	1.0	1.5	2.0	Independent of ν
	λ	31.0	34.5	55.0	87.0	120.0	
			3. Edges a are C; edge b is S				
	a/b	0.1340	0.2887	0.5	0.866	1.866	Independent of ν
	λ	20.36	26.30	36.80	61.20	166.8	
			4. Edges a are C; edge b is C				
	a/b	0.5	0.866	1.866			Independent of ν
	λ	46.8	74.4	186.0			

5. Right triangle

a/b	1.0	1.732
λ	49.35	92.11

Independent of ν

6. Asymmetric triangle

	β (deg of arc)			
a/b	10	20	30	45
0.5	24.78	25.06	25.64	27.78
1.0	46.28	47.71	50.57	60.22
1.5	74.64	77.85	84.21	105.1

Independent of ν

Cases 7–17 Point-supported equilateral triangles of side a

$\nu = 0.12$ for Cases 7–17

7.

$\lambda = 6.49$

TABLE 2-14 *(Continued)*

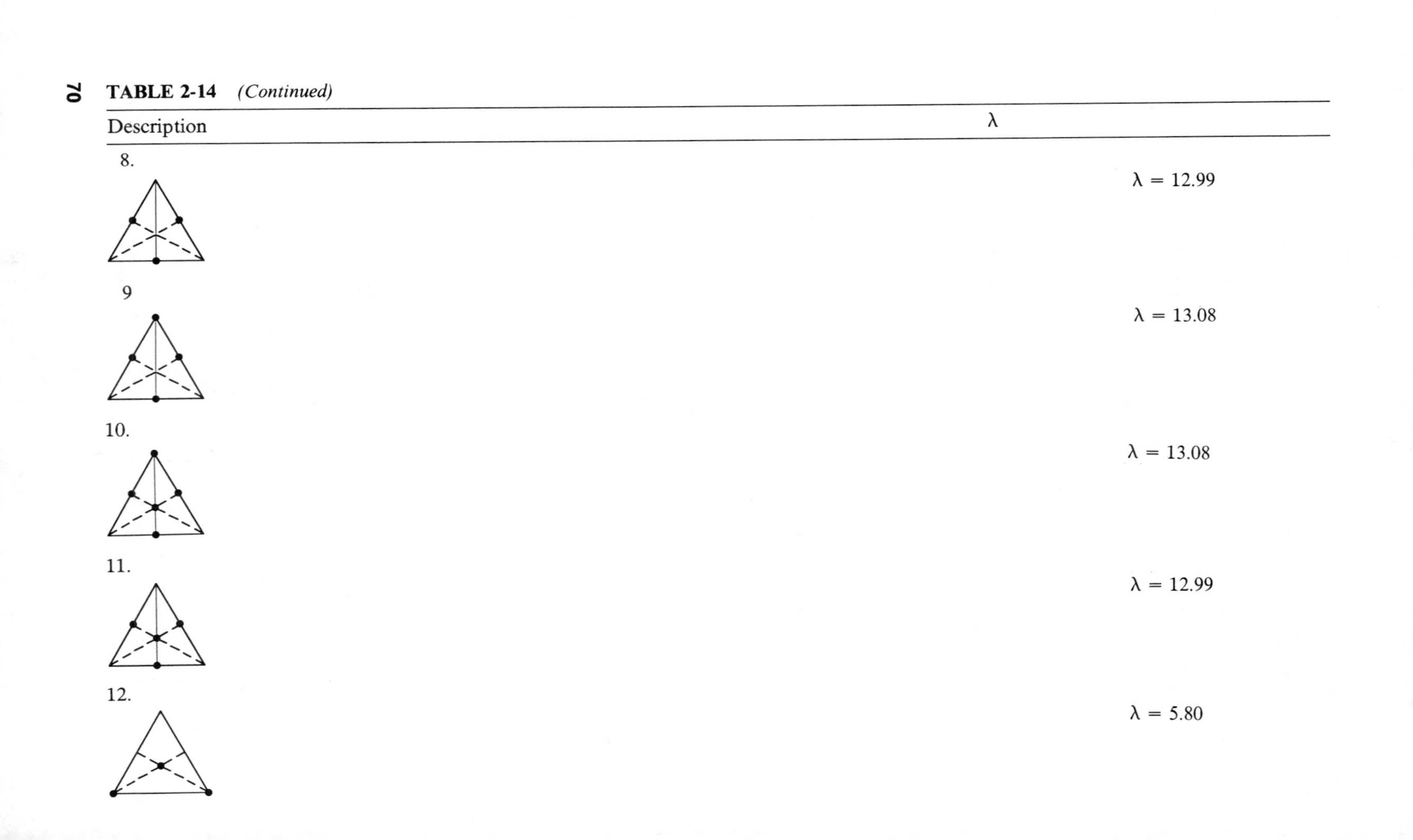

Description	λ
8.	$\lambda = 12.99$
9	$\lambda = 13.08$
10.	$\lambda = 13.08$
11.	$\lambda = 12.99$
12.	$\lambda = 5.80$

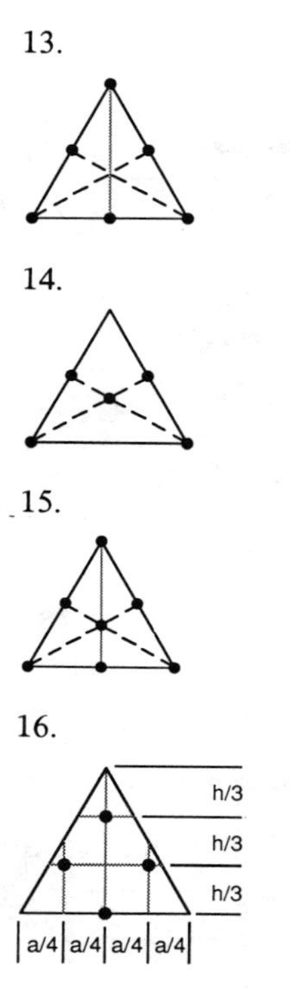

13. $\lambda = 33.46$

14. $\lambda = 14.08$

15. $\lambda = 33.55$

16. $\lambda = 17.04$

TABLE 2-14 *(Continued)*

Description	λ
17.	λ = 18.06

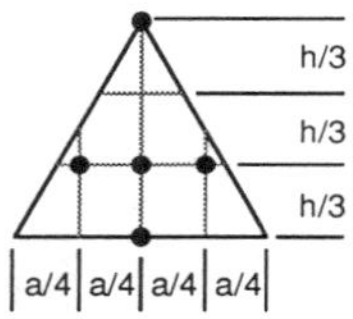

18. Symmetric trapezoid

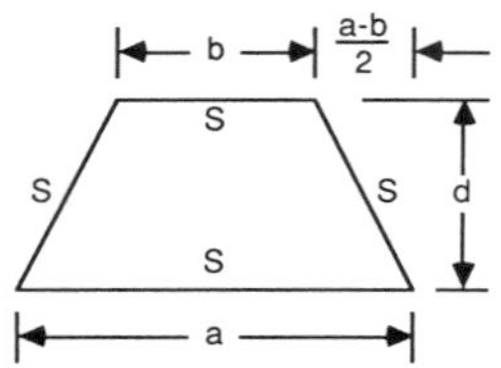

	b/a						
d/a	0.0	0.2	0.4	0.6	0.8	1.0	Independent of ν
0.5	98.78	76.50	63.18	55.97	51.85	49.35	
2/3	69.70	55.09	44.70	38.38	35.54	32.08	
1.0	45.85	37.75	30.79	25.64	22.13	19.74	
1.5	32.74	28.04	23.64	19.72	16.58	14.26	

19. Asymmetric trapezoid

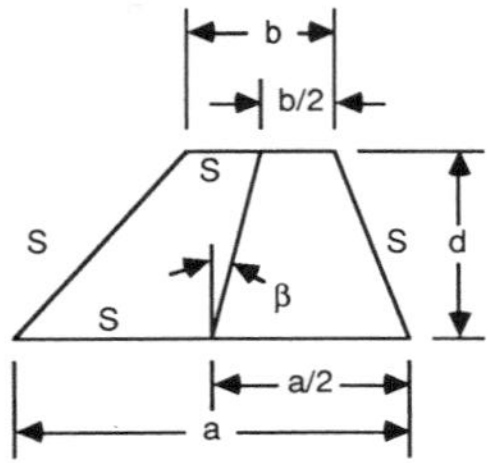

	d/a						
β	0.5		1.0		1.5		
(deg)	$b/a = 0.4$	$b/a = 0.8$	$b/a = 0.4$	$b/a = 0.8$	$b/a = 0.4$	$b/a = 0.8$	Independent of ν
10	63.42	52.00	31.20	22.37	24.05	16.88	
20	64.26	52.47	32.50	23.18	25.39	17.86	
30	66.02	53.44	35.14	24.87	28.02	19.85	
45	72.81	56.94	44.06	30.97	36.59	26.81	

20–21. Regular polygon with n sides of length a

Edge condition	Number of sides, n				
	4	5	6	7	8
20. S	19.74	11.01	7.152	5.068	3.794
21. C	35.08	19.71	12.81	9.081	6.787

Independent of ν

Cases 22 and 23. Arbitrarily shaped edge

Characteristic length = $a = (A/\pi)^{1/2}$
i = mode number.
22. $\lambda_i = 4.977i$ for S edge condition, $\nu = 0.3$
23. $\lambda_i = 10.22i$ for C edge condition, independent of ν

Cases 24–34 Point-supported regular hexagons of side a

$\nu = 0.12$ for Cases 24–34

$\lambda = 2.376$

TABLE 2-14 *(Continued)*

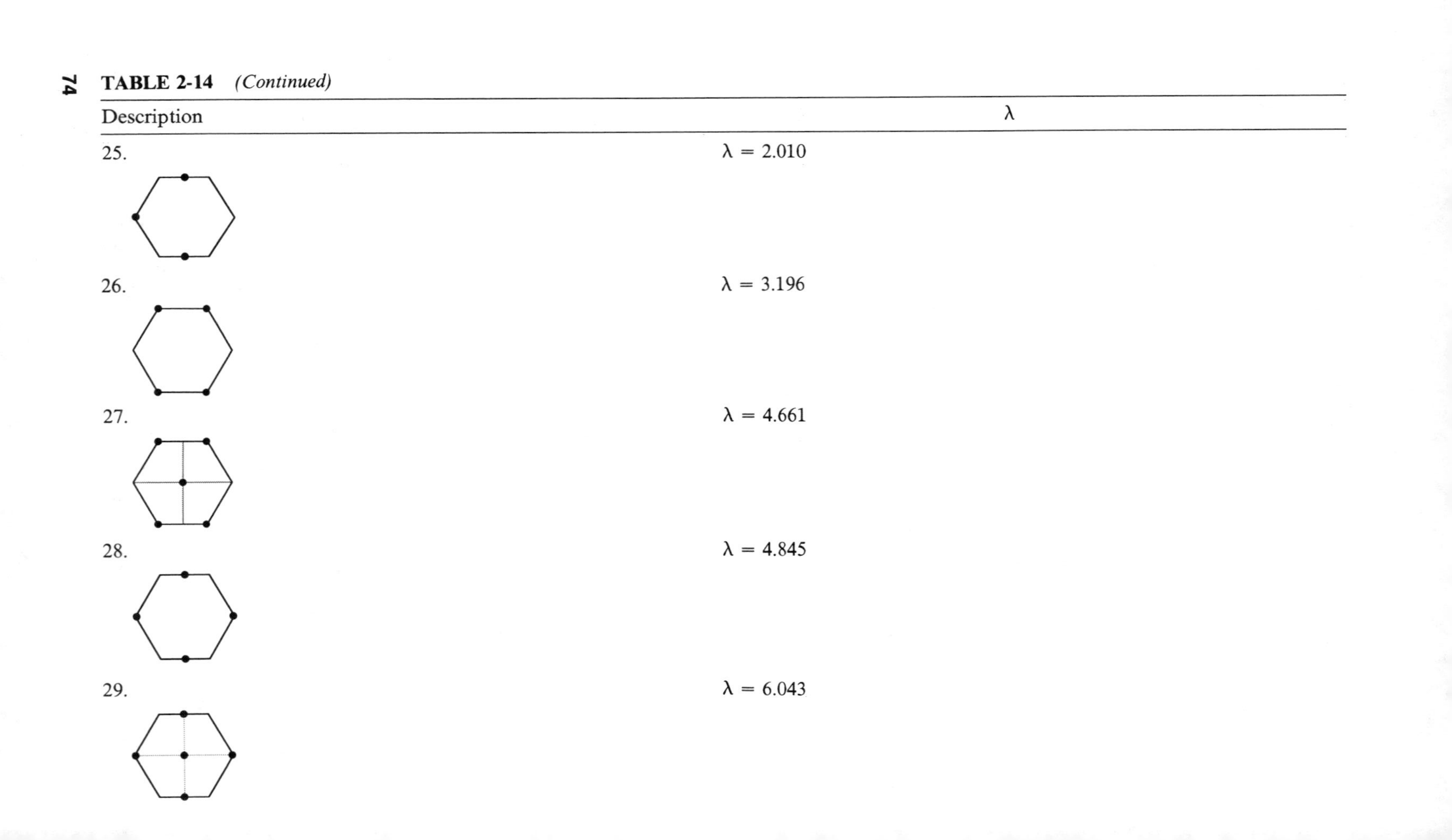

Description	λ
25.	$\lambda = 2.010$
26.	$\lambda = 3.196$
27.	$\lambda = 4.661$
28.	$\lambda = 4.845$
29.	$\lambda = 6.043$

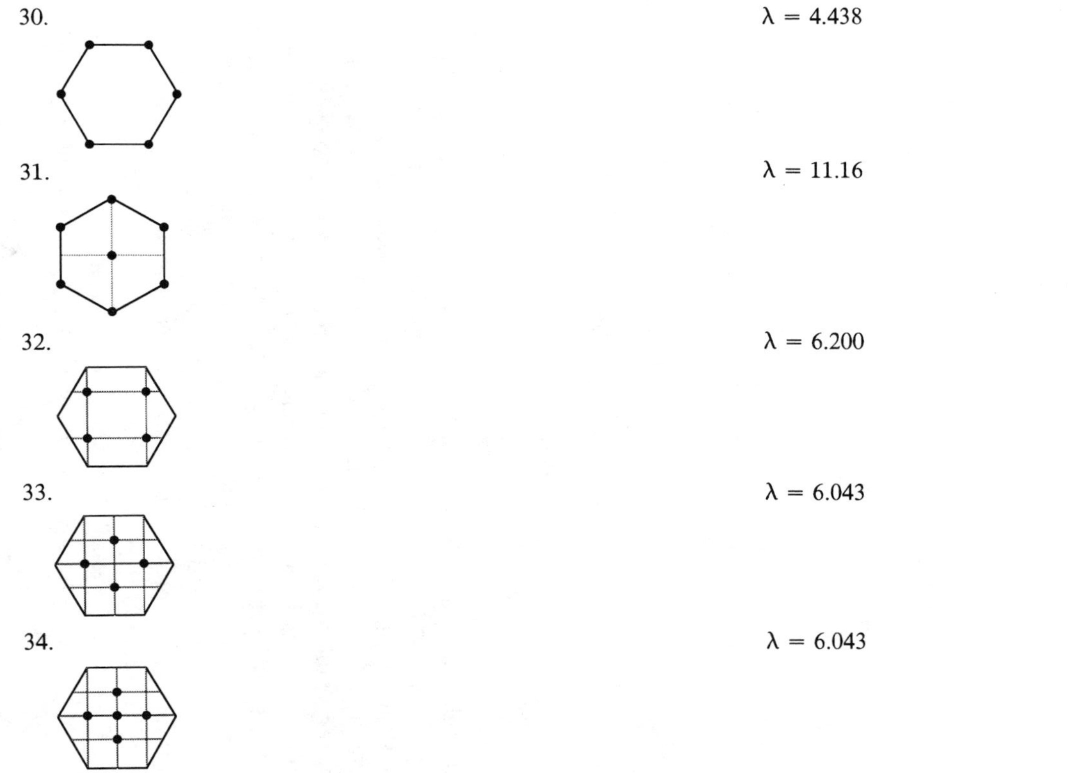
30.
λ = 4.438
31.
λ = 11.16
32.
λ = 6.200
33.
λ = 6.043
34.
λ = 6.043

2.7 ORTHOTROPIC PLATES

Orthotropic plates have material properties that are dependent on direction. The thin, rectangular orthotropic plates addressed in this section have their orthotropic axes aligned with the plate axes, which are aligned with the edges of the rectangular plate. The elastic constants for the thin orthotropic plate are two elastic moduli, E_x and E_y; two Poisson's ratios, ν_x and ν_y; and the shear modulus, G. However,[64]

$$\nu_y E_x = \nu_x E_y \tag{2-28}$$

so there are only four independent orthotropic constants for a thin orthotropic plate. For analysis purposes, the following four orthotropic constants are defined:

$$\begin{aligned} D_x &= E_x h^3/12(1 - \nu_x\nu_y) \quad \text{lbf} \cdot \text{in.} \\ D_y &= E_y h^3/12(1 - \nu_x\nu_y) \quad \text{lbf} \cdot \text{in.} \\ D_k &= Gh^3/12 \quad \text{lbf} \cdot \text{in.} \\ D_{xy} &= D_x\nu_y + 2D_k \quad \text{lbf} \cdot \text{in.} \end{aligned} \tag{2-29}$$

For a uniform, rectangular plate of side a parallel to the X axis and side b parallel to the Y axis, the fundamental natural frequency is[65]

$$F = (\pi/2)(g/\rho)^{1/2}\left[G_1^4 D_x/a^4 + G_2^4 D_y/b^4 + 2H_1H_2D_{xy}/a^2b^2 + 4D_k(J_1J_2 - H_1H_2)/a^2b^2\right]^{1/2} \tag{2-30}$$

where ρ = weight per unit area of plate (lbf/in.^{2})
g = gravitational acceleration on earth = 386 in./sec^2

The dimensionless constants G, H, and J are given in Table 2-15.[65]

The direction-dependent properties of orthotropic plates may also be due to the use of stiffeners made of isotropic materials. A rectangular network of stiffeners rigidly connected at their intersections (a grillage), either by itself or joined to a rectangular plate, can be treated as an orthotropic plate when the bending rigidity differs between orthogonal axes. Equation (2-30) and Table 2-15 may be applied when the stiffeners are uniform and evenly and closely spaced compared with the distance between vibration nodes, all parallel stiffeners are identical, and the plate contains two orthogonal planes of symmetry normal to the plane of the plate. In this case the constant ρ in Eq. (2-30) is the average weight per unit area of the entire assembly. Table 2-16 contains values of the equivalent orthotropic constants for grillages and stiffened plates.[66] These constants may be used in Eq. (2-30) in place of those from Eq. (2-29).

TABLE 2-15 Orthotropic Plate Constants

Description	Constants for Use in Eq. (2-30)[a]					
Edged supported: start at left edge and read clockwise:	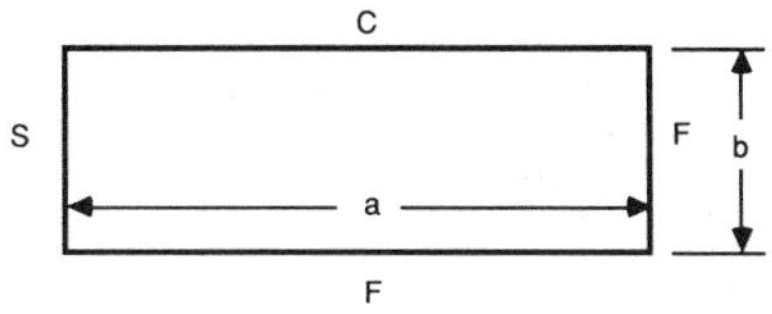					
Reads: *SCFF*	G_1	G_2	H_1	H_2	J_1	J_2
1. *SFFF*	0	0	0	0	0.3040	0
2. *CFFF*	0.597	0	−0.0870	0	0.471	0
3. *SFFS*	0	0	0	0	0.3040	0.3040
4. *SFSF*	1	0	1	0	1	0
5. *CFSF*	1.25	0	1.165	0	1.165	0
6. *CFFS*	0.597	0	−0.0870	0	0.471	0.3040
7. *CFCF*	1.506	0	1.248	0	1.248	0
8. *CFFC*	0.597	0.597	−0.0870	−0.0870	0.471	0.471
9. *SFSS*	1	0	1	0	1	0.3040
10. *SFSC*	1	0.597	1	−0.0870	1	0.471
11. *CFCS*	1.506	0	1.248	0	1.248	0.3040
12. *CFSS*	1.25	0	1.165	0	1.165	0.3040
13. *CFSC*	1.25	0.597	1.165	−0.0870	1.165	0.471
14. *CFCC*	1.506	0.597	1.248	−0.0870	1.248	0.471
15. *SSSS*	1	1	1	1	1	1
16. *SSSC*	1	1.25	1	1.165	1	1.165
17. *CSSC*	1.25	1.25	1.165	1.165	1.165	1.165
18. *SCSC*	1	1.506	1	1.248	1	1.248
19. *CSCC*	1.506	1.25	1.248	1.165	1.248	1.165
20. *CCCC*	1.506	1.506	1.248	1.248	1.248	1.248

[a] From Ref. 65.

2.8 SHELLS

A shell is a curved plate. The curvature is defined with respect to the midsurface of the shell and may be simple, like a cylinder, or complex. The shell may be closed, like a tube, or open like a hemispherical cap. Because the curvature of the shell couples its flexural vibrations with its extensional vibrations, the analysis of the shell's natural frequencies and mode shapes becomes much more complicated than for flat plates. Also, the frequency does not necessarily increase monotonically with increasing values of the modal indices; the lowest-frequency mode may be associated with high values of the modal indices, that is, with many nodes.

A cylindrical shell will have frequencies and mode shapes that are the same as previously obtained in this chapter for the transverse, longitudinal, and

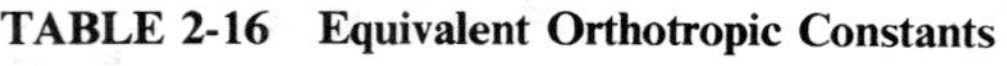

TABLE 2-16 Equivalent Orthotropic Constants

a = plate width (in.)
b = plate length (in.)
h = plate thickness without grillage (in.)
C = torsion constant (in.^{4}), Table 2-2

I = moment of inertia about neutral axis (in.^{4}), Table 2-2
H = total plate thickness including grillage (in.)
E = modulus of elasticity (lbf/in.^{2})
ν = Poisson's ratio

Description	D_x (lbf · in.)	D_y (lbf · in.)	D_{xy} (lbf · in.)
1. Grillage	$E_a I_a / b_1$	$E_b I_b / a_1$	$E_a C_a / 2b_1 + E_b C_b / 2a_1$

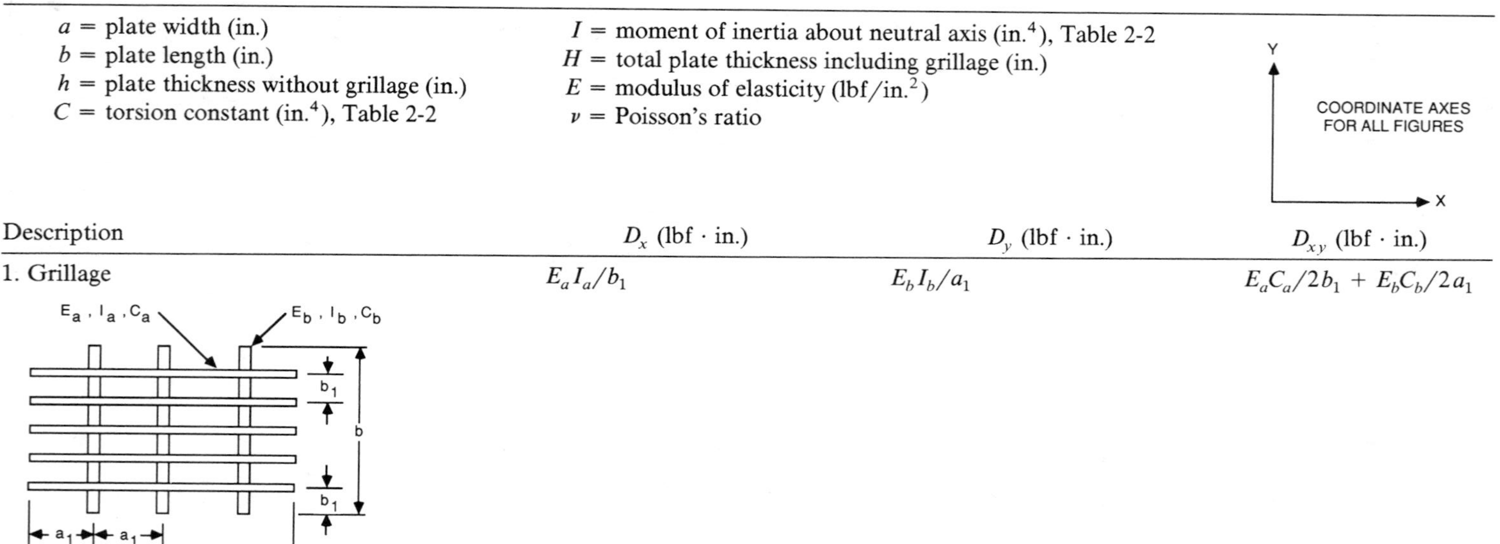

2. Plate with stiffeners in one direction 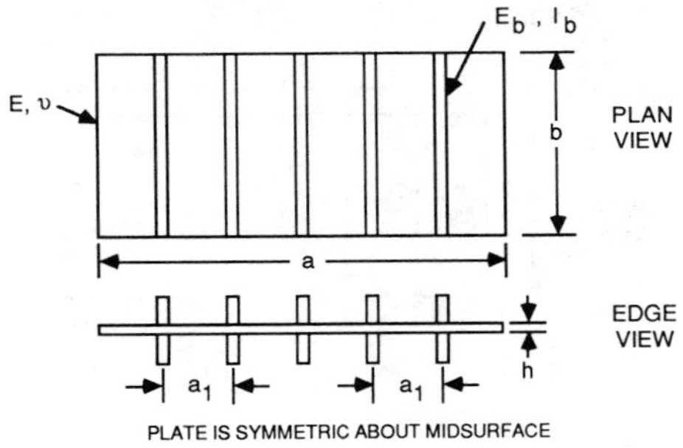	$Eh^3/12(1 - \nu^2)$	$Eh^3/12(1 - \nu^2) + E_b I_b/a_1$ I_b = moment of inertia of stiffener alone with respect to midsurface	$Eh^3/12(1 - \nu^2)$
3. Plate with stiffeners in two directions 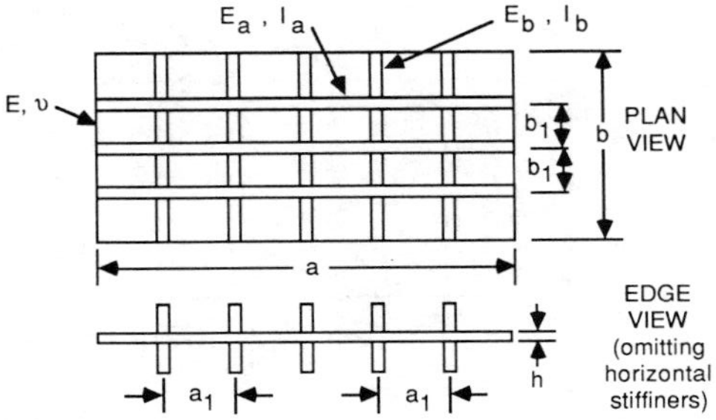	$Eh^3/12(1 - \nu^2) + E_a I_a/b_1$ I_a = moment of inertia of stiffener along with respect to midsurface	$Eh^3/12(1 - \nu^2) + E_b I_b/a_1$ I_b = moment of inertia of stiffener alone with respect to midsurface	$Eh^3/12(1 - \nu^2)$

TABLE 2-16 *(Continued)*

Description	D_x (lbf · in.)	D_y (lbf · in.)	D_{xy} (lbf · in.)
4. Plate with rectangular ribs in one direction	$\dfrac{Ea_1h^3}{12\left[a_1 - t + (h/H)^3 t\right]}$	EI_r/a_1 See first column for I_r	$Eh^3/12(1+\nu) + C_r E/a_1$ C_r = torsion constant of rib alone
5. Corrugated plate	$\dfrac{Eh^3}{12(1-\nu^2)\left(1 + \dfrac{\pi^2H^2}{16L^2}\right)}$	$\dfrac{EHh^2}{4}\left(1 - \dfrac{0.81}{1 + 0.156H^2/L^2}\right)$	$\dfrac{Eh^3(1 + \pi^2H^2/16L^2)}{12(1+\nu)}$

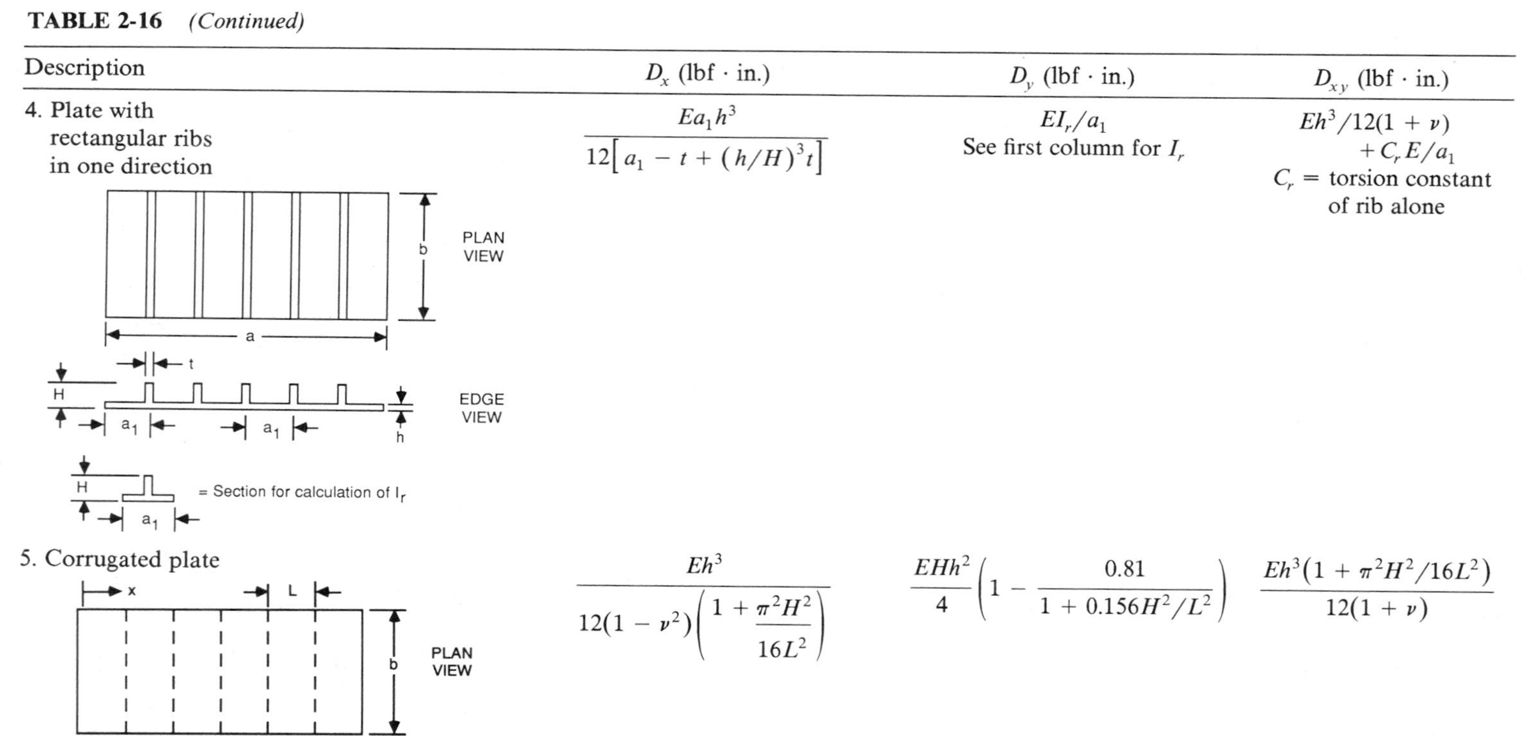

6. Fiber-reinforced plate 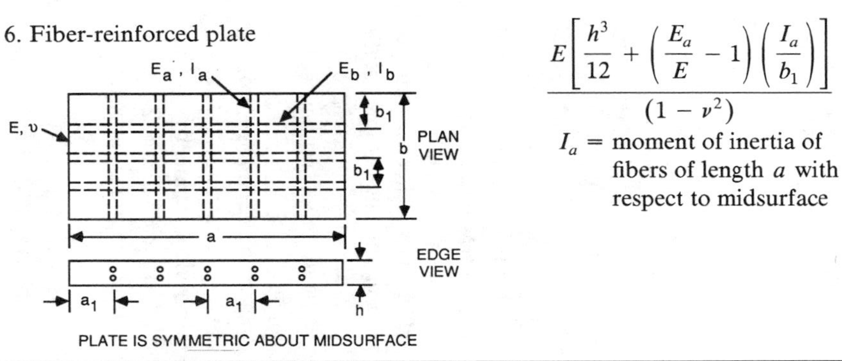	$\dfrac{E\left[\dfrac{h^3}{12} + \left(\dfrac{E_a}{E} - 1\right)\left(\dfrac{I_a}{b_1}\right)\right]}{(1 - \nu^2)}$ I_a = moment of inertia of fibers of length a with respect to midsurface	$\dfrac{E\left[\dfrac{h^3}{12} + \left(\dfrac{E_b}{E} - 1\right)\left(\dfrac{I_b}{a_1}\right)\right]}{(1 - \nu^2)}$ I_b = moment of inertia of fibers of length b with respect to midsurface	$(D_x D_y)^{1/2}$

*Subscript a refers to stiffener of length a
Subscript b refers to stiffener of length b
The equivalent orthotropic constants in this table are for use in Eq. (2-30), with $D_k = D_{xy}/2$ lbf · in.

TABLE 2-17 Circular, Cylindrical Shells

$$F_{ij} = (\lambda_{ij}/2\pi R)[E/\mu(1 - \nu^2)]^{1/2} \text{ Hz}$$

where i = number of circumferential waves in the mode shape
j = number of longitudinal half-waves in the mode shape
λ_{ij} = dimensionless parameter provided in the table
R = cylinder radius to midsurface (in.)
L = length of cylinder (in.)
h = thickness of cylinder wall (in.)
μ = mass density of shell material (lbm/in.^{3})
ν = Poisson's ratio of shell material, dimensionless
E = elastic modulus of shell material (lbf/in.^{2})

Description	λ_{ij}, Simply Supported Ends without Axial Constraint
1. Bending modes	$i = 1;\ j = 1, 2, 3, \ldots$ $\lambda_{ij} = j^2\pi^2(1 - \nu^2)^{1/2}R^2/2^{1/2}L^2$ for $L/jR > 8$ Same as Case 2 of Table 2-1 for a cylindrical beam with $I = \pi R^3 h$

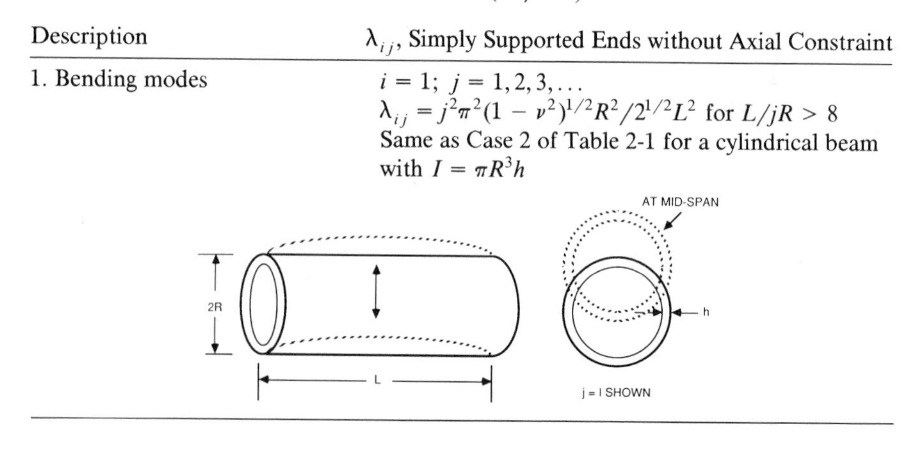

torsional vibrations of tubular beams and the flexural in-plane and extensional in-plane vibrations of rings. In addition, the cylindrical shell will have vibration modes unique to shells. Because of the coupling due to curvature, the shell deformation can vary from purely extensional to purely flexural.

In this section it is assumed that the shell walls are of constant thickness less than 10% of the shell radius and are composed of linear, elastic, homogeneous isotropic material. There are no loads applied to the shells, and rotary inertia and shear deformation are neglected. The results presented here are approximate values of the dimensionless constant λ, valid only for a particular range of L/R for the specified type of vibration mode.

Table 2-17 presents some of the natural frequencies of circular, cylindrical shells, simply supported at their ends and without axial constraints.[67] Cases 1 and 2 provide the same frequency values as for thin-walled beams of circular cross section (Compare Case 1 of Table 2-17 with Case 2 of Table 2-1 with $I = \pi R^3 h$). Case 5 illustrates that the lowest-frequency mode is not always associated with the lowest modal indices.

When the cylindrical shell of Table 2-17 is subjected to a uniform axial load T_x per unit length of circular edge and a uniform circumferential load T_θ per

TABLE 2-17 *(Continued)*

Description	λ_{ij}, Simply Supported Ends without Axial Constraint
2. Axial modes	$i = 0;\ j = 1, 2, 3, \ldots$ $\lambda_{0j} = j\pi(1 - \nu^2)^{1/2}R/L$ for $L/jR > 8$
3. Torsion modes	$i = 0;\ j = 1, 2, 3 \ldots$ $\lambda_{0j} = j\pi(1 - \nu^2)^{1/2}R/2^{1/2}L$

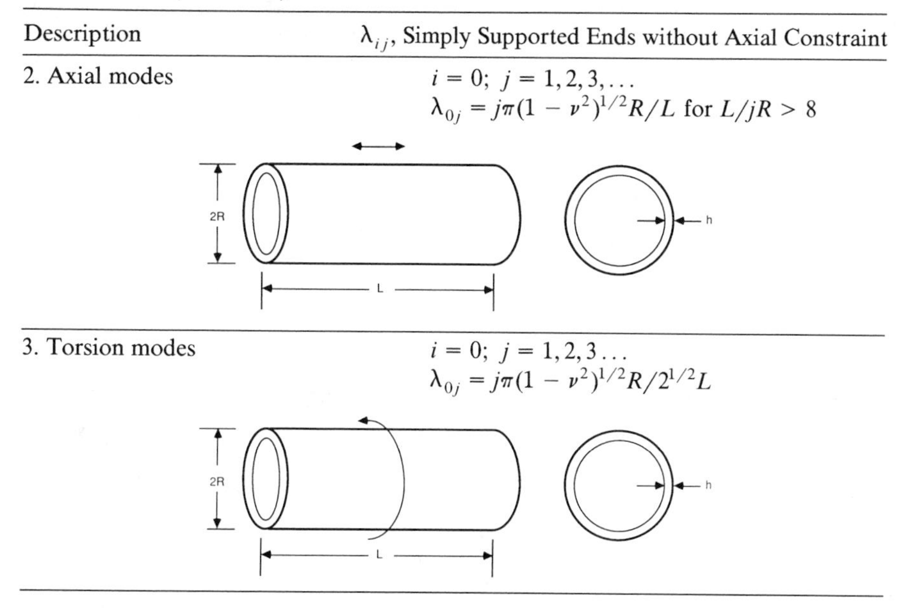

unit axial length, the natural frequencies are given by[68]

$$F_{ij} = \left[F_{0ij}^2 + j^2T_x/4\mu hL^2 + i^2T_\theta/4\theta^2\mu hR^2\right]^{1/2} \text{ Hz} \tag{2-31}$$

where F_{0ij} is given in Table 2-17 ($T_x = T_\theta = 0$); i, j, μ, h, L, and R are defined in Table 2-17; T_x = axial load per unit length of cylinder edge (lbf/in.); T_θ = circumferential load per unit axial length (lbf/in.); T_x and T_θ are positive if they produce tensile stress in the shell; and T_x and T_θ are negative if they produce compressive stress in the shell.

For the cylindrically curved panel shown in Fig. 2-17 with all four edges clamped, the fundamental natural frequency is given by[69, 70]

$$F = \left[F_0^2 + \alpha Egh/4\pi^2R^2\rho(1 - \nu^2)\right]^{1/2} \text{ Hz} \tag{2-32}$$

where F_0 is given in Table 2-12 (Case 20) with $a = L$ and $b = R\theta$, R, L, h, E, and ν are as defined in Table 2-17 and Fig. 2.17, ρ = panel weight per unit area (lbf/in.^{2}); g = gravitational acceleration on earth = 386 in./sec^2, and

TABLE 2-17 *(Continued)*

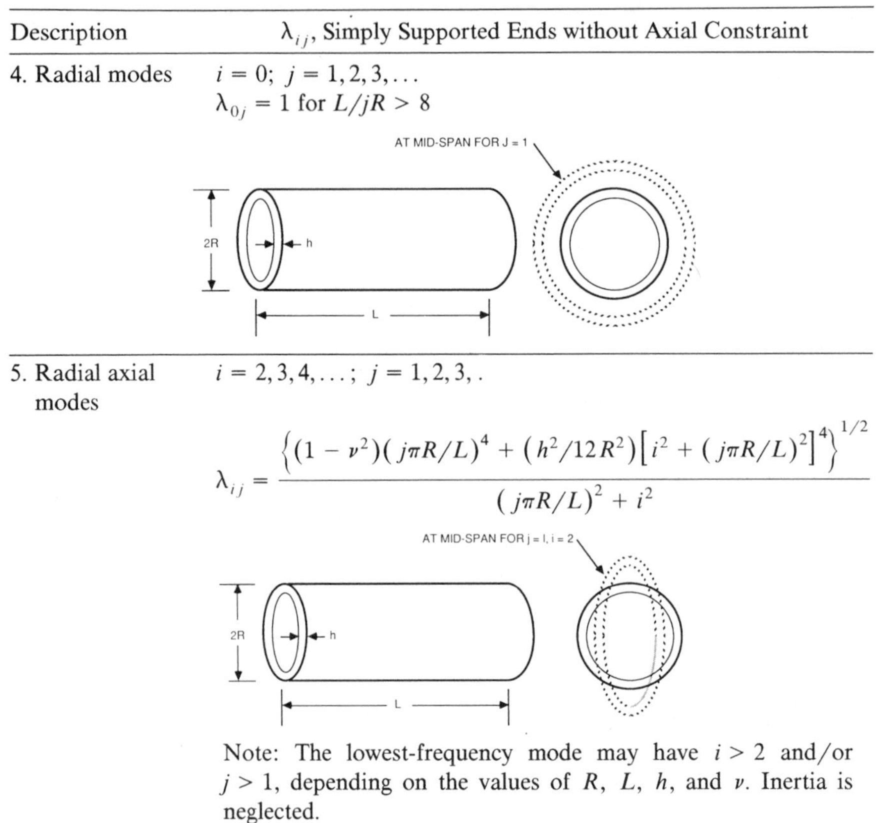

Description	λ_{ij}, Simply Supported Ends without Axial Constraint
4. Radial modes	$i = 0;\ j = 1, 2, 3, \ldots$ $\lambda_{0j} = 1$ for $L/jR > 8$
5. Radial axial modes	$i = 2, 3, 4, \ldots;\ j = 1, 2, 3, .$ $\lambda_{ij} = \dfrac{\left\{(1 - \nu^2)(j\pi R/L)^4 + (h^2/12R^2)\left[i^2 + (j\pi R/L)^2\right]^4\right\}^{1/2}}{(j\pi R/L)^2 + i^2}$ Note: The lowest-frequency mode may have $i > 2$ and/or $j > 1$, depending on the values of R, L, h, and ν. Inertia is neglected.

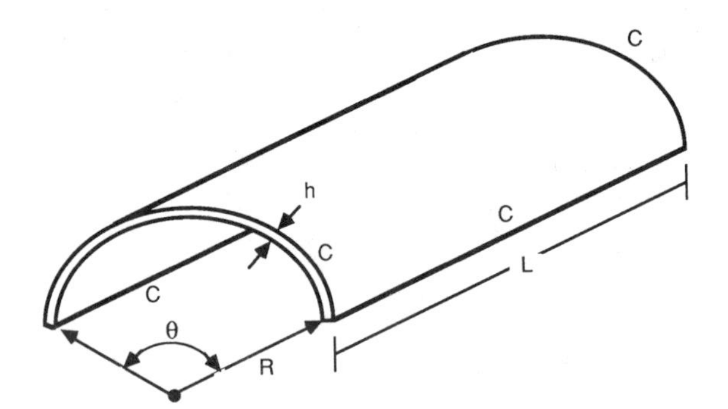

FIGURE 2.17 Cylindrically curved panel.

TABLE 2-18 Conical Shells

$$F = (\lambda/2\pi R)(E/\mu)^{1/2}$$

where λ = dimensionless parameter provided in the table
R = radius of cone base (in.)
E = elastic modulus of shell material (lbf/in.^{2})
μ = mass density of shell material (lbm/in.^{3})
h = thickness of shell wall (in.)
α = half angle of cone (deg. of arc)
ν = Poisson's ratio of shell material, dimensionless
R_1 = minor radius of the frustrum of a cone (in.)
L = slant height of the frustrum of a cone (in.)

Description	λ for Lowest Natural Frequency[a]								
1. Cone, clamped base	For $\nu = 0.3$								
	$12(1-\nu^2)(R/h)^2/\tan^4\alpha$	0.1	0.4	1.0	4.0	10	10^2	10^3	10^5
	λ	1,049	266.9	110.4	32.07	16.33	6.096	3.574	1.802

TABLE 2-18 *(Continued)*

Description	λ for Lowest Natural Frequency[a]
2. Cone frustrum, simply supported base and top	For $\nu = 0.3$ $\lambda = (R/L)/(1-\nu^2)^{1/2} \times$ (values in table)
3. Cone frustrum, clamped base and top	$\lambda = \dfrac{\sin\alpha}{(1 - R_1/R)(2+2\nu)^{1/2}} \times$ (values in table)

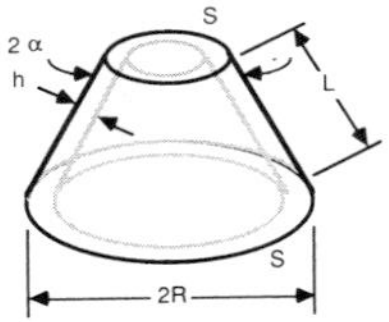

Case 2 (simply supported base and top):

α (deg)	h/R = 0.03	0.01	0.005	0.001
70	0.479	0.287	0.199	0.0950
50	0.652	0.386	0.282	0.138
30	0.776	0.479	0.350	0.172
10	0.891	0.553	0.432	0.229

Case 3 (clamped base and top):

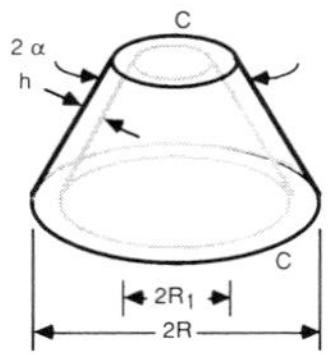

R/R_1 =	1	2	5	10	20	50
Values =	3.142	3.197	3.389	3.547	3.667	3.760

[a] From Ref. 68.

$\alpha = A/B$; where

$$A = 250{,}564 - 75{,}762.6(1 + \nu^2) + 75.6772(1 - \nu) \cdot \left[(500.564 - 151.354\nu^2)(b^2/L^2) + 349.209(L^2/b^2)\right] + 22{,}908.2\nu^2$$

and

$$B = 250{,}564 + 37{,}881(1 - \nu)(b^2/L^2 + L^2/b^2) - 22{,}908.2\nu$$

A segment of a shallow, spherical shell will have a fundamental natural frequency given by[71]

$$F = \left(F_0^2 + Eg/4\pi^2\rho R^2\right)^{1/2} \text{ Hz} \tag{2-33}$$

where F_0 is given in Table 2-12 for a flat plate that has the same projected shape and size (the same shadow on a plane surface) as does the spherical segment, and the same thickness, materials, and boundary conditions. The parameters E, ρ, and R have been previously defined, as has the constant g.

Table 2-18 presents some of the natural frequencies of thin, conical shells.[68, 72–74]

REFERENCES

1 D. S. Steinberg, *Vibration Analysis for Electronic Equipment*, New York: Wiley-Interscience, 1973, pp. 36–37.

2 R. J. Roark and W. C. Young, *Formulas for Stress and Strain*, 5th ed., New York: McGraw-Hill, 1975, p. 89.

3 R. D. Blevins, *Formulas for Natural Frequency and Mode Shape*, New York; Van Nostrand Reinhold, 1979, p. 175.

4 T. C. Chang and R. R. Craig, *J. Eng. Mech. Div., ASCE* **95**, 1027–1031 (1969).

5 R. J. Roark and W. C. Young, in Ref. 2, pp. 290–296.

6 A. V. K. Murty and R. P. Shimpi, *J. Sound Vib.* **36**, 273–284 (1974).

7 F. J. Shaker, "Effect of Axial Load on Mode Shapes and Frequencies of Beams," Lewis Research Center Report NASA-TN-9109, December, 1975.

8 S. Timoshenko, D. H. Young, and W. Weaver, Jr., *Vibration Problems in Engineering*, 4th ed., New York: Wiley, 1974, pp. 31–41.

9 P. A. A. Laura, J. L. Pombo, and E. A. Susemihl, *J. Sound Vib.* **37**, 161–168 (1974).

10 B. R. Bhat and H. Wagner, *J. Sound Vib.* **45**, 304–307 (1976).

11 J. M. Gere, *J. Appl. Mech.* **21**, 381–387 (1954).

12 J. B. Carr, *Aeron. J., Royal Aeron. Soc.* **73** (704), 672–674 (1969).

13 S. Timoshenko et al., in Ref. 8, pp. 401–405.

14 R. D. Blevins, in Ref. 3, pp. 183–185.

15 G. R. Cowper, *J. Appl. Mech.* **33**, 335–340 (1966).

16 R. D. Blevins, in Ref. 3, pp. 176–178.

17 D. S. Steinberg, in Ref. 1, pp. 63–64.

18 H. D. Conway and J. F. Dubil, *J. Appl. Mech.* **32**, 932–934 (1965).

19 H. H. Mabie and C. B. Rogers, *J. Acoust. Soc. Am.* **44**, 1739–1741 (1968).

20 H. H. Mabie and C. B. Rogers, *J. Acoust. Soc. Am.* **51**, 1771–1774 (1972).

21 H. C. Wang and W. J. Worley, "Tables of Natural Frequencies and Nodes for Transverse Vibration of Tapered Beams," NASA-CR-443, University of Illinois, April 1966.

22 H. H. Mabie and C. B. Rogers, *J. Eng. Industry* **98**, 1335–1341 (1976).

23 B. Downs, *J. Appl. Mech.* **44**, 737–742 (1978).

24 R. P. Goel, *J. Sound Vib.* **47**, 1–7 (1976).

25 A. E. H. Love, *A Treatise on the Mathematical Theory of Elasticity*, 4th ed., New York: Dover, 1944, pp. 452–453.

26 K. Sato, *J. Acoust. Soc. Am.* **57**, 113–115 (1975).

27 C. G. Culver, *J. Struct.* **83**, 189–203 (1967).

28 A. S. Veletsos et al., *J. Eng. Mech. Div., ASCE* **98**, 311–329 (1972).

29 R. D. Blevins, in Ref. 3, p. 209.

30 J. Kirkhope, *J. Acoust. Soc. Am.* **59**, 86–88 (1976).

31 J. Kirkhope, *J. Eng. Mech. Div., ASCE* **102**, 239–247 (1976).

32 W. B. Bickford and S. P. Maganty, *J. Sound Vib.* **108**, 503–507 (1986).

33 W. J. Austin and A. S. Veletsos, *J. Eng. Mech. Div., ASCE* **98**, 735–753 (1973).

34 W. C. Hurty and M. F. Rubinstein, *J. Franklin Inst.* **278**, 124–132 (1964).

35 S. Venkatesan and V. X. Kunukkasseril, *J. Sound Vib.* **60**, 511–534 (1978).

36 A. W. Leissa, *J. Sound Vib.* **31**, 257–293 (1973).

37 D. S. Steinberg, *Machine Design* **48**, 116–119 (1976).

38 D. J. Johns and R. Nataroja, *J. Sound Vib.* **25**, 75–82 (1972).

39 R. S. Srinivasan and K. Munaswamy, *J. Sound Vib.* **39**, 207–216 (1975).

40 M. Petyt and W. H. Mirza, *J. Sound Vib.* **21**, 355–364 (1972).

41 W. Nowacki, *Dynamics of Elastic Systems*, New York: Wiley, 1963, p. 228.

42 H. L. Cox, *Quart. J. Mech. Appl. Math.* **8**, 454–456 (1955).

43 G. V. Rao, *J. Sound Vib.* **38**, 271 (1975).

44 F. C. Appl and N. R. Byers, *J. Appl. Mech.* **32**, 163–167 (1965).

45 J. E. Ashton, *J. Struct. Div., ASCE* **95**, 787–790 (1969).

46 J. E. Ashton, *J. Eng. Mech. Div., ASCE* **95**, 497–500 (1969).

47 R. F. Hegarty and T. Ariman, *Int. J. Solids Structures* **11**, 895–906 (1975).

48 P. Paramasiram, *J. Sound Vib.* **30**, 173–178 (1973).

49 R. C. Colwell and H. C. Hardy, *Phil. Mag.* **24**, 1041–1055 (1937).

50 H. Carrington, *Phil. Mag.* **50**, 1261–1264 (1925).

51 R. Y. Bodine, *J. Appl. Mech.* **26**, 666–668 (1959).

52 R. E. Roberson, *J. Appl. Mech.* **18**, 349–352 (1951).

53 S. S. Rao and A. S. Prasad, *J. Sound Vib.* **42**, 305–324 (1975).

54 S. M. Vogel and D. W. Skinner, *J. Appl. Mech.* **32**, 926–931 (1965).

55 A. W. Leissa, *J. Sound Vib.* **6**, 145–148 (1967).

56 R. P. McNitt, *J. Aerospace Sci.* **29**, 1124–1125 (1962).

57 H. Cox and B. Klein, *Aeron. Quart.* **7**, 221–224 (1956).

58 I. Chopra and S. Durvasula, *J. Sound Vib.* **19**, 379–392 (1971).

59 I. Chopra and S. Durvasula, *J. Sound Vib.* **20**, 125–134 (1972).

60 T. Ota et al., *Bull. JSME* **4**, 478–481 (1961).

61 P. A. Shahady et al., *J. Acoust. Soc. Am.* **42**, 806–809 (1967).

62 P. A. A. Laura and R. Gutierrez, *J. Sound Vib.* **48**, 327–332 (1976).

63 D. Pneuli, *J. Appl. Mech.* **42**, 815–820 (1975).

64 A. E. H. Love, in Ref. 25, pp. 105–107.

65 S. M. Dickinson, *J. Sound Vib.* **61**, 1–8 (1978).

66 S. Timoshenko and W. Woinowski-Krieger, *Theory of Plates and Shells*, 2nd ed., New York: McGraw-Hill, 1959, pp. 366–371; and R.D. Blevins, in Ref. 3, pp. 283–285.

67 R. D. Blevins, in Ref. 3, pp. 303–305.

68 A. W. Leissa, "Vibrations of Shells," NASA-SP-288, Ohio State University, 1973, pp. 331–388.

69 J. J. Webster, *Int. J. Mech. Sci.* **10**, 571–582 (1968).

70 J. L. Sewall, "Vibration Analysis of Cylindrically Curved Panels with Simply Supported or Clamped Edges and Comparison with Some Experiments," Langley Research Center, NASA TN D-3791, 1967.

71 W. Soedel, *J. Sound Vib.* **29**, 457–461 (1973).

72 J. F. Dreher and A. W. Leissa, "Axisymmetric Vibration of Thin Conical Shells," *Proc. 4th Southwestern Conference on Theoretical and Appl. Mech.* (*New Orleans, LA*), *Feb. 29–Mar. 1, 1968*, pp. 163–181.

73 E. I. Grigolywk, *Izv. Akad. Nank. SSR, O.T.D.*, No. 6, 1956 NASA T T F-25.

74 H. Garnet et al., *J. Appl. Mech.* **28**, 571–573 (1961).

3

NATURAL FREQUENCY OF SIMPLE STRUCTURES

Structures are often made up of simple components such as the beams, rings, arcs, plates, and shells addressed in Chapter 2. The natural frequencies of such a structure cannot usually be found from the frequencies of these components. However, the stiffness (Chapter 2, page 22), damping, and mass of these components; the stiffness and damping of the connections between components; and the type of attachment of the structure to mounting surfaces will determine the natural frequencies of the structure. Estimates of natural frequencies can only be made for simple structures without developing an FEA (finite-element-analysis) model and utilizing an FEA computer program. But even rough estimates of natural frequency can provide a relatively rapid means of comparing maximum acceleration, stress, and fatigue in different design approaches, and identifying potential problem areas in a structure. This type of information can help to avoid excessive modification of the FEA model when a comprehensive computer analysis is done.

The methods presented in this chapter for evaluating natural frequencies of simple structures will typically have a frequency error on the low side, which will result in a conservative (larger than actual) estimate of stress.

3.1 COMPOSITE BEAMS

Beams may be made of two or more layers of different materials adhered to one another, with each layer running the length of the beam. In this section, the layers are assumed to have constant, rectangular cross sections. The layers

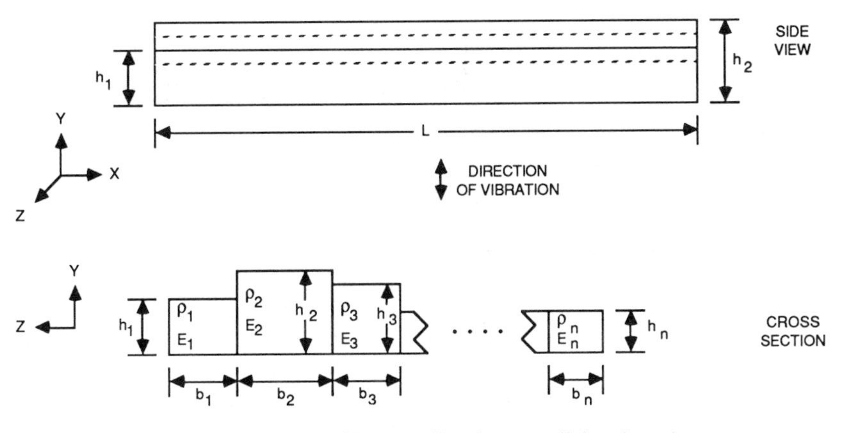

FIGURE 3.1 Layered beam, vibration parallel to interfaces.

may be oriented so that the direction of vibration is parallel to the layer interfaces (Section 3.1.1) or normal to the layer interfaces (Section 3.1.2).

3.1.1 Vibration Parallel to Layer Interfaces

Figure 3.1 shows the case where the direction of vibration is parallel to the layer interfaces. In Section 2.3 the term EI/L is referred to as the stiffness of a beam. Defining EI as the stiffness factor of a beam, the stiffness factor of the composite beam of Fig. 3.1 is

$$EI = \sum_{i=1}^{n} E_i I_i = \frac{1}{12} \sum_{i=1}^{n} E_i b_i h_i^3 \text{ lbf} \cdot \text{in.}^2 \tag{3-1}$$

where E_i = modulus of elasticity of layer i (lbf/in.^{2})
b_i = width of layer i (in.)
h_i = height of layer i (in.)
I_i = area moment of inertia of layer i about neutral (Z) axis (in.^{4})

The weight per unit length of the composite beam is

$$W = \sum_{i=1}^{n} \rho_i b_i h_i \text{ lbf/in.} \tag{3-2}$$

where ρ_i is the weight density of layer i (lbf/in.^{3}).

Equations (3-1) and (3-2) may be used in the frequency formulas of Section 2.4.1 when the composite beam is uniformly loaded and the neutral axis through the beam cross section, parallel to the Z axis, remains undeflected (no bending along the Z axis).

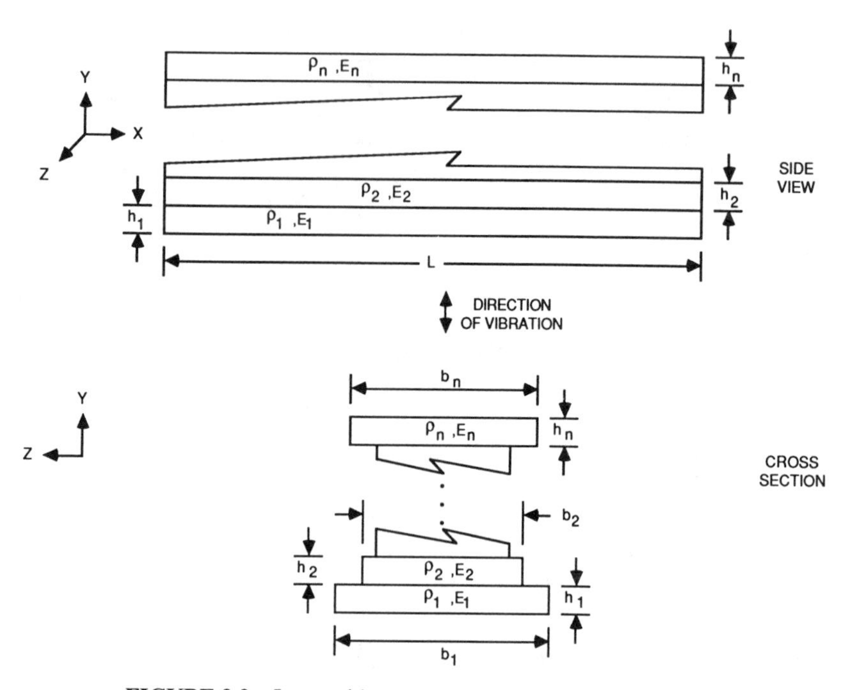

FIGURE 3.2 Layered beam, vibration normal to interfaces.

3.1.2 Vibration Normal to Layer Interfaces

Figure 3-2 shows the case where the vibration is normal to the layer interfaces. The stiffness factor of the composite beam of Fig. 3-2 is

$$EI = \sum_{i=1}^{n} b_i h_i E_i \left[\left(\overline{Y} - Y_i \right)^2 + \tfrac{1}{12} h_i^2 \right] \text{ lbf} \cdot \text{in.}^2 \tag{3-3}$$

where

$$Y_i = \sum_{j=0}^{i-1} h_j + \tfrac{1}{2} h_i \text{ in.,} \quad \text{where } h_0 = 0$$

and

$$\overline{Y} = \sum_{i=1}^{n} b_i h_i E_i Y_i \bigg/ \sum_{i=1}^{n} b_i h_i E_i \text{ in.}$$

See Section 3.1.1 for definitions.

The weight per unit length of the composite beam W is given by Eq. (3-2). (See also Section 2.4.1, Sandwich Beams.) Equations (3-2) and (3-3) may be used in the slender-beam frequency formulas of Section 2.4.1 to obtain approximate values of natural frequency. Accuracy is improved when the layer widths b_i approach equality with one another.

3.2 STEPPED BEAMS

Stepped beams have two or more different cross sections along their span, resulting in two or more different moments of inertia. Figure 3.3 shows two examples of stepped cantilever beams. In Fig. 3.3*a* the beam has two different cross sections and the average moment of inertia for the beam is[1]

$$I_A = L^3 I_1 I_2 / \left[3\left(a^2 b + b^2 a + b^3/3\right) I_1 + a^3 I_2\right] \text{ in.}^4 \tag{3-4}$$

In Fig. 3.3*b* the beam has three different cross sections and the average moment of inertia for the beam is

$$I_A = L^3 I_1 I_2 I_3 / \left\{3\left[(a+b)^2 c + (a+b)c^2 + c^3/3\right] I_1 I_2 + 3\left(a^2 b + ab^2 + b^3/3\right) I_1 I_3 + a^3 I_2 I_3\right\} \text{ in.}^4 \tag{3-5}$$

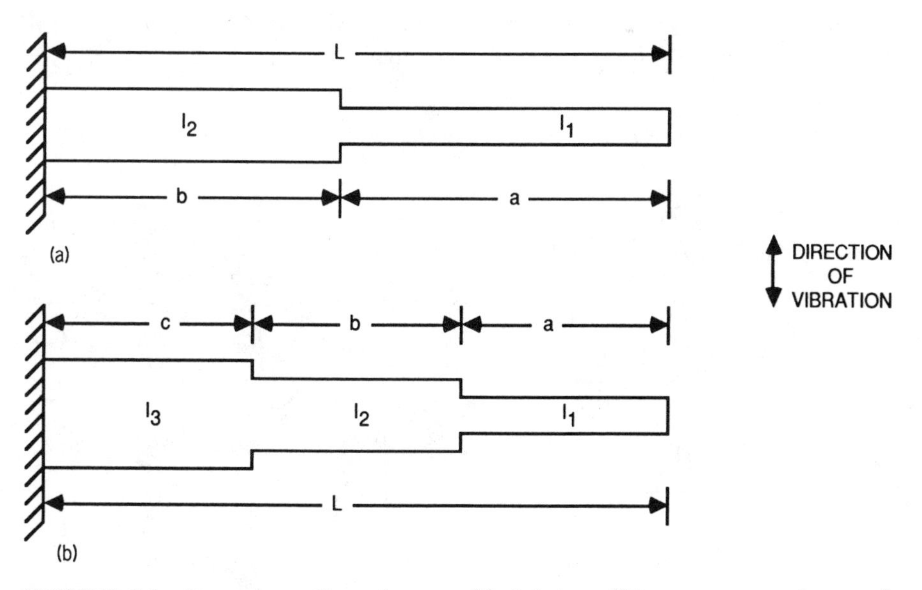

FIGURE 3.3 Stepped cantilever beams with (*a*) two different cross sections and (*b*) three different cross sections.

In general, for a beam with n different cross sections, an approximate value for the average moment of inertia is

$$I_A = \frac{1}{L} \sum_{i=1}^{n} x_i I_i \text{ in.}^4 \tag{3-6}$$

where x_i = spanwise length of cross section i (in.)
I_i = moment of inertia of cross section i (in.^{4})
$L = \Sigma_i x_i$ = full span of beam (in.)

Equations (3-4), (3-5), and (3-6) may be used in the frequency formulas of Section 2.4.1 to obtain values of natural flexural frequencies of stepped beams. Equations (3-4) and (3-5) are only for cantilever beams and should provide accurate results for slender beams. Equation (3-6) may be used for any end-support conditions, and will usually yield a natural frequency roughly 5–10% lower than the correct value.

3.3 SLENDER RIGHT ANGLES AND U BENDS

Figure 3.4 shows a right angle and a U bend with intermediate supports. The ends E may have any combination of pinned P or clamped C boundary conditions. The intermediate supports S prevent transverse motion (perpendicular to the beam axis) at the support, but allow the beam to move parallel to its own axis and to rotate about any axis. The fundamental natural frequency for vibration in the plane of the figures (in-plane vibration) is[2]

$$F = (\lambda/2\pi R^2)(EI_y g/W)^{1/2} \text{ Hz} \tag{3-7}$$

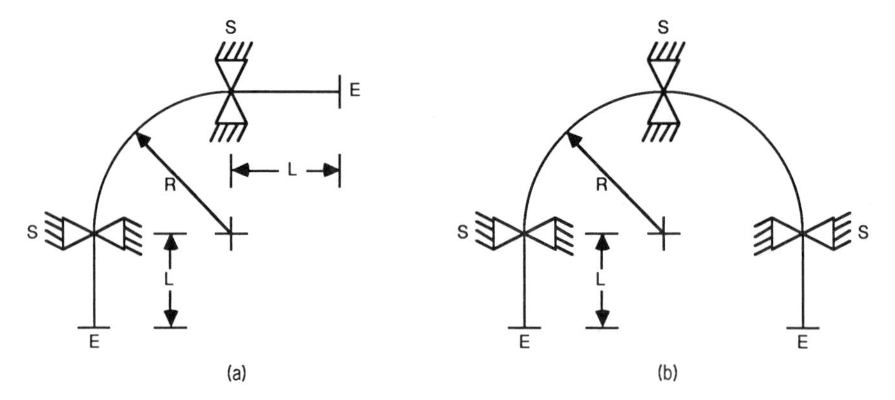

FIGURE 3.4 (*a*) Slender right angle bend; (*b*) slender U bend.

TABLE 3-1 In-Plane Vibration of Right Angles and U Bends[a]

	λ for Use in Eq. (3-7)					
	Right Angles (Fig. 3.4a)			U Bends (Fig. 3.4b)		
L/R	P–P	P–C	C–C	P–P	P–C	C–C
0	22.8	22.8	22.8	4.5	4.5	4.5
0.4	18.3	18.5	19.0	3.7	3.8	3.8
0.8	14.5	15.5	16.8	3.4	3.5	3.5
1.2	8.3	8.3	11.8	3.2	3.3	3.3
1.6	5.0	5.0	7.5	2.8	2.9	3.1
2.0	3.5	3.5	5.0	2.4	2.6	2.9

[a] L = length of legs (in.), R = radius of curvature, (in.), P = pinned end condition, C = clamped end condition. Vibration in the plane of Figs. 3.4a and 3.4b.

where λ = dimensionless frequency parameter in Table 3-1,
R = radius of curvature shown in Fig. 3.4 (in.)
E = modulus of elasticity of beam material (lbf/in.^{2})
I_y = area moment of inertia about axis perpendicular to the plane of the figures (in.^{4})
g = gravitational acceleration at surface of earth = 386 in./sec^2
W = weight per unit length of beam (lbf/in.)

The fundamental natural frequency for vibration perpendicular to the plane of the figures (out-of-plane vibration) is

$$F = (\lambda/2\pi R^2)(GI_p g/W)^{1/2} \text{ Hz} \tag{3-8}$$

where I_y, R, E, g, and W are as defined for Eq. (3-7), and
λ = dimensionless frequency parameter in Table 3-2
G = shear modulus [$E/2(1 + \nu)$ lbf/in.^{2}]
ν = Poisson's ratio, dimensionless
I_p = polar area moment of inertia = $I_x + I_y$ (in.^{4})
I_x = area moment of inertia about axis in the plane of the figure and perpendicular to the local beam axis (in.^{4})

Equations (3-7) and (3-8) do not take into account shear deformation, cross-sectional distortion due to torsion, or coupling of rotation and displacement. The rotary inertia of the beam twisting about its own axis is not included in Eq. (3-7) but is included in Eq. (3-8). However, the values of λ given in Table 3-2 are only valid for circular beams or tubes with a value of $\nu = 0.3$.

TABLE 3-2 Out-of-Plane Vibration of Right Angles and U Bends[a]

	λ for Use in Eq. (3-8)					
	Right Angles (Fig. 3.4*a*)			U Bends (Fig. 3.4*b*)		
L/R	$P–P$	$P–C$	$C–C$	$P–P$	$P–C$	$C–C$
0	9.5	9.5	9.5	5.8	5.9	5.9
0.4	7.5	7.6	7.8	5.1	5.3	5.3
0.8	6.0	6.4	6.8	4.7	4.8	4.9
1.2	5.0	5.4	5.9	4.3	4.4	4.6
1.6	3.5	4.1	5.0	3.6	3.8	4.2
2.0	2.6	3.0	3.8	2.7	2.9	3.6

[a]See footnote *a* of Table 3.1. Vibration perpendicular to the plane of Figs. 3.4*a*and 3.4*b*. λ values for $\nu = 0.3$.

3.4 SIMPLE FRAMES

The simple frames shown in Figure 3.5 are also called portal frames in structural applications or bents in electronic applications. The following formulas provide approximate values for the fundamental natural frequencies in the specified vibration modes.[3]

For in-plane vertical vibration with legs hinged at the supports, Fig. 3.5*a*,

$$F = (1/2\pi)\{48EI_1 g/WL^3[1 - 2.25/(2K + 3)]\}^{1/2}\ \text{Hz} \qquad (3\text{-}9)$$

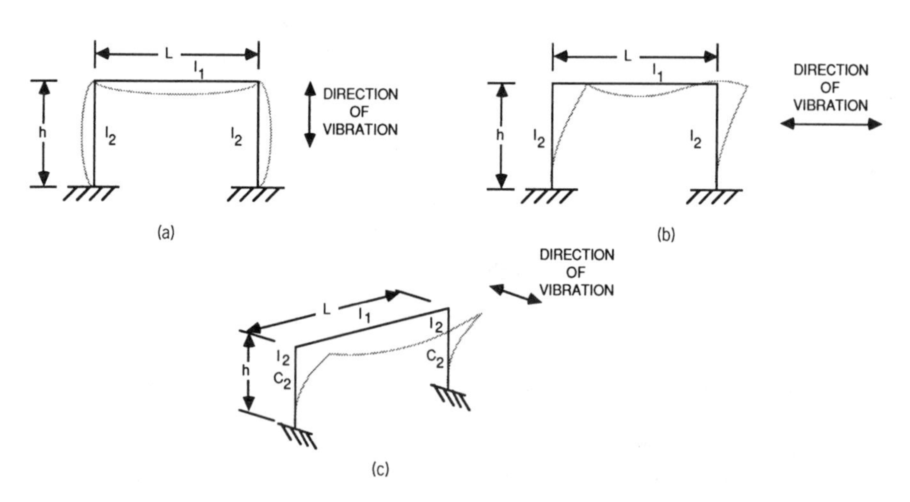

FIGURE 3.5 A simple frame in (*a*) in-plane vertical vibration, (*b*) in-plane lateral vibration, and (*c*) out-of-plane transverse vibration.

where E = modulus of elasticity of frame material (lbf/in.^{2})
I_1 = area moment of inertia of top of frame about neutral axis (in.^{4})
g = gravitational acceleration at surface of earth = 386 in./sec^2
W = total weight of frame (lbf)
L = length of top (span of frame) (in.)
$K = hI_1/LI_2$, dimensionless
h = height of frame (length of leg) (in.)
I_2 = area moment of inertia of frame legs about neutral axis (in.^{4})

For in-plane vertical vibration with legs fixed at the supports, Fig. 3.5*a*,

$$F = (1/2\pi)\{48EI_1g/WL^3[1 - 3/(2K + 4)]\}^{1/2} \text{ Hz} \qquad (3\text{-}10)$$

For in-plane lateral vibration with legs fixed at the supports, Fig. 3.5*b*,

$$F = (1/2\pi)\{24EI_2g/Wh^3[1 + 3/(6K + 1)]\}^{1/2} \text{ Hz} \qquad (3\text{-}11)$$

For out-of-plane transverse vibration with legs fixed at the supports, Fig. 3.5*c*,

$$F = (g^{1/2}/2\pi)\{(W/2)[L^3/24EI_1 + h^3/3EI_2 - L^4GC_2/32EI_1(2hEI_1 + LGC_2)]\}^{-1/2} \text{ Hz} \quad (3\text{-}12)$$

where C_2 is the torsional constant (in.^{4}).

The approximate fundamental natural frequency for a rigid body of mass M_0 supported by n slender, uniform legs of length L, all in the same plane, clamped at their feet and at the rigid body, as shown in Fig. 3.6, for vibration in the plane of the legs, is given by[4]

$$F = (1/2\pi)[(12\textstyle\sum E_iI_i)/L^3(M_0 + 0.37\sum M_i)]^{1/2} \text{ Hz} \qquad (3\text{-}13)$$

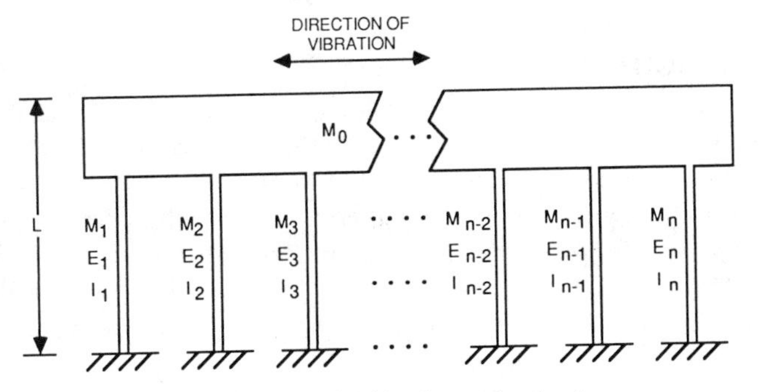

FIGURE 3.6 Rigid body on slender legs.

where M_0 = rigid body mass (lbm)
M_i = mass of leg i (lbm)
E_i = modulus of elasticity of leg i (lbf/in.^{2})
I_i = area moment of inertia of leg i about its neutral axis (in.^{4})
Σ = sum over all legs, $i = 1, 2, 3, \ldots, n$
n = number of legs ≥ 2

3.5 STIFFENED PLATES

A plate may be made stiffer by the addition of ribs, thereby increasing the natural frequency and decreasing the deflections and stresses. In Section 2.7, the orthotropic plate constants from Table 2-15 and the equivalent orthotropic constants from Table 2-16 may be used in Eq. (2-30) to obtain the natural frequencies of stiffened rectangular plates with various boundary conditions. For example, Case 4 of Table 2-16 represents a plate with parallel ribs running in one direction, added to one side of the plate. These equivalent orthotropic constants (Table 2-16) are based on the assumption that the stiffeners are an integral part of the plate with no weakness at the joints. If the stiffeners are attached to the plate by screws or bolts or other means, the stiffness of the joint and of the stiffened plate may be reduced.[5] The efficiency factor η for attachment of the stiffeners to the plate may vary from 10% ($\eta = 0.1$) for quick-disconnect fasteners to 25% or 50% ($\eta = 0.25$–0.50) for screwed or bolted joints to 100% ($\eta = 1.0$) for welded joints. This efficiency factor will also be a function of the vibration frequency and vibration acceleration for screwed or bolted joints. The efficiency factor may be taken into account in calculating the natural frequency of a stiffened plate by reducing the effective thickness of the stiffener by a factor η. For example, in Case 4 of Table 2-16, the rib thickness t would be replaced by an equivalent thickness ηt in the expression for D_x. In Case 2 of Table 2-16, the reduced rib thickness would be taken into account by replacing the area moment of inertia I_b by ηI_b in the expression for D_y.

3.6 HOUSINGS

Housings may be analyzed to estimate their fundamental natural frequencies. These frequencies may include flexural vibration along one or more axes of the structure, torsional vibration, and coupled modes of vibration. The frequencies will depend on the geometry and material properties of the structure, the attachment efficiency factor between parts of the structure (Section 3.5), and the connection of the structure to mounting surfaces (the boundary conditions).

3.6.1 Flexure

Figure 3.7 shows a housing composed of several structural elements. It is mounted by means of brackets attached at each end of the longer dimension, near the bottom (Fig. 3.7*b*). It will be analyzed as a simply supported beam that can vibrate in flexure in the *X* direction and in the *Y* direction and that can vibrate in torsion about the *Z* axis. Evaluation of the flexural frequencies requires an estimate of the stiffness factors of the structure, $E_x I_x$ and $E_y I_y$.

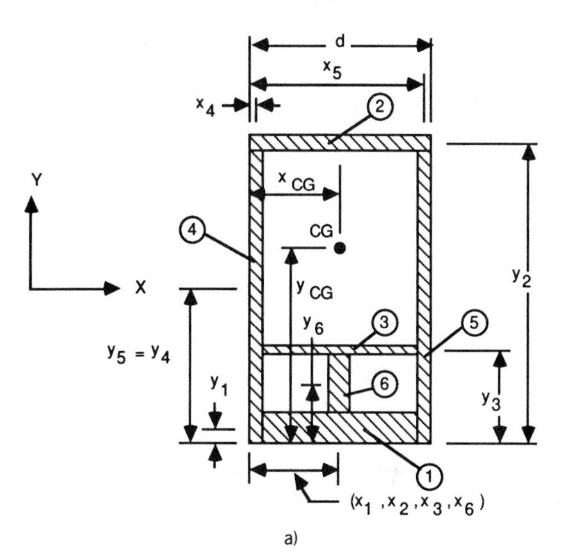

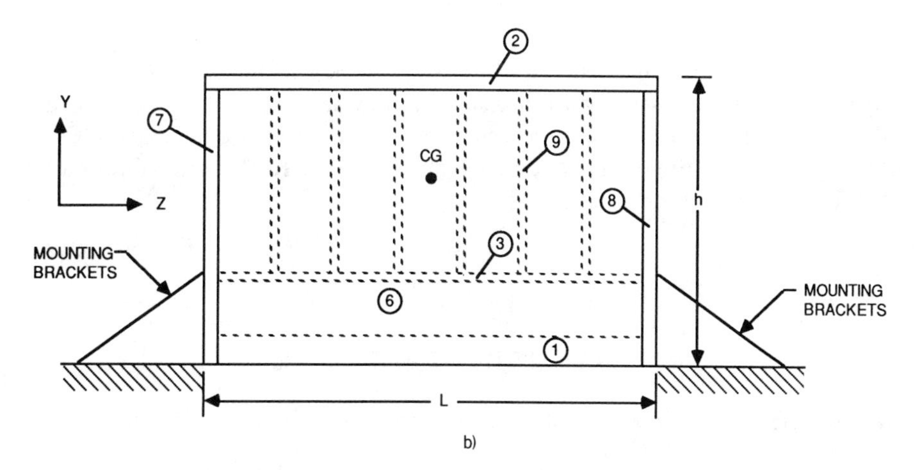

FIGURE 3.7 Housing: (*a*) side view cross section; (*b*) front view cross section.

These are determined as follows:

$$E_x I_x = \sum_{i=1}^{6} \eta_i E_i \left[A_i (\overline{X} - X_i)^2 + I_{x,i} \right] \text{ lbf} \cdot \text{in.}^2 \tag{3-14}$$

where η_i = attachment efficiency factor for element i, dimensionless (See Section 3.5)
E_i = modulus of elasticity of element i (lbf/in.^{2})
A_i = cross-sectional area of element i in the x–y plane (in.^{2})
X_i = distance from left edge of structure to neutral axis (or midpoint) of element i (in.)
$I_{x,i}$ = area moment of inertia of element i about the neutral axis parallel to the Y direction at X_i (in.^{4})

and

$$\overline{X} = \sum_{i=1}^{6} \eta_i A_i E_i X_i \bigg/ \sum_{i=1}^{6} \eta_i A_i E_i \text{ in.} \tag{3-15}$$

$$E_y I_y = \sum_{i=1}^{6} \eta_i E_i \left[A_i (\overline{Y} - Y_i)^2 + I_{y,i} \right] \text{ lbf} \cdot \text{in.}^2 \tag{3-16}$$

where η_i, E_i, and A_i are defined following Eq. (3-14), and
Y_i = distance from bottom of structure to neutral axis (or midpoint) of element i (in.)
$I_{y,i}$ = area moment of inertia of element i about the neutral axis parallel to the X axis at y_i (in.^{4})

and

$$\overline{Y} = \sum_{i=1}^{6} \eta_i A_i E_i Y_i \bigg/ \sum_{i=1}^{6} \eta_i A_i E_i \text{ in.} \tag{3-17}$$

Note that structural elements 7, 8, and 9 are not included in Eqs. (3-14) through (3-17) because they are not subjected to bending, but are either fixed to the mounting brackets or displaced parallel to their own plane. Using Case 2 from Table 2-1, the fundamental natural frequencies for flexural vibration in the X and Y directions are

$$F_x = (\pi / 2L^2)(E_x I_x g / W)^{1/2} \text{ Hz} \tag{3-18}$$

and

$$F_y = (\pi / 2L^2)(E_y I_y g / W)^{1/2} \text{ Hz}$$

where $E_x I_x$ and $E_y I_y$ are found from Equations (3-14) and (3-16), and
L = length of housing, Fig. 3.7b (in.)
g = acceleration of gravity at surface of earth = 386 in./sec^2
W = weight of housing per unit length (lbf/in.) (total weight = WL)

If the ends of the housing, structural elements 7 and 8, were mated with and fixed to mounting surfaces, then Case 3 of Table 2-1 would apply and the constant in Eq. (3-18) would be $(22.373/\pi)$ instead of π.

3.6.2 Torsion

Acceleration in the X direction will produce torsion as well as bending, since the housing is supported near its bottom and the center of gravity (CG) is located above the support (Fig. 3.7*b*). The axis of rotation for torsion will be at the bottom of the housing parallel to the Z axis. See Fig. 3.8. Since the CG is not on the axis of rotation, the torsional natural frequency of the housing will be coupled to the flexural natural frequency F_x.

The frequency formula for torsional vibration is taken from Table 2-3:

$$F_\theta = (\lambda/2\pi L)\left(CGg/\mu I_\mathrm{p}\right)^{1/2}\ \mathrm{Hz} \tag{3-19}$$

where C = torsional constant of beam cross section (in.^{4})
G = shear modulus of beam material (lbf/in.^{2})
μ = weight density of beam material (lbf/in.^{3})
I_p = polar area moment of inertia of beam cross section about the beam axis of torsion (in.^{4})

Cases 1 and 3 of Table 2-3 do not take into account the polar mass moment of inertia J about the axis of rotation, which is an important factor in the present case. The half-housing shown in Fig. 3.8 may be analyzed using Case 2 of Table 2-3 with the frequency constant λ found by solving the transcendental

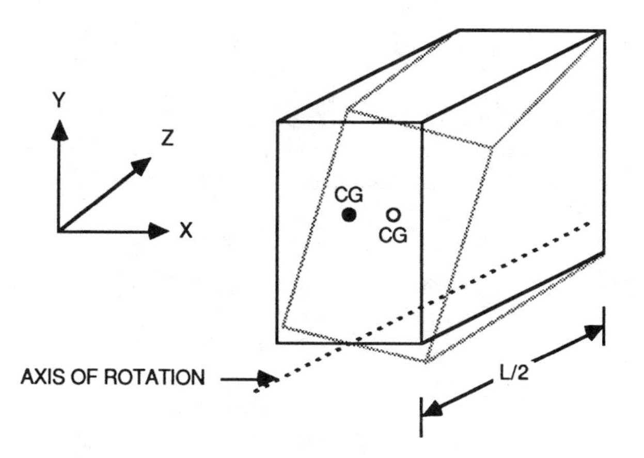

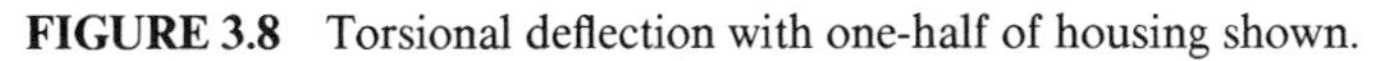
FIGURE 3.8 Torsional deflection with one-half of housing shown.

equation that includes the parameter J:

$$\cot \lambda = (Jg/\mu LC)\lambda \tag{3-20}$$

For the half-housing, L must be replaced by $L/2$ in Eqs. (3-19) and (3-20), and J must be the polar mass moment of inertia about the axis of rotation for the half-housing. The same results may be obtained by using the formula

$$F_\theta = (1/2\pi)(4CG/LJ)^{1/2} \text{ Hz} \tag{3-21}$$

instead of Eq. (3-19) when $(\mu LI_p/gJ)^{1/2} \ll 1$. In Eq. (3-21), L is the full length of the housing and J is the polar mass moment of inertia about the axis of rotation for the full housing. C and G have the same values as in Eqs. (3-19) and (3-20). Using the relation $\mu = Mg/LA$ lbf/in.^{3}, where M is the total mass of the housing and A is the housing cross-sectional area in the X–Y plane, and equating Eqs. (3-19) and (3-21) yields:

$$I_p/A = \lambda^2 J/M \text{ in.}^2 \tag{3-22}$$

where the radius of gyration is $(I_p/A)^{1/2}$. In arriving at Eq. (3-22), L was replaced by $L/2$ in Eq. (3-19).

The value of the polar area moment of inertia about the axis of rotation I_p may be estimated by

$$I_p = (E_x I_x + E_y I_y)/E_a \text{ in.}^4 \tag{3-23}$$

where E_a is an appropriate average value for the structure. In the case where the structure is composed primarily of a single structural material, E_a is the modulus of elasticity for that material.

The value of J is given by

$$J = \sum_{i=1}^{9} \left[J_{z,i} + m_i \left(x_i^2 + y_i^2 \right) \right] \text{ lbm} \cdot \text{in.}^2 \tag{3-24}$$

where x_i, y_i are as previously defined, and

$J_{z,i}$ = polar mass moment of inertia about axis through the neutral axis of element i and parallel to the axis of rotation (lbm · in.^{2})

m_i = mass of element i (lbm)

A more rapid but less accurate approximation for J is

$$J = (M/12)(4h^2 + d^2) \text{ lbm} \cdot \text{in.}^2 \tag{3-25}$$

where M is the total mass of the housing. (See Fig. 3.7.)

If J is calculated for the entire housing, then only half of its value must be used in Eq. (3-20).

The torsional constant C is much more difficult to estimate accurately, even for a simple structure. Table 2-2 may be used, supported by other material,[6] or the following rough approximation may be used:

$$C \simeq \tfrac{1}{2} I_p \text{ in.}^4 \tag{3-26}$$

3.6.3 Coupled Modes

As pointed out previously, acceleration in the X direction produces both bending at a frequency F_x and torsion at a frequency F_θ. These vibration modes will be coupled to produce a fundamental natural mode of the structure, which can be approximated by Dunkerley's method[7]:

$$F_c = \left(F_x^{-2} + F_\theta^{-2}\right)^{-1/2} \text{ Hz} \tag{3-27}$$

where F_x and F_θ are found from Eqs. (3-18) and (3-19). The coupling of modes of vibration always results in a natural frequency lower than the coupled frequencies, and a consequent increase in deflection and stress. In the example given in the preceding two sections, this coupling of modes may be avoided by mounting the housing so that the center of gravity (CG) lies on the mounting plane. Torsional modes may still occur, but they will not be coupled with the bending mode. Figure 3.9 is an example of a CG mount.

3.6.4 Other Housing Configurations

There are other housing geometries and mounting configurations where the housing may be modeled as a beam or a plate with boundary conditions that approximate the mounting attachments. In these cases the frequency formulas of Sections 2.4, 2.6, and 2.7 may be used to estimate the fundamental natural frequency of the housing. The flexural stiffness factors EI and the torsional frequency parameters C, I_p, and J must be estimated as in Section 3.6.3.

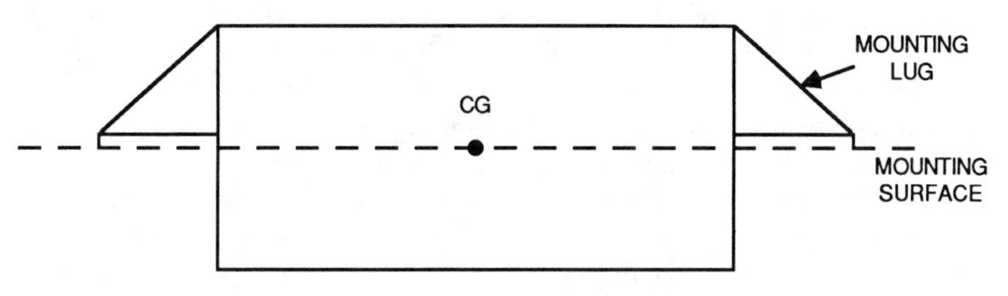

FIGURE 3.9 A center-of-gravity (CG) mount.

The examples shown in Fig. 3.10 could be analyzed as follows:
Figure 3.10*a*:

1. For $L > h, t$: A beam with simply supported ends. Flexural vibration in X and Z directions. Torsional vibration about the axis through centroid, parallel to Y axis. Flexural and torsional modes not coupled.
2. For $L \simeq h > t$: A plate with two opposite sides simply supported, other two sides free. Flexural vibration in Z direction.
3. For $L \simeq t > h$: A plate with four corner supports. Flexural vibration in X direction.

Figure 3.10*b*: For $L \simeq h > t$: A plate with three simply supported sides. Flexural vibration in Z direction.
Figure 3.10*c*: Same as Fig. 3.10*b*.
Figure 3.10*d*:

1. For $h > L, t$: A cantilever beam. Flexural vibration in Y and Z directions. Torsional vibration about the axis through centroid, parallel to X axis. Flexural and torsional modes coupled if CG is not at the midpoint of the L and the t dimensions.
2. For $L \simeq h > t$: A plate simply supported on one side with other three sides free. Flexural vibration in Z direction. Torsional vibration about the axis through centroid parallel to X axis. Flexural and torsional modes coupled if CG is not at the midpoint of the L dimension.

3.7 LUMPED ELEMENTS

One type of model that may be used to represent structures is the lumped-element model. In this approach, parts of the structures are treated as masses and other parts as springs. Rigid, heavy components may be treated as masses, while flexible, lightweight components may be treated as springs. The spring elements may be some of the same structural components that make up the masses, even though they are treated as massless in the analysis. The combined weight of the masses must add up to the total weight of the complete structure.

Figure 3.11 shows a three-degree-of-freedom structure composed of a transformer mounted on a bracket that is attached to a PWB mounted in a housing. The bottom of the housing is fixed to a mounting plate. When the acceleration is parallel to the mounting plate and normal to the PWB, the housing will vibrate as a cantilever, the PWB will vibrate as a loaded plate, and the bracket will vibrate in the direction shown. Figure 3.12 is a lumped-element model of the structure shown in Fig. 3.11. The values of the spring

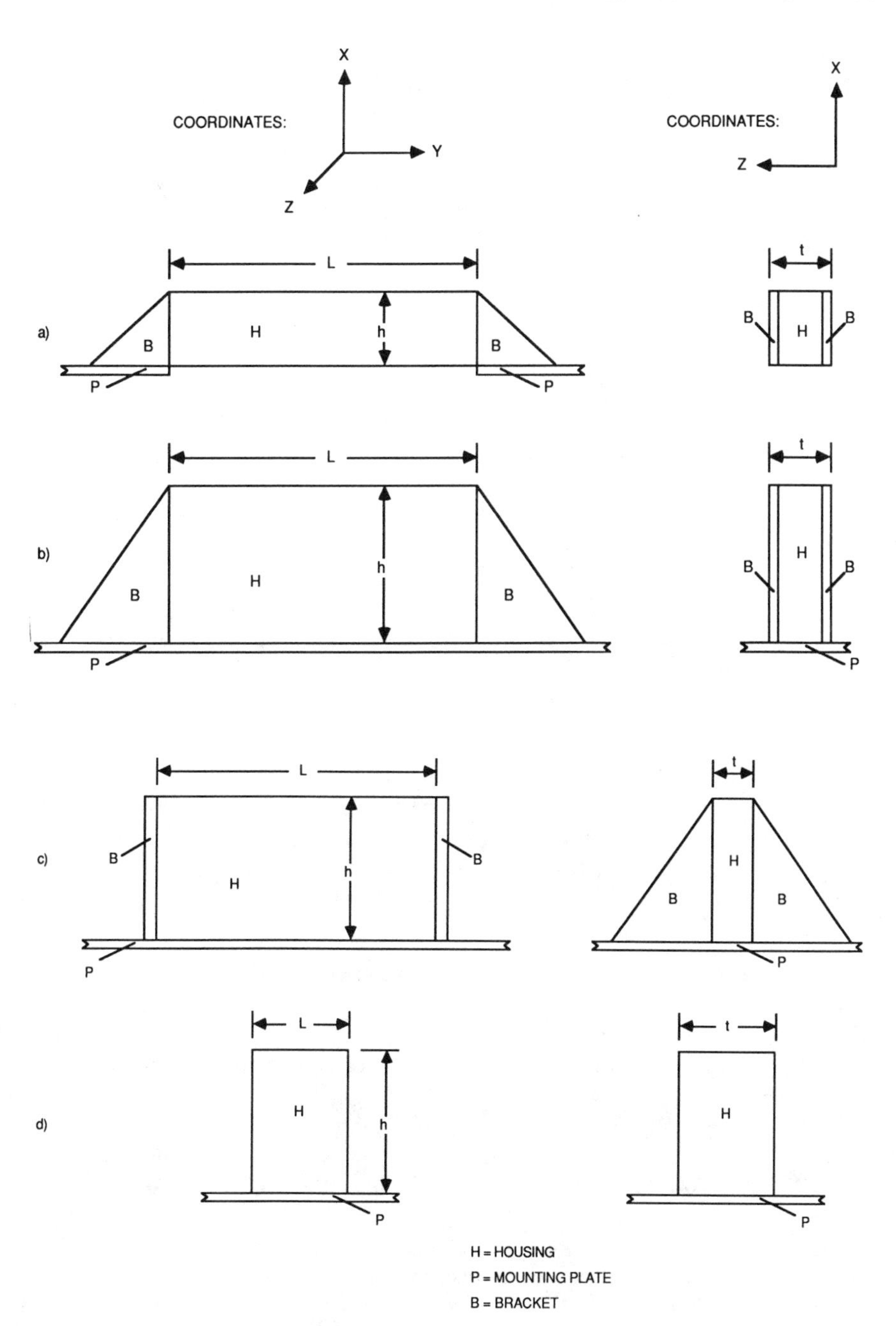

FIGURE 3.10 Examples of housing and mounting configurations.

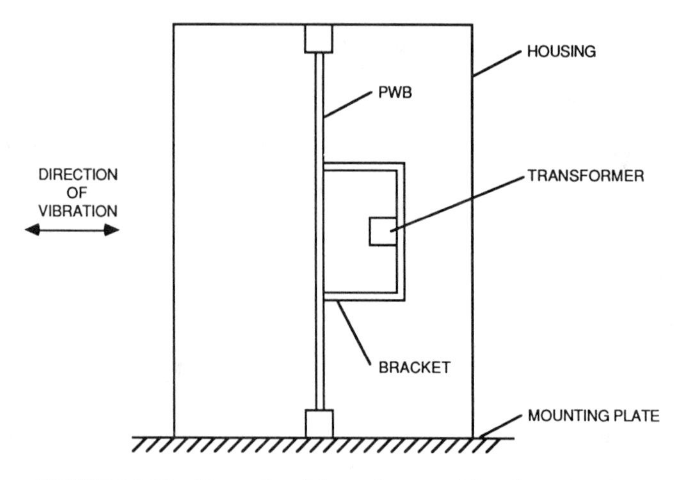

FIGURE 3.11 Example of three-degree-of-freedom structure.

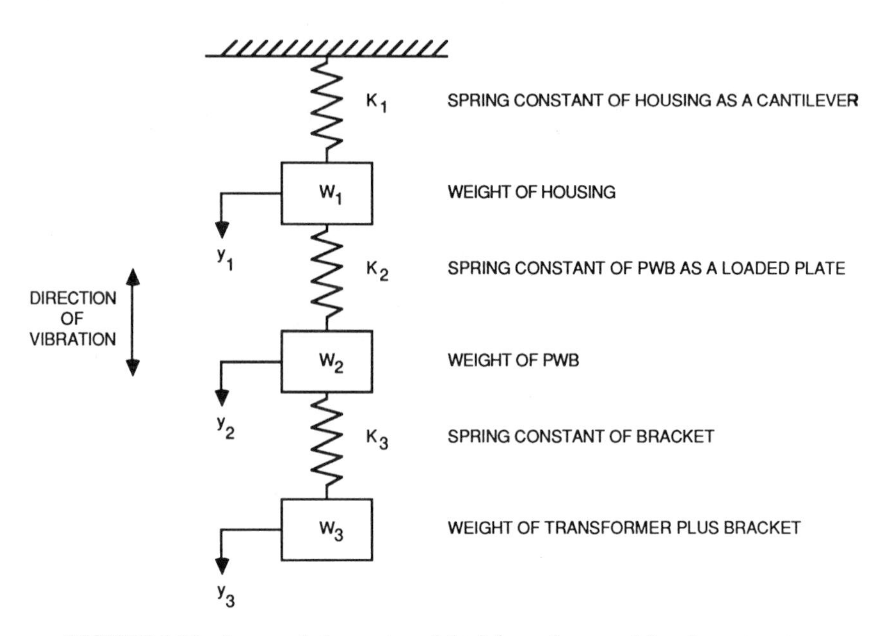

FIGURE 3.12 Lumped-element model of three-degree-of-freedom structure.

constant K may be determined by use of the following equation:

$$K = (W/g)(2\pi F)^2 \text{ lbf/in.} \tag{3-28}$$

where W = weight of element (lbf)
g = acceleration of gravity at surface of earth = 386 in./sec^2
F = fundamental natural frequency of the element (Hz)

The frequency formulas for finding the values of F in Eq. (3-28) are given by Eq. (3-9) for the bracket, by Table 2-12 for the PWB and by Table 2-1 for the cantilever housing. The static deflections, for a 1-g acceleration, are

$$
\begin{aligned}
y_1 &= (W_1 + W_2 + W_3)/K_1 \text{ in.} \\
y_2 &= y_1 + (W_2 + W_3)/K_2 \text{ in.} \\
y_3 &= y_1 + y_2 + W_3/K_3 \text{ in.}
\end{aligned}
\tag{3-29}
$$

The fundamental natural frequency of the structure shown in Figs. 3.11 and 3.12 is[8]

$$F = \frac{1}{2\pi}\left[g \sum_{i=1}^{3} W_i y_i \Big/ \sum_{i=1}^{3} W_i y_i^2\right]^{1/2} \text{ Hz} \tag{3-30}$$

For a structure with n degrees of freedom, which can be represented by n spring/mass elements in series,

$$y_i = \sum_{j=1}^{i-1} y_j + \sum_{j=i}^{n} W_j/K_i \text{ in.} \tag{3-31}$$

and

$$F = \frac{1}{2\pi}\left[g \sum_{i=1}^{n} W_i y_i \Big/ \sum_{i=1}^{n} W_i y_i^2\right]^{1/2} \text{ Hz} \tag{3-32}$$

REFERENCES

1 D. S. Steinberg, *Vibration Analysis For Electronic Equipment*, New York; Wiley-Interscience, 1973, pp. 127–136.

2 L. S. S. Lee, *J. Eng. Industry* **97**, 23–32 (1975).

3 D. S. Steinberg, in Ref. 1, pp. 246–249.

4 R. D. Blevins, *Formulas for Natural Frequency and Mode Shape*, New York: Van Nostrand Reinhold, 1979, p. 221.

5 D. S. Steinberg, in Ref. 1, pp. 304–306.

6 R. J. Roark and W. C. Young, *Formulas for Stress and Strain*, 5th ed., New York: McGraw-Hill; 1975 pp. 290–303.

7 D. S. Steinberg, in Ref. 1, pp. 63–64.

8 D. S. Steinberg, in Ref. 1, p. 50.

4

RANDOM VIBRATION

The imposition of a random-vibration environment on a structure or a component is one of the most severe sources of acceleration loads that the structure must survive without a sacrifice of mechanical integrity.

Three different types of vibration were discussed in Chapter 2: flexural (bending) vibration in which there is no stretching or compression of the structural elements; extensional vibration in which structural elements undergo stretching and compression; and torsional vibration in which elements twist about their long axis. It was pointed out in Section 2.3 that bending modes of vibration were of greatest interest since they usually have the lowest frequencies, largest displacements, and largest stresses. This chapter primarily addresses flexural vibrations, and, to a lesser extent, torsional vibrations, and reviews the processes for estimating vibration-induced loads and their resultant deflections and stresses. The discussion is limited to ductile materials and to strains that are in the elastic range of the material.

4.1 OVERVIEW OF VIBRATION ANALYSIS

Figure 4.1 is a schematic that illustrates the analysis process used to estimate the maximum material stresses and fatigue life of a structural element in a random-vibration environment. The vibration environment is specified by the power spectral density (PSD) and the duration of the vibration input t_0. The structural element is characterized by its fundamental natural frequency F_0, its

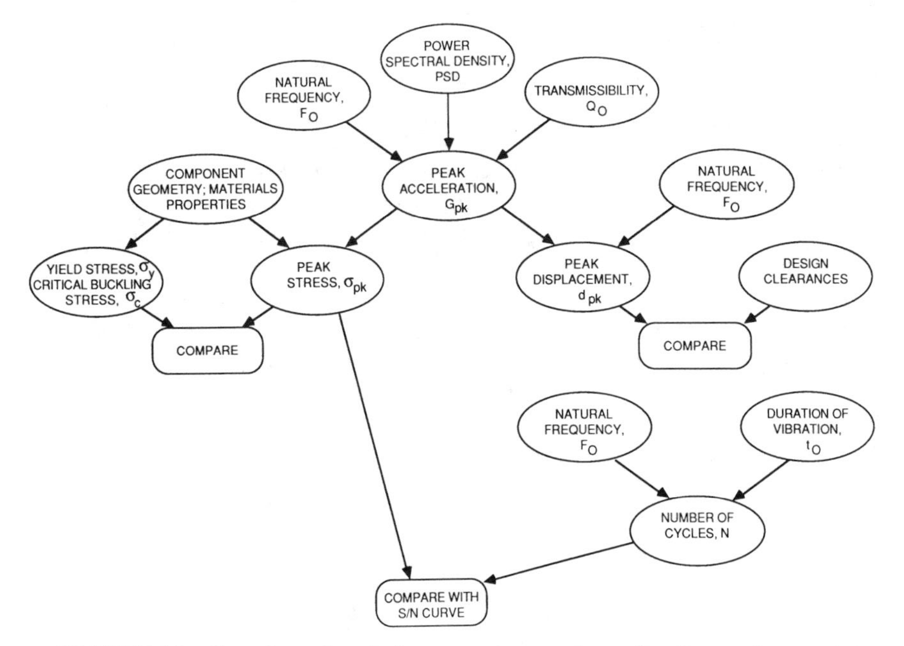

FIGURE 4.1 Overview of analysis process in a random-vibration environment.

transmissibility Q_0, its geometry, and its material properties. The three parameters F_0, PSD, and Q_0 determine the peak acceleration G_{pk}. F_0 and G_{pk} determine the peak displacement d_{pk}, which may be compared with the design clearance to identify undesirable contacts between structural parts during vibration. The component geometry, material properties, and G_{pk} are used to estimate the peak material stress σ_{pk}, which may be compared with the material yield strength σ_y or the critical buckling stress σ_c if elastic instability can occur. The product of F_0 and the specified duration of the random-vibration input t_0 will yield the number of vibration cycles N. The values of σ_{pk} and N may then be compared to the S/N curve characteristic of the structural material in order to estimate fatigue life margin.

4.2 PEAK ACCELERATION

The value of the peak acceleration of a structure G_{pk} in a random-vibration environment is found from the frequency F_0 and the transmissibility Q of the structure and the specified PSD of the random-vibration environment. See Eq. (4-8). The value of F_0 is found as outlined in Chapters 2 and 3; the parameters PSD and Q are discussed in the following subsections.

4.2.1 Power Spectral Density

Figure 4.2 illustrates a typical random-vibration PSD to which, for example, a space structure would be exposed during launch. The vertical axis is the PSD in units of g^2/Hz, where $g = 386\ \text{in./sec}^2$ is the gravitational acceleration at the earth's surface. The horizontal axis is the frequency in Hz. Both axes are usually on a log scale, as shown in the figure, although linear scales may be used.[1] In the following, the log–log format is addressed. The PSD is usually specified in terms of initial and final values at the low- and high-frequency ends of the spectrum; plateau values where the PSD is constant over a frequency range; and slopes in units of dB per octave over frequency ranges where the PSD is increasing or decreasing with increasing frequency. In addition, a value of the root-mean-square acceleration over the entire specified frequency range G_{rms} is usually provided. The dimensionless slope m between two points on the log–log PSD plot is given by

$$m = \ln(\text{PSD}_2/\text{PSD}_1)/\ln(F_2/F_1) \tag{4-1}$$

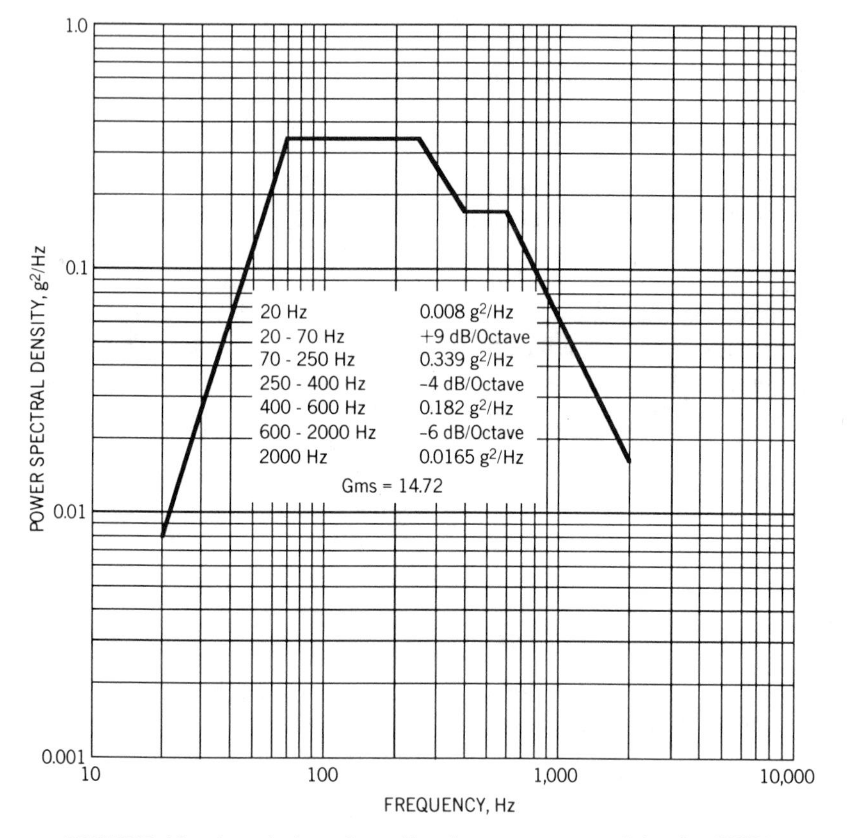

FIGURE 4.2 A typical random-vibration power spectral density (PSD).

where the logarithm may be to base e, base 10, or any base, and

F_2 = the higher-frequency point on the PSD plot (Hz)
F_1 = the lower-frequency point on the PSD plot (Hz)
PSD_2 = the PSD value at frequency F_2
PSD_1 = the PSD value at frequency F_1

The slope m' in units of dB per octave is given by

$$m' = 3m \text{ dB/octave} \tag{4-2}$$

The slopes m and m' may be either positive or negative. The mean PSD value $\overline{PSD}$ between F_1 and F_2 is

$$\overline{PSD} = [(PSD_1)/(1+m)]\left[F_2(F_2/F_1)^m - F_1\right] \tag{4-3}$$

In the case of a plateau of constant PSD value, $m = 0$ and Eq. (4-3) becomes

$$\overline{PSD} = PSD_1(F_2 - F_1) \tag{4-4}$$

The value of G_{rms} over the entire PSD spectrum is

$$G_{rms} = \left(\sum \overline{PSD}\right)^{1/2} \tag{4-5}$$

where Σ is the sum of all of the PSD segments in the spectrum. G_{rms} is a measure of the overall acceleration level of the PSD spectrum.

4.2.2 Transmissibility

When a harmonic input force acts on a structure, the structure will undergo forced vibration at the same frequency as the input force. The transmissibility Q is the dimensionless ratio of the maximum output force (or acceleration) of the structure P_{out} to the maximum input force (or acceleration), P_{in}:

$$Q = P_{out}/P_{in} = \left\{\left[1 + (2R_F R_C)^2\right] \Big/ \left[\left(1 - R_F^2\right)^2 + (2R_F R_C)^2\right]\right\}^{1/2} \tag{4-6}$$

where $R_F = F/F_0$ is the frequency ratio
F = frequency of harmonic input force (Hz)
F_0 = fundamental natural frequency of the structure (Hz)
$R_C = C/C_c$
C = damping of the structure (1bf · sec/in.)
C_c = critical damping (1bf · sec/in.)

Equation (4-6) is for a parallel spring and viscous dashpot combination commonly used to represent mechanical structures. See Chapter 6 for a more-detailed explanation and for other types of damping.

When the structure is critically damped, a displacement from its equilibrium configuration and subsequent removal of external forces to allow free vibration will result in the structure just returning to its equilibrium configuration without any "overshoot" or oscillatory motion. When the structure is not critically damped, there will be oscillatory motion of decreasing amplitude with time for the case where this actual damping, C, is less than critical, $R_C < 1$. For overdamping, $R_C > 1$, there is no oscillatory motion. For the case where $F = F_0$, or $R_F = 1$, Eq. (4-6) becomes

$$Q_0 = \left[1 + (2R_C)^2\right]^{1/2} \Big/ (2R_C) \tag{4-7}$$

where Q_0 is the transmissibility at the structure's natural resonance frequency and Q_0 is a maximum.

When R_C or Q_0 cannot be measured or estimated, rules of thumb may be used:

- For printed circuit boards (PCB):
  - Generally, $5 < Q_0 < 15$.
  - $Q_0 \approx F_0^{1/2}$, where F_0 is the natural frequency of the PCB.
  - $Q_0 \approx \frac{1}{2}F_0^{1/2}$ for spring edge guide supports or for a large PCB with stiffening ribs and $F_0 \approx 100$ Hz.
  - $Q_0 \approx 2F_0^{1/2}$ for a small PCB with no stiffening ribs and $F_0 \approx 400$ Hz.
  - For a given PCB, Q_0 is larger for lower values of input acceleration.
- For cantilevered beams supporting electronic components, $Q_0 \approx 1.8F_0^{1/2}$.
- For a rectangular chassis supported at the ends of the long dimension, with removable top and bottom covers, $Q_0 \approx 8$.
- For a large ($6 \times 6 \times 20$ in.^{3}) electronic chassis with a complex internal structure and different natural frequencies in different parts of the system, $3 < Q_0 < 17$.
- For "light" damping, $C < 0.2$, the value of Q will be proportional to
  - the square root of the system stiffness;
  - the square root of the system mass;
  - the inverse of the system damping.

4.2.3 Estimation of Peak Acceleration

The peak acceleration response of a system to a random vibration environment can be approximated by

$$G_{\text{pk}} = 3\left[(\pi/2)(\text{PSD})Q_0F_0\right]^{1/2} g \tag{4-8}$$

where PSD = value of PSD at F_0, from PSD spectrum, in g^2/Hz
Q_0 = transmissibility at F_0, dimensionless
F_0 = natural frequency of system (Hz)
g = gravitational acceleration at surface of earth (386 in./sec^2)

Equation (4-8) is a good estimate of G_{pk} when there is a fairly uniform value of PSD over a frequency range of $F_{min} \leq 0.5F_0$ to $F_{max} \geq 1.5F_0$. This value of G_{pk} is a three-sigma value, which means that an acceleration response greater than that given by Eq. (4-8) will occur only 0.28% of the time.

4.3 PEAK DEFORMATION

If the deformation of a component or structure is known for a 1-g acceleration applied in a specified direction with relation to the structure, then the deformation due to an acceleration of G_{pk} in the same direction will be

$$d_{pk} = (G_{pk}/g)\,d_0 \text{ in.} \tag{4-9}$$

where d_0 is static deformation (in inches) due to a 1-g acceleration provided d_{pk} is within the linear elastic range of the structural materials and elastic instability (buckling) does not occur.

Accurate values of 1-g deformations have been calculated for many structural elements,[2] and some of these are presented in Table 4-1. For complex structures, sine-sweep and random-vibration tests yield values of accelerations as a function of frequency, and analysis procedures also provide estimates of acceleration as a function of frequency for vibration environments (see Section 4.2). An approximation of the single-amplitude peak deformation is given by

$$d_{pk} = 9.8G_{pk}/F_0^2 \text{ in. for linear deformations} \tag{4-10}$$

where G_{pk} is estimated as in Section 4.2 and F_0 is the structure's natural frequency.

The estimate of Eq. (4-10) is a three-sigma value at the location of maximum structural deformation, for example, in the center of a uniform plate with uniform edge supports or at the free end of a cantilever beam. Deformations greater than d_{pk} will occur only 0.28% of the time.

The value of d_{pk} from Eqs. (4-9) or (4-10) may be compared to the design clearance between parts where the peak deformation occurs. If the clearance is not greater than d_{pk}, dynamic contact may occur during vibration and result in damage or loss of the mechanical integrity of the structure. For PCBs the peak deformation may also affect the fatigue life of wire leads and solder joints. Extensive testing of PCBs indicates that a fatigue life of at least 10^7 cycles can be expected if the peak deformation at the center of the PCB is limited to[3]

$$d_{pk} \leq 0.003b/L^2 \text{ in.} \tag{4-11}$$

TABLE 4-1 Maximum Stress and Displacement in Transversly Loaded Uniform, Slender Beams

$\sigma_{pk} = G_{pk}\sigma_m$ = peak material stress (psi)
$d_{pk} = G_{pk}d_m$ = peak displacement from equilibrium (in.)

where G_{pk} = peak acceleration in g units (Section 4.2)
σ_m = maximum material stress with 1-g acceleration normal to beam (psi)
d_m = maximum displacement from equilibrium with 1-g acceleration normal to beam (in.)
g = gravitational acceleration at surface of earth = 386 in./sec^2
F = free end of beam
S = simply supported (pinned) end of beam
C = clamped (fixed) end of beam
x = distance along beam axis from left end of beam (in.)
L = length (span) of beam (in.)
a = distance from left end of beam to concentrated load, W (in.)
$R = a/L$
t = thickness (depth) of beam (in.)
b = width of beam (in.) (Table 4-2).
c = distance from neutral axis to extreme point on beam section (in.)
e = effective width of beam (in.)
W = concentrated load at $x = a$ (lbf)
w = uniform unit load (lbf/in. of longitudinal axis)
E = modulus of elasticity of beam material (lbf/in.^{2})
I = area moment of inertial of beam cross section about centroidal axis (in.^{4})

Cases 1–4. Concentrated load at $x = a$; $R = a/L$

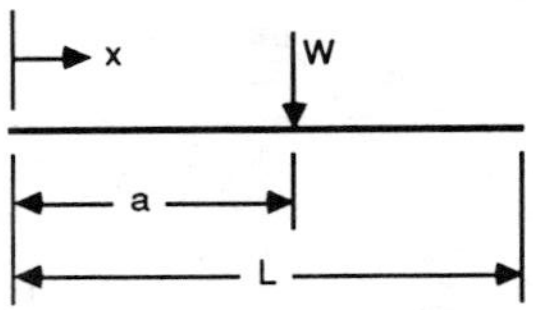

Case	Maximum stress	Maximum deflection
1. Left end C; right end F; Eq. (2-9)	$\sigma_m = (WLc/I)R$ at $x = 0$	$d_m = (W/6EI)L^3R^2(3 - R)$ at $x = L$
2. Left end C; right end S (frequency formula intermediate to Cases 3 and 4)	$\sigma_m = (WLc/I)R(2 - 3R + R^2)/2$ at $x = 0$	For $R \leq 0.5858$ $d_m = (W/6EI)L^3R^2(1 - R)$ $[(1 - R)/(3 - R)]^{1/2}$ at $x = [(1 - R)/(3 - R)]^{1/2}L$
	$\sigma_m = (WLc/I)$ $R^2(3 - 4R + R^2)/2$ at $x = a$	For $R \geq 0.5858$ $d_m = (W/3EI)L^3(1 - R)(2R - R^2)^3/$ $(2 + 2R - R^2)^2$ at $x = [(2 - 2R + R^2)/(2 + 2R - R^2)]L$
3. Left end C; right end C; Eqs. (2-12) and (2-13)	$\sigma_m = (WLc/I)R^2(1 - R)$ at $x = 0$ or $x = L$, where $R = \max[(a/L), (1 - a/L)]$	$d_m = (W/1.5EI)L^3R^2(1 - R)^3/(3 - 2R)^2$ at $x = [2(1 - R)/(3 - 2R)]L$, where $R = \min[(a/L), (1 - a/L)]$

TABLE 4-1 *(Continued)*

Description of Loading and End Supports[a]	Magnitude and Location of Maximum Stress σ_m (psi)	Magnitude and Location of Maximum Deflection d_m (in.)
4. Left end S; right end S; Eqs. (2-10) and (2-11)	$\sigma_m = (WLc/I)R(1 - R)$ at $x = a$	$d_m = (W/3EI)L^3(1 - R)[(2R - R^2)/3]^{3/2}$ at $x = \{1 - [(2R - R^2)/3]^{1/2}\}L$ where $R = \max[(a/L), (1 - a/L)]$
Cases 5–8. Uniform load on entire span.		
5. Left end C; right end F; Table 2-1, Case 1	$\sigma_m = (wL^2c/I)(1/2)$ at $x = 0$	$d_m = (w/EI)(L^4/8)$ at $x = L$
6. Left end C; right end S; Table 2-1, Case 3	$\sigma_m = (wL^2c/I)(1/8)$ at $x = 0$	$d_m = (w/EI)(0.0054L^4)$ at $x = 0.5785L$
7. Left end C; right end C; Table 2-1, Case 4	$\sigma_m = (wL^2c/I)/12$ at $x = 0$ and $x = L$	$d_m = (w/EI)(L^4/384)$ at $x = L/2$
8. Left end S; right end S; Table 2-1, Case 2.	$\sigma_m = (wL^2c/I)/8$ at $x = L/2$	$d_m = (w/EI)L^4(5/384)$ at $x = L/2$

[a] The equation number or the table number and Case number are of the applicable frequency formula from Chapter 2, for use in finding G_{pk}.

where b is the short side of a rectangular PWB (in.) and L is the maximum length of a component mounted on the PWB (in.) Combining Eqs. (4-10) and (4-11) yields a limit on component length for achievement of a long random-vibration fatigue life:

$$L \leq 0.0175\left(b/G_{\text{pk}}\right)^{1/2} F_0 \text{ in.} \tag{4-12}$$

4.4 STRESS

The stress in the structure will be determined by the structure's geometry and material properties and by the loads and displacements due to the random-vibration environment (Section 4.3). The maximum stresses in the structure must be compared to the yield stresses and ultimate stresses of the materials to determine the safety factors protecting structural integrity. In designs where elastic instability (buckling) is possible, the structural stresses must be compared to the critical buckling stresses to establish a safety factor against buckling.

4.4.1 General, Principal, and Equivalent Stresses

The most general case for the stress at a point in the material consists of tensile or compressive stresses along three orthogonal axes (triaxial stress) and orthogonal shear stresses, normal to the axes, on each of the three planes defined by these axes. In Fig. 4.3, this case is broken down into separate figures for clarity. Figure 4.3*a* shows the three tensile or compressive stresses σ_x, σ_y, and σ_z, acting along the three axes x, y, and z. Figure 4.3*b* shows the shear stress τ_{xy} in the YZ plane acting in the Y direction, and the equal shear stress τ_{yx} in the XZ plane acting in the X direction. The first subscript refers to the normal to the plane in which the stress acts; the second subscript refers to the direction of the stress. Figure 4.3*c* shows the shear stress τ_{xz} in the YZ plane acting in the Z direction, and the equal shear stress τ_{zx} in the XY plane acting in the X direction. Figure 4.3*d* shows the shear stress τ_{yz} in the XZ plane acting in the Z direction and the equal shear stress τ_{zy} in the XY plane acting in the Y direction. The equalities

$$\tau_{xy} = \tau_{yx}, \quad \tau_{xz} = \tau_{zx}, \quad \tau_{yz} = \tau_{zy} \tag{4-13}$$

are necessary for equilibrium. In the most general case, all of the stresses shown in Fig. 4.3 are combined.

A rotation of the coordinate axes shown in Fig. 4.3 can reduce the general-stress case to one in which the three orthogonal planes defined by the new, rotated axes are acted on by purely normal stresses, tension or compression. These new planes are called principal planes and have no shear stress on

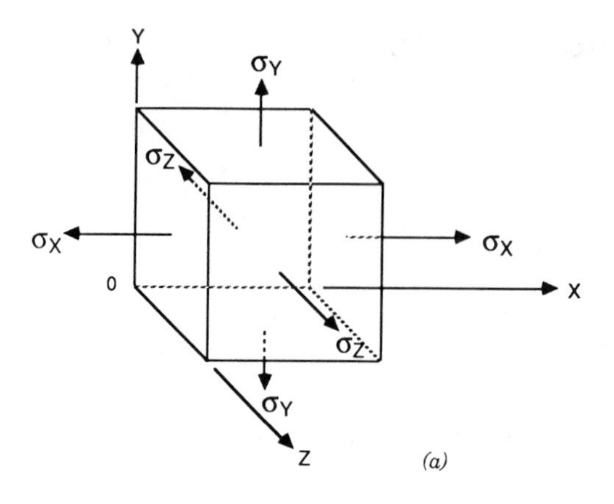

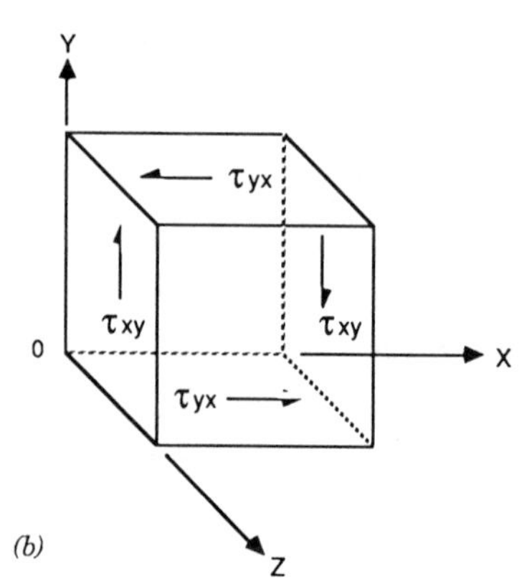

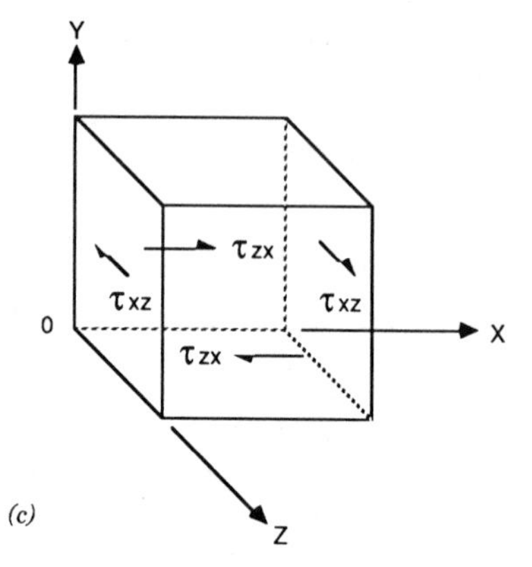

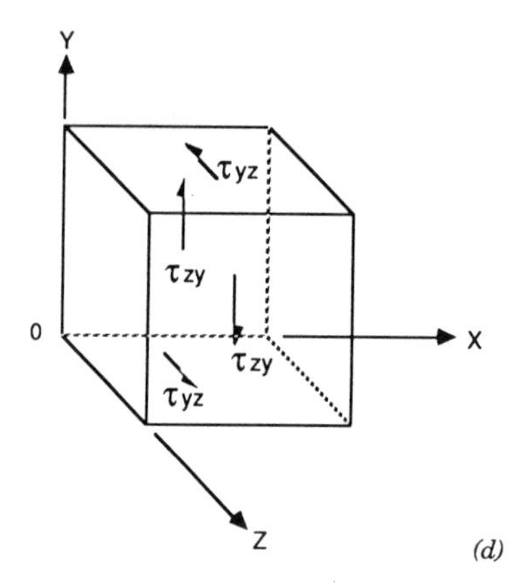

FIGURE 4.3 General stress case. In the general case, all of these stresses are superimposed.

them. The stresses normal to the principal planes are the principal stresses. If the values of σ_x, σ_y, σ_z, τ_{xy}, τ_{xz}, and τ_{yz} are given for the general-stress case, then the magnitudes and directions (with respect to the original axes) of the principal stresses and the maximum stress may be determined. Also, the magnitude and directions of the maximum shear stresses may be determined.

The principal stresses are[4]

$$\sigma_n = a_1/3 + 2\{-(H/3)\cos[(\gamma + 2n\pi - 2\pi)/3]\}^{1/2} \quad (n = 1, 2, 3) \quad (4\text{-}14)$$

where $\gamma = \cos^{-1}[(G/2)(-H/3)^{-3/2}]$

$G = (a_3 - a_1 a_2/3 + 2a_1^3/27)$

$H = a_2 - a_1^2/3$

$a_1 = \sigma_x + \sigma_y + \sigma_z$

$a_2 = \sigma_x\sigma_y + \sigma_y\sigma_z + \sigma_z\sigma_x - \tau_{xy}^2 - \tau_{yz}^2 - \tau_{zx}^2$

$a_3 = \sigma_x\sigma_y\sigma_z + 2\tau_{xy}\tau_{yz}\tau_{zx} - \sigma_x\tau_{yz}^2 - \sigma_y\tau_{zx}^2 - \sigma_z\tau_{xy}^2$

In Eq. (4-14), σ_1 is the maximum principal stress and σ_3 is the minimum principal stress. The equivalent stress is defined as[5]

$$\sigma_e = \tfrac{1}{2}^{1/2}\left[(\sigma_1 - \sigma_2)^2 + (\sigma_2 - \sigma_3)^2 + (\sigma_3 - \sigma_1)^2\right]^{1/2} \quad (4\text{-}15)$$

and is used in von Mises' criterion, which requires

$$\sigma_e \leq \sigma_y \quad (4\text{-}16)$$

to avoid yielding of the material, where σ_y is the yield stress of the material in simple tension. When the three principal stresses are equal in magnitude and sign, it is a state of hydrostatic stress; when one principal stress is zero, it is a state of plane stress; and when two principal stresses are zero it is a state of either simple tension or simple compression.

The maximum shear stresses are given by

$$\tau = \pm(\sigma_1 - \sigma_2)/2; \quad \pm(\sigma_2 - \sigma_3)/2; \quad \pm(\sigma_3 - \sigma_1)/2 \quad (4\text{-}17)$$

These shear stresses are on planes that are parallel to one of the principal axes and inclined at 45° to the other two.

4.4.2 Stress Concentration

In identifying regions of maximum stress in a structure and in estimating the magnitudes of these stresses it is necessary to take into account any stress concentrations that may be present. Stress concentrations are increased stress that can be caused by application of a load over a small area of a component or by flaws in the component material. However, the stress concentrations that are generally of design importance are those due to the geometric features of the component, such as abrupt changes in the cross section (e.g., a stepped beam), holes, grooves, and notches. These increased stresses are accounted for by the the use of a stress-concentration factor[5] given by

$$K = \sigma_e'/\sigma_e \tag{4-18}$$

where σ_e is the equivalent stress, Eq. (4-15), in the absence of the stress concentration geometric feature, and σ_e' is the equivalent stress with the geometric feature present. Stress-concentration factors are tabulated in Chapter 7, Section 7.4.

4.5 STRESSES AND DISPLACEMENTS IN STRUCTURAL ELEMENTS

The maximum acceleration-induced material stress in a structural element may occur where the deflection is a maximum, as in the case of a simply supported uniform rectangular plate; where the deflection is a minimum, as in the case of a uniform cantilever beam; or where the deflection has an intermediate value between the two extremes. As in the case of peak deformations (Section 4.3) accurate values of 1-*g* stresses have been calculated for many structural elements.[2] Some structural elements may be approximately represented as uniform beams with a combination of free, simply supported, or clamped ends. In many of these cases the beams may have a small (< 3) span to depth ratio or a relatively large width to span ratio (≥ 1). Other structural elements may be represented as uniform plates with straight or curvilinear boundaries and a combination of free, simply supported, or clamped edges. Peak stresses for these structural elements are tabulated in the following subsections.

4.5.1 Stresses and Displacements in Beams

The maximum stress in any transverse section of a beam (any cross section normal to the beam longitudinal axis) is usually expressed in terms of the bending moment M at that section:

$$\max \sigma = Mc/I \text{ psi} \tag{4-19}$$

where M = bending moment at the section (in.· lbf)
I = moment of inertia of the beam cross section about the neutral axis (in.^{4})
c = distance from the neutral axis to the most remote point of the section (in.)

The maximum stress in the entire beam σ_m will usually occur at that longitudinal position where M is maximum. In Table 4-1, the values of max M have been used in Eq. (4-19) to obtain values of σ_m.

Table 4-1 presents the locations and magnitudes of the maximum stress σ_m and maximum displacement d_m in a uniform, slender beam subjected to a 1-g acceleration normal to its longitudinal axis.[6] Peak stresses and displacements are obtained by multiplying the 1-g values by G_{pk} as found in Section 4.2. In this table, the absolute values of σ_m and d_m are given, with σ_m always occurring at that point on the beam cross section (on the surface) farthest from the neutral axis. The x coordiante of the beam cross section at which σ_m or d_m occurs is given beneath the formulas for σ_m and d_m, respectively. Also provided, in the "Description" column, are the equation number, or the table number and case number, of the applicable frequency formula from Chapter 2, for use in finding G_{pk}. When the x locations of σ_m are the same for two or more different loading cases on the same beam, the σ_m values may be added to obtain a total σ_m, provided the total stress is within the linear elastic region. In adding σ_m values, the algebraic signs must be observed, so that loads in the same direction add and loads in opposite directions subtract. A similar comment applies to the addition of d_m values for different loading cases. For example, the σ_m values may be added together and the d_m values may be added together for Cases 1 and 5, since in both cases the σ_m occurs at $x = 0$ and the d_m occurs at $x = L$.

For a simply supported, uniformly loaded beam of small span to depth ratio L/t, the values of σ_m obtained from Case 8 of Table 4-1 must be multiplied by the factors provided in Table 4.2[6]:

$$\sigma_m(L/t) = K(L/t)(WL^2c/I)/8 \text{ psi} \tag{4-20}$$

where all of the quantities are defined in Table 4-1, and σ_m occurs at $x = L/2$, midspan. It should be noted that in this case the frequency formula (Table 2-1, Case 2) cannot be expected to be accurate, and, consequently, G_{pk} may be in error. For the same beam with clamped edges, the frequency formula from Table 2-9, Case 3, may be used to obtain G_{pk}, and the σ_m value from Table 4-1, Case 7 may be used.

For a simply supported, uniformly loaded beam of large width to span ratio b/L, the values of σ_m obtained from Table 4-1, Case 8, may be used. However, in calculating the load-bearing capacity of the beam, an effective beam width e must be used[6]:

$$e \simeq 0.95b \text{ in.} \tag{4-21}$$

where b is the actual width of beam (in.).

Equation (4-21) is accurate in the range $1 \leq b/L \leq 2$. For the same beam with clamped edges, Table 4-1, Case 7 may be used, with E replaced by $E/(1 - \nu^2)$ in the frequency formula used in Table 2-1, Case 4. Other cases of beams with large b/L ratios are better represented by rectangular flat plates, as in the following section.

TABLE 4-2 Stress Multiplication Factors for Beams of Small Span / Depth Ratio[a]

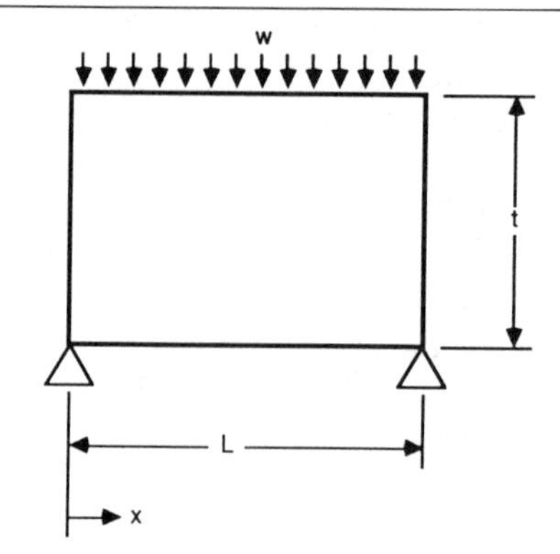

L/t	$K(L/t)$, Tension[b,c]	$K(L/t)$, Compression[b,c]
3.0	1.025	1.030
2.5	1.046	1.035
2.0	1.116	1.022
1.5	1.401	0.879
1.0	2.725	0.600
0.5	10.95	2.365
0.33	24.70	5.160

[a] For a uniformly loaded, simply supported beam of length (span) L in. and thickness (depth) t in.
[b] $K(L, t)$ is for use in Eq. (4-20).
[c] $K(L, t)$, Tension, applies to the tensile stress at the underside of the beam. $K(L, t)$, Compression, applies to the compression stress at the topside of the beam. Both of these maximum stresses occur at $x = L/2$.

For a slender beam under simultaneous axial and transverse loading, the values of σ_{m} and d_{m} may be found by applying the multiplication factors of Table 4-3[7] to the values of stress and displacement found from Table 4-1:

$$\begin{aligned} \sigma_{\mathrm{m}} &= K_{\sigma}\left(PL^2/EI\right)^{1/2}\sigma'_{\mathrm{m}} \text{ psi} \\ d_{\mathrm{m}} &= K_{d}\left(PL^2/EI\right)^{1/2} d'_{\mathrm{m}} \text{ in.} \end{aligned} \tag{4-22}$$

where the primed quantities on the right-hand side of Eq. (4-22) are the values from Table 4-1, the K_{σ} and K_d are provided in Table 4-3, and the unprimed stress and displacement on the left-hand side of Eq. (4-22) are for the simultaneous axial and transverse loading. Provided in the "Description" column of Table 4-3 are the case number from Table 4-1 for use in Eq. (4-22) and the frequency formula from Chapter 2 for use in finding G_{pk} as described in Section 4.2.

TABLE 4-3 Stress and Displacement Multiplication Factors for Beams under Simultaneous Axial and Transverse Loading

$\sigma_m = G_{pk}\sigma_m$ = peak material stress (psi)
$d_{pk} = G_{pk} d_m$ = peak displacement from equilibrium (in.)

where
G_{pk} = peak acceleration in g units, Section 4.2
σ_m = maximum material stress with 1-g acceleration normal to beam (psi), from Eq. (4-22)
d_m = maximum displacement from equilibrium with 1-g acceleration normal to beam (in.) from Eq. (4-22)
g = gravitational acceleration at surface of earth = 386 in./sec^2
$K_\sigma(PL^2/EI)$ = multiplication factor for stress, provided in the table, for use in Eq. (4-22), dimensionless
$K_d(PL^2/EI)$ = multiplication factor for displacement, provided in the table, for use in Eq. (4-22), dimensionless
P = axial load (lbf)
L = length of beam (in.)
E = modulus of elasticity of beam material (lbf/in.^{2})
I = area moment of inertia beam cross section about centroid axis (in.^{4})
F = free end of beam
S = simply supported (pinned) end of beam
C = clamped (fixed) end of beam
a = distance from left end of beam to concentrated load, W (in.)

Description of Transverse Loading, End Supports, Case Numbers, and Frequency Formulas	Description of Axial Loading, Multiplication Factors	Values of Axial Compression and Tension Are in Units of $(PL^2/EI)^{1/2}$, Dimensionless. K_σ and K_d Are Dimensionless Multiplication Factors.				
1. Uniform transverse load;	Axial Compression	0.6	1.2	1.8	2.4	3.0
left end C; right end S.	K_σ	1.0122	1.0515	1.1273	1.2635	1.5243
Table 4-1, Case 6;[b]	K_d	1.0176	1.0742	1.1846	1.3848	1.7736
Eqs. (2-6) and (2-7).	Axial Tension	1.0	2.0	4.0	8.0	12.0
for frequency	K_σ	0.9681	0.8874	0.6900	0.4287	0.3033
	K_d	0.9543	0.8397	0.5694	0.2524	0.1323

TABLE 4-3 *(Continued)*

Description of Transverse Loading, End Supports, Case Numbers, and Frequency Formulas	Description of Axial Loading, Multiplication Factors	Values of Axial Compression and Tension Are in Units of $(PL^2/EI)^{1/2}$, Dimensionless. K_σ and K_d Are Dimensionless Multiplication Factors.				
2. Uniform transverse load; left end C; right end C. Table 4-1, Case 7; Eqs. (2-6) and (2-8) for frequency	Axial Compression	0.8	1.6	2.4	3.2	4.0
	K_σ	1.0190	1.0801	1.1979	1.4078	1.7993
	K_d	1.0163	1.0684	1.1686	1.3455	1.6722
	Axial Tension	1.0	2.0	4.0	8.0	12.0
	K_σ	0.9716	0.8945	0.6728	0.3200	0.1617
	K_d	0.9756	0.9092	0.7152	0.3885	0.2228
3. Concentrated transverse load at free end of beam ($a = L$); left end C; right end F; Table 4-1, Case 1, with $R = 1.0$; Eq. (2-9) for frequency	Axial Compression	0.2	0.4	0.6	0.8	1.0
	K_σ	1.0136	1.0570	1.1402	1.2870	1.5574
	K_d	1.0163	1.0684	1.1686	1.3455	1.6722
	Axial Tension	0.5	1.0	2.0	4.0	8.0
	K_σ	0.9242	0.7616	0.4820	0.2498	0.1250
	K_d	0.9092	0.7152	0.3885	0.1407	0.0410
4. Concentrated transverse load at middle of beam ($a = L/2$); left end S; right end S; Table 4-1, Case 4, with $R = 0.5$; Eq. (2-11) for frequency	Axial Compression	0.4	0.8	1.2	1.6	2.0
	K_σ	1.0136	1.0570	1.1402	1.2870	1.5574
	K_d	1.0163	1.0684	1.1686	1.3455	1.6772
	Axial Tension	1.0	2.0	4.0	8.0	12.0
	K_σ	0.9242	0.7616	0.4820	0.2498	0.1667
	K_d	0.9092	0.7152	0.3885	0.1407	0.0694
5. Concentrated transverse load at middle of beam ($a = L/2$); left end C right end C; Table 4-1, Case 3, with $R = 0.5$; Eq. (2-13) for frequency	Axial Compression	0.8	1.6	2.4	3.2	4.0
	K_σ	1.0136	1.0570	1.1402	1.2870	1.5574
	K_d	1.0163	1.0684	1.1686	1.3455	1.6722
	Axial Tension	1.0	2.0	4.0	8.0	12.0
	K_σ	0.9797	0.9242	0.7616	0.4820	0.3317
	K_d	0.9756	0.9092	0.7152	0.3885	0.2228

[a] The equation number provided is of the applicable frequency formula from Chapter 2, for use in finding G_{pk}. Also given is the case number from Table 4.1 for finding the values of σ'_m and d'_m for use in Equation (4-22).

[b] The location of d_m is at $x = 0.5785L$ for Table 4-1, Case 6, while the location of K_d in Case 1 is at $x = 0.5$ for both the axial compression and axial tension cases. This will result in an error in the use of Eq. (4-22).

For a slender, tapered beam of uniform density under transverse loading due to its own weight, the values of σ_m and d_m may be found by applying the multiplication factors of Table 4-4 to the values of stress and displacement found from Table 4-1[8]:

$$\begin{aligned} \sigma_m &= K_\sigma(\alpha, \beta)\sigma'_m \text{ psi} \\ d_m &= K_d(\alpha, \beta) d'_m \text{ in.} \end{aligned} \tag{4-23}$$

where the primed quantities on the right-hand side of Eq. (4-23) are the values from Table 4-1, the K_σ and K_d are provided in Table 4-4, and the unprimed stress and displacement on the left-hand side of Eq. (4-23) are for the tapered beam. Provided in Table 4-4 are the case numbers from Table 4-1 for use in Eq. (4-23) and the case numbers from Table 2-10 (Chapter 2) for use in finding G_{pk} as described in Section 4.2, with an indication of whether α and β are equal to unity or not equal to unity. The figure in Table 4-4 shows the tapered beam geometry and defines the parameters α and β. These parameters are combined, and values of K_σ and K_d are provided for five specified values of $\alpha^3\beta = 0.25$, 0.50, 2.0, 4.0, and 8.0. Also, for each case, only K_σ or K_d is available, never both. The values of σ'_m and d'_m are calculated from the formulas in Table 4-1 using the beam dimensions at the right end of the beam, as shown in the figures in Table 4-4.

For a slender beam in torsion, the measure of displacement is angle of twist θ expressed as

$$\theta = ML/CG \text{ rad} \tag{4-24}$$

where M = torsional moment (lbf · in.)
L = length of beam (in.)
C = torsional constant (in.^{4})
G = shear modulus (lbf/in.^{2})

(See Figure 2.10.)

For beams of circular cross section, $C = I_p$, the polar area moment of inertia about the centroid of the beam cross section. Values of C are provided in Table 2-2. As described in Section 3.6.2, acceleration will produce torsion when the center of gravity does not lie on the axis of rotation (the support axis) of the component or structure. Let M_0 represent the torsional moment for a 1-g acceleration normal to the moment arm between the center of gravity and axis of rotation. Using M_0 in Eq. (4-24) yields θ_0, which is the torsional equivalent of the 1-g displacement, d_m, in the transverse loading of a beam. Let τ_0 represent the maximum shear stress (lbf/in.^{2}) due to the twist θ_0. Table 4-5[9] presents values and locations of τ_0 corresponding to some of the torsional constants included in Table 2-2, with the case number from Table 2-2 included. See Table 2-3 for frequency formulas for calculation of G_{pk}.

TABLE 4-4 Stress and Displacement Multiplication Factors for Transversly Loaded, Tapered Beams[a]

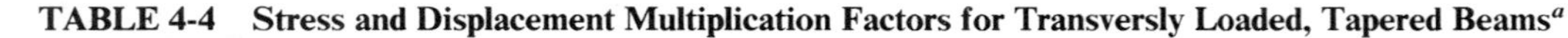

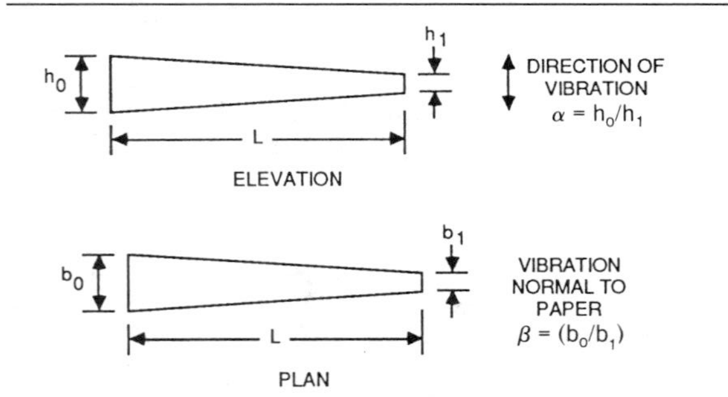

$\sigma_{pk} = G_{pk}\sigma_m$ = peak material stress (psi)
$d_{pk} = G_{pk}d_m$ = peak displacement from equilibrium (in.)
where G_{pk} = peak acceleration in g units, Section 4.2

σ_m = maximum material stress with 1-g acceleration normal to beam (psi), from Eq. (4-23)

d_m = maximum displacement from equilibrium with 1-g acceleration normal to beam (in.), from Eq. (4-23)

g = gravitational acceleration at surface of earth = 386 in./sec^2

K_σ = multiplier for stress, provided in the table for use in Eq. (4-23), dimensionless

K_d = multiplier for displacement, provided in the table, for use in Eq. (4-23), dimensionless

F = free end of beam

S = simply supported (pinned) end of beam
C = clamped (fixed) end of beam
h_0 = depth of beam at left end (in.)
h_1 = depth of beam at right end (in.)
b_0 = width of beam at left end (in.)
b_1 = width of beam at right end (in.)
L = length of beams (in.)
$\alpha = h_0/h_1$; $\beta = b_0/b_1$

Case Numbers	End Conditions: Left end	End Conditions: Right End	Table 4.1 Case Number for Eq. (4-23)	Table 2-10 Case Number for Frequency	Multiplier is	Column Headings are Values of $\alpha^3\beta$; Table Contents are Values of K_σ or K_d: 0.25	0.50	2.0	4.0	8.0
1.	C	F	5	10 ($\alpha = 1$; $\beta \neq 1$)	K_d	2.711	1.695	0.561	0.302	0.158
2.	C	F	5	3, 7 ($\alpha \neq 1$; $\beta = 1$)	K_d	2.981	1.734	0.572	0.324	0.182
3.	C	F	5	11, 12, 13, 14 ($\alpha \neq 1$; $\beta \neq 1$)	K_d	3.013	1.738	0.573	0.328	0.187
4.	C	C	7	1 ($\alpha \neq 1$; $\beta = 1$)	K_σ	1.297	1.144	0.867	0.745	0.635
5.	S	S	8	5 ($\alpha \neq 1$; $\beta = 1$)	K_d	1.919	1.400	0.700	0.480	0.322

[a] In using Table 4-1 to calculate values of σ'_m and d'_m for use in Eq. (4-23), the values of I, c, and w at the right end of the beam should be used: $I = b_1 h_1^3/12$ in.^{4}; $c = h_1/2$ in.; $w = \rho b_1 h_1$ lbf/in.; where ρ is the weight density of beam material (lbf/in.^{3}).

TABLE 4-5 Maximum Shear Stress for a Slender Beam in Torsion

$\theta_{pk} = G_{pk}\theta_0$ = peak angle of twist (rad)
$\tau_{pk} = G_{pk}\tau_0$ = peak shear stress (lbf/in.^{2})

where G_{pk} = peak acceleration in g units, Section 4.2; See Table 2-3 for frequency formulas
θ_0 = obtained from Eq. (4-24) with $M = M_0$
τ_0 = provided in the table
M_0 = torsional moment for a 1-g acceleration normal to the moment arm between the center of gravity and the axis of rotation
L = length of beam (in.)
C = torsional constant (in.^{4}), Table 2-2
G = shear modulus (lbf/in.^{2})
g = acceleration of gravity at surface at earth = 386 in./sec^2

For a beam composed of thin walls of uniform thickness,

$\max \tau_0 = M_0(3S + 1.8t)/S^2t^2$ lbf/in.^{2}

where M_0 = defined above
S = length of midwall perimeter (in.)
t = uniform wall thickness (in.)

Beam Cross Section	Formula for Shear Stress, τ_0 (lbf/in.^{2})
1. Annulus	$\max \tau_0 = 2M_0R_1/\pi(R_1^4 - R_2^4)$ at R_1 (For a solid section, $R_2 = 0$) Table 2-2, Case 1
2. Ellipse	$\max \tau_0 = 2M_0/\pi ab^2$ at ends of minor axis Table 2-2, Case 2

4.5.2 Stresses and Displacements in Plates

Table 4-6 presents the locations and magnitudes of the maximum stress σ_m and the maximum displacement d_m in uniformly loaded flat plates of constant thickness with straight edges, subjected to a 1-g acceleration normal to the plate.[10] Peak stresses and displacements are obtained by multiplying the 1-g values by G_{pk} as found in Section 4.2. In this table the absolute values of σ_m and d_m are given, with σ_m always occurring at the plate surfaces. The location at which σ_m or d_m occurs is given beneath the data for σ_m and d_m, respectively. In some cases the location or values of σ_m or d_m are not provided. In

TABLE 4-5 *(Continued)*

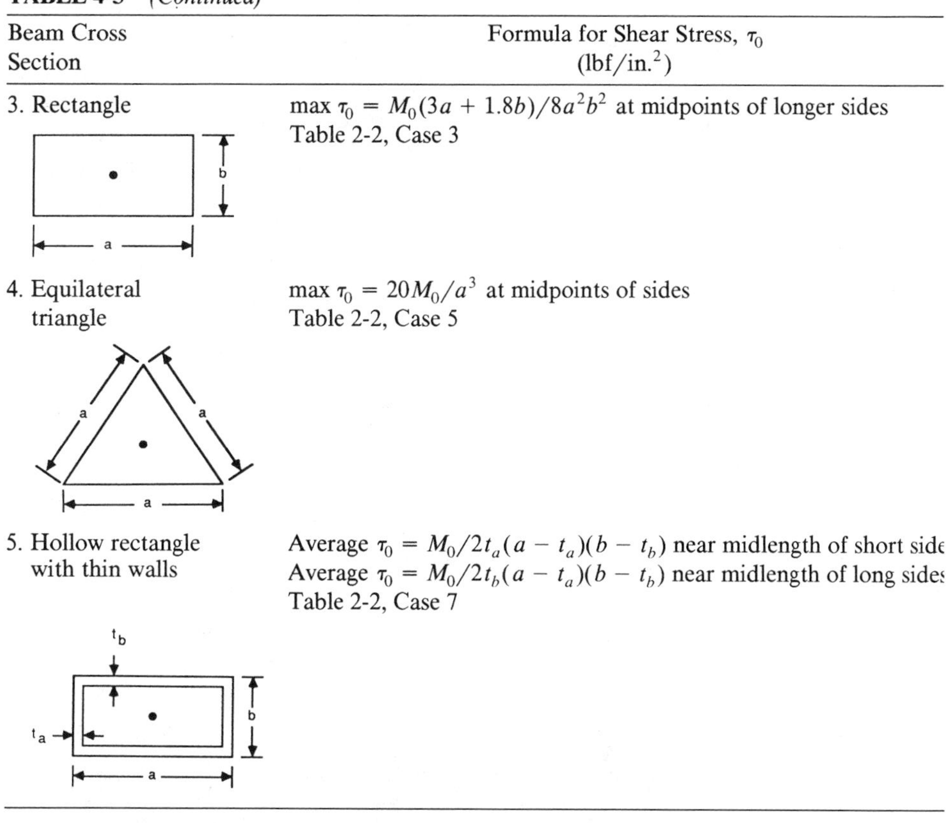

Beam Cross Section	Formula for Shear Stress, τ_0 (lbf/in.^{2})
3. Rectangle	max $\tau_0 = M_0(3a + 1.8b)/8a^2b^2$ at midpoints of longer sides Table 2-2, Case 3
4. Equilateral triangle	max $\tau_0 = 20M_0/a^3$ at midpoints of sides Table 2-2, Case 5
5. Hollow rectangle with thin walls	Average $\tau_0 = M_0/2t_a(a - t_a)(b - t_b)$ near midlength of short side Average $\tau_0 = M_0/2t_b(a - t_a)(b - t_b)$ near midlength of long sides Table 2-2, Case 7

the "Description" column are the table number and Case number of the applicable frequency formula from Chapter 2, for use in finding G_{pk}, and, where applicable, the value of Poisson's ratio ν used in obtaining the numerical values of the constants α and β in the table. For values of ν in the range 0.15–0.30 errors in σ_m should not exceed 15% and errors in d_m should not exceed 8%. Note that in Cases 1, 3, and 4 the values of ν used in Table 4-6 are not the same as those used in the frequency formula of Chapter 2, which may cause some error.

For rectangular plates with simply supported edges calculations were based on the assumption that the corners of the plate were held down. If the corners are permitted to rise, the stress at the center of the plate will increase, becoming 35% larger for a square plate. The boundary condition notation for rectangular plates is the same as used in Table 2-12.

Table 4-7 presents the stresses σ and the displacements d in uniformly loaded flat plates of constant thickness with circular or elliptical boundaries, subjected to a 1-g acceleration normal to the plate.[11] Peak stresses and displacements are obtained by multiplying the 1-g values by G_{pk} as found in

TABLE 4-6 Maximum Stress and Displacement in Uniformly Loaded Flat Plates with Straight Edges[a]

$\sigma_m = \beta q b^2 / t^2$ psi
$d_m = \alpha q b^4 / (E t^3)$ in.

$\sigma_{pk} = G_{pk} \sigma_m$ = peak material stress (psi)
$d_{pk} = G_{pk} d_m$ = peak displacement from equilibrium (in.)

where G_{pk} = peak acceleration in g units, Section 4.2
σ_m = maximum material stress with 1-g acceleration normal to plate (psi)
d_m = maximum displacement from equilibrium with 1-g acceleration normal to plate (in.)
g = gravitational acceleration at surface of earth = 386 in./sec^2
F = free edge
C = clamped edge
S = simply supported edge
a = length of an edge of the plate (in.) (long edge of rectangle)
b = length of an edge of the plate (in.) (short edge of rectangle)
t = thickness of plate (in.)
q = load per unit area of plate (lbf/in.^{2}) = ρt
ρ = weight density of plate material (lbf/in.^{3})
E = modulus of elasticity of plate material (lbf/in.^{2})
ν = Poisson's ratio
α, β = numerical constants provided in the table.

Description of Geometry and Edge Supports	Values of α and β
Cases 1–9. Rectangular; start at left edge and read clockwise	

Reads: *SCFF*

1. *CFFC*
Table 2-12, Case 8
$\nu = 0.2$

a/b	1.00	1.33	2.00	2.67	4.00	8.00
β	1.769	1.246	0.632	0.398	0.188	0.050

σ_m at or near a C–S corner

2. *SFSS*
Table 2-12, Case 9
$\nu = 0.3$

a/b	0.25	0.50	0.667	1.00	1.50	2.00
β	0.80	0.79	0.77	0.67	0.45	0.36
α	0.167	0.165	0.160	0.140	0.106	0.080

3. *SFSC*
Table 2-12, Case 10
$\nu = 0.2$

a/b	0.25	0.50	0.75	1.00	1.50	2.00	3.00
β	0.048	0.190	0.386	0.665	1.282	1.804	2.450

σ_m at center of clamped edge for $a/b \geq 1.0$ and at center of free edge for $a/b \leq 1.0$.

4. *CFCC*
Table 2-12, Case 14
$\nu = 0.2$

a/b	0.25	0.50	0.75	1.00	1.50	2.00	3.00
β	0.031	0.126	0.286	0.511	1.073	1.568	2.105

σ_m at a C–F corner for $a/b \leq 2.0$ and at center of clamped edge opposite free edge for $a/b \geq 3.0$

5. *SSSS*
Table 2-12, Case 15
$\nu = 0.3$

a/b	1.0	1.2	1.4	1.6	1.8	2.0	3.0	4.0	5.0	∞
β	0.2874	0.3762	0.4530	0.5172	0.5688	0.6102	0.7134	0.7410	0.7476	0.7500
α	0.0444	0.0616	0.0770	0.0906	0.1017	0.1110	0.1335	0.1400	0.1417	0.1421

σ_m at center of plate; d_m at center of plate.

6. *SSSC*
Table 2-12, Case 16
$\nu = 0.3$

a/b	0.25	0.286	0.333	0.400	0.500	0.667	1.000	1.500	2.000	2.500	3.000	3.500	4.000
β	0.75	0.75	0.75	0.74	0.73	0.67	0.50	0.66	0.73	0.74	0.74	0.75	0.75
α	0.139	0.137	0.132	0.122	0.101	0.071	0.030	0.046	0.054	0.056	0.057	0.058	0.058

TABLE 4-6 *(Continued)*

Description of Geometry and Edge Supports	Values of α and β													
7. *SCSC*	a/b	0	0.500	0.556	0.625	0.714	0.833	1.000	1.200	1.400	1.600	1.800	2.000	∞
Table 2-12, Case 18	β	0.750	0.7146	0.6912	0.6540	0.5988	0.5208	0.4182	0.4626	0.4860	0.4968	0.4971	0.4973	0.5000
	α	—	0.0922	0.0800	0.0658	0.0502	0.0349	0.0210	0.0243	0.0262	0.0273	0.0280	0.0283	0.0285
	σ_m at center of clamped edge; d_m at center of plate													
8. *CSCC*	a/b	0.25	0.50	0.75	1.00	1.50	2.00	3.00						
Table 2-12, Case 19	β	0.031	0.121	0.242	0.343	0.539	0.657	0.718						
$\nu = 0.2$	For $a/b \leq 1.0$, σ_m at $0.4b$ from S edge, on clamped edges of length b; for $a/b > 1.0$, σ_m at center of clamped edge of length a.													
9. *CCCC*	a/b	1.0	1.2	1.4	1.6	1.8	2.0	∞						
Table 2-12, Case 20	β	0.3078	0.3834	0.4356	0.4680	0.4872	0.4974	0.5000						
$\nu = 0.3$	α	0.0138	0.0188	0.0226	0.0251	0.0267	0.0277	0.0284						
	σ_m at center of long edge; d_m at center of plate													
10. Equilateral triangle, all edges S, $\nu = 0.3$, Table 2-14, Case 1	$\beta = 0.1085$ at center of plate; σ_m not at center of plate; $\alpha = 0.00694(1 - \nu^2)$; d_m at center of plate; $b = a =$ length of side of triangle													

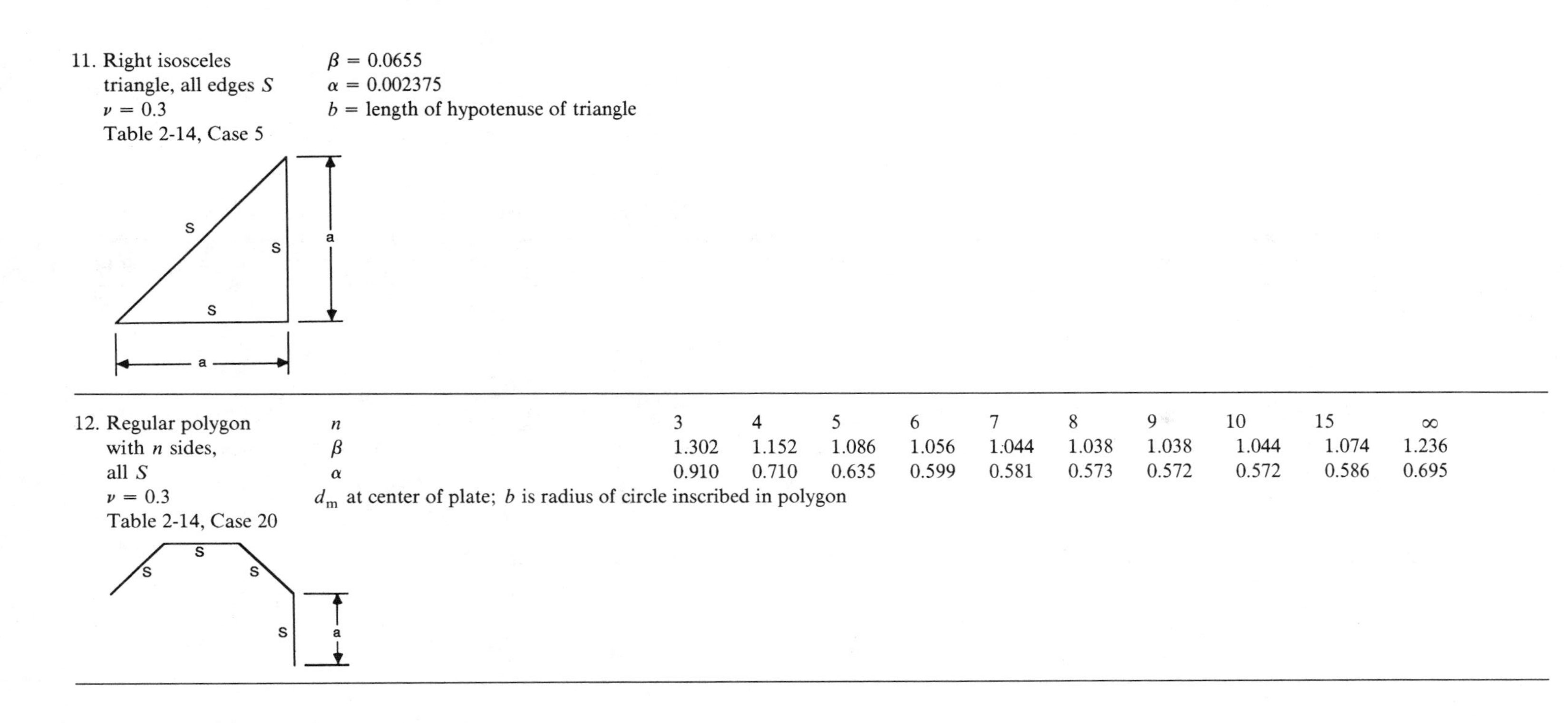

Case											
11. Right isosceles triangle, all edges S $\nu = 0.3$ Table 2-14, Case 5	$\beta = 0.0655$ $\alpha = 0.002375$ b = length of hypotenuse of triangle										
12. Regular polygon with n sides, all S $\nu = 0.3$ Table 2-14, Case 20	n	3	4	5	6	7	8	9	10	15	∞
	β	1.302	1.152	1.086	1.056	1.044	1.038	1.038	1.044	1.074	1.236
	α	0.910	0.710	0.635	0.599	0.581	0.573	0.572	0.572	0.586	0.695
	d_m at center of plate; b is radius of circle inscribed in polygon										

TABLE 4-6 *(Continued)*

Description of Geometry and Edge Supports	Values of α and β									
13. Regular polygon with n sides, all C	n	3	4	5	6	7	8	9	10	∞
$\nu = 0.3$	β	1.423	1.232	1.132	1.068	1.023	0.990	0.964	0.944	0.750
Table 2-14, Case 21	α	0.264	0.221	0.203	0.194	0.188	0.184	0.182	0.180	0.171
	σ_m at center of sides, d_m at center of the plate; b is radius of circle inscribed in polygon									

C C C C a

[a] In the "Description" column is the table number and case number of the applicable frequency formula from Chapter 2, for use in finding G_{pk}. The boundary condition notation for rectangular plates is the same as used in Table 2-12. The values of ν used to calculate the constants α and β in Cases 1, 3, and 4 are not the same as those used in the frequency formula of Chapter 2, which may cause some error.

TABLE 4-7 Maximum Stress and Displacement in Flat Plates with Circular or Elliptical Boundaries

σ = material stress with 1-g acceleration normal to plate (psi)
d = displacement from equilibrium with 1-g acceleration normal to plate, (in.)
$\sigma_{pk} = G_{pk}\sigma$ = peak material stress (psi)
$d_{pk} = G_{pk}d$ = peak displacement (in.)
G_{pk} = peak acceleration in g units (Section 4.2)
g = gravitational acceleration at surface of earth = 386 in./sec^2
q = load per unit area of plate (lbf/in.^{2}) = ρt
ρ = weight density of plate material (lbf/in.^{3})
t = thickness of plate (in.)
M = concentrated weight at center of plate (lbf)
r_0 = radius of concentrated weight at center of plate (in.)
r_1 = effective radius of concentrated weight at center of plate (in.)
a = radius of plate; outer radius of annulus, semiminor axis of ellipse (in.)
b = semimajor axis of ellipse (in.)
$D = Et^3/12(1 - \nu^2)$(lbf · in.)
E = modulus of elasticity of plate material (lbf/in.^{2})
ν = Poisson's ratio
β, α = numerical constants provided in the table
F = free edge
S = simply supported edge
C = clamped edge
Subscript m indicates a maximum absolute value of parameter
Subscript a indicates the stress or displacement at the outer edge, Cases 5–12
Subscript b indicates the stress or displacement at the inner edge, Cases 5–12
Subscript r indicates the stress in the radial direction, Cases 5–12
Subscript t indicates the stress in the tangential direction, Cases 5–12

TABLE 4-7 *(Continued)*

Description of Geometry, Supports and Loading	Values of σ and d
1. Circle, S edge Table 2-13, Case 1 S a	$\sigma_m = 0.375(3 + \nu)qa^2/t^2$, at center of plate $d_m = [(5 + \nu)/64(1 + \nu)]qa^4/D$, at center of plate
2. Circle, C edge Table 2-13, Case 2 C a	$\sigma_m = 0.75qa^2/t^2$, at edge of plate $d_m = qa^4/64D$, at center of plate
3. Circle, S edge, weight M of radius r_0 at center, $q = 0$ Table 2-13, Case 7 S r_0 M a	$\sigma_m = (1.5M/\pi t^2)[(1 + \nu)\ln(a/r_1) + 1]$, at center of plate $d_m = [(3 + \nu)/(1 + \nu)](Ma^2/16\pi D)$, at center of plate where $r_1 = (1.6r_0^2 + t^2)^{1/2} - 0.675t$ for $r_0 < 0.5t$ $r_1 = r_0$ for $r_0 > 0.5t$

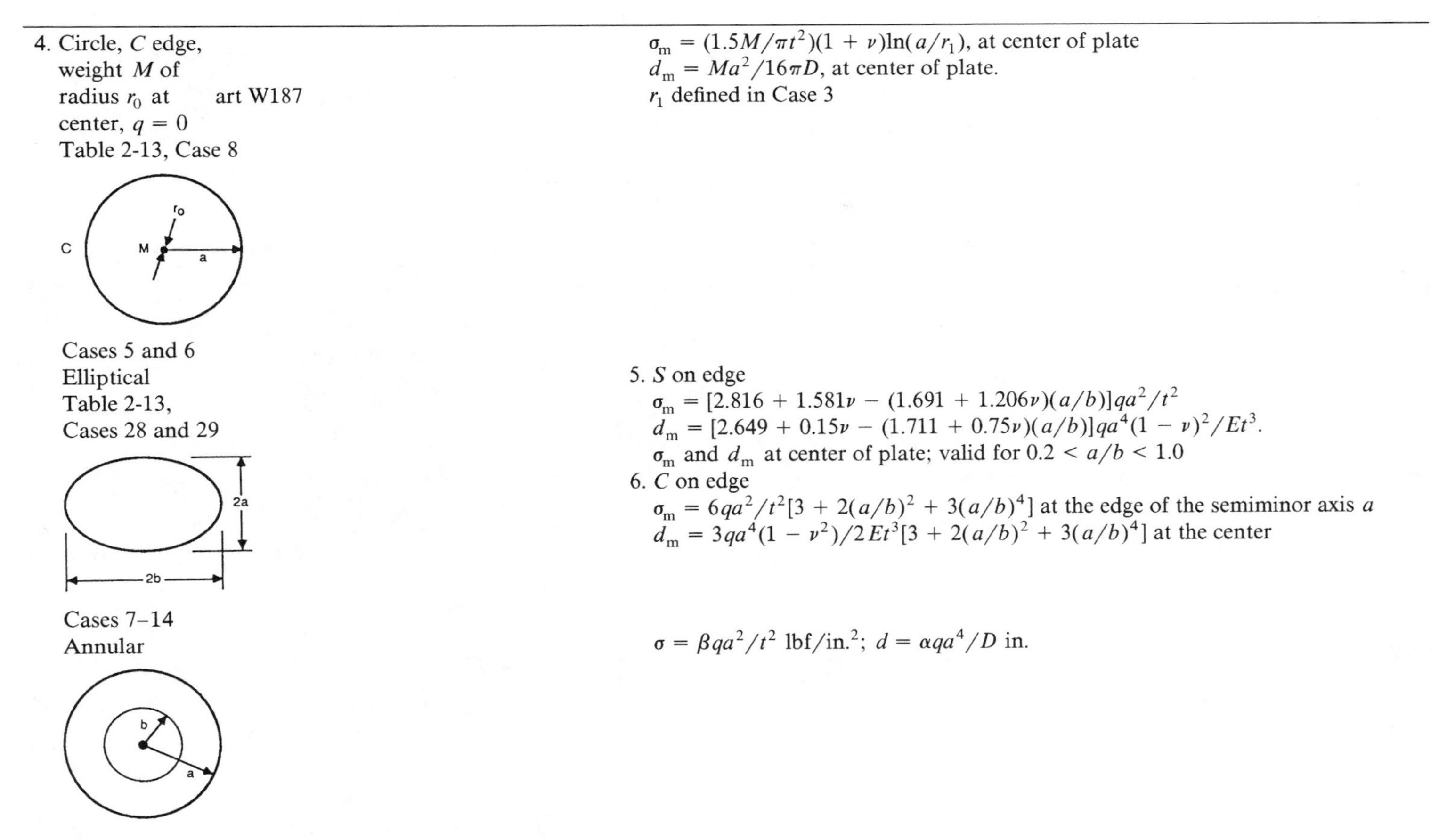

4. Circle, C edge, weight M of radius r_0 at center, $q = 0$ Table 2-13, Case 8 art W187	$\sigma_m = (1.5M/\pi t^2)(1 + \nu)\ln(a/r_1)$, at center of plate $d_m = Ma^2/16\pi D$, at center of plate. r_1 defined in Case 3
Cases 5 and 6 Elliptical Table 2-13, Cases 28 and 29	5. S on edge $\sigma_m = [2.816 + 1.581\nu - (1.691 + 1.206\nu)(a/b)]qa^2/t^2$ $d_m = [2.649 + 0.15\nu - (1.711 + 0.75\nu)(a/b)]qa^4(1 - \nu)^2/Et^3$. σ_m and d_m at center of plate; valid for $0.2 < a/b < 1.0$ 6. C on edge $\sigma_m = 6qa^2/t^2[3 + 2(a/b)^2 + 3(a/b)^4]$ at the edge of the semiminor axis a $d_m = 3qa^4(1 - \nu^2)/2Et^3[3 + 2(a/b)^2 + 3(a/b)^4]$ at the center
Cases 7–14 Annular	$\sigma = \beta qa^2/t^2$ lbf/in.^{2}; $d = \alpha qa^4/D$ in.

TABLE 4-7 *(Continued)*

Case Number	Inside Edge	Outside Edge	Parameter	a/b				
				0.1	0.3	0.5	0.7	0.9
7	S	F	$\beta_{b,t}$	7.6404	3.6876	2.0484	1.0452	0.3126
(Case 20)			$\alpha_{m,a}$	0.1115	0.1158	0.0826	0.0378	0.0051
8	F	S	$\beta_{m,b,t}$	2.3790	1.9632	1.4424	0.8814	0.2982
(Case 21)			$\alpha_{m,b}$	0.0687	0.0761	0.0624	0.0325	0.0048
9	C	F	$\beta_{b,r}$	5.7876	2.4618	1.0416	0.3246	0.0318
(Case 22)			$\alpha_{m,a}$	0.0757	0.0318	0.0086	0.0011	—
10	F	C	$\beta_{a,r}$	0.7476	0.6810	0.4800	0.2166	0.0282
(Case 23)			$\beta_{b,t}$	0.8688	0.4668	0.1626	0.0312	0.00096
			$\alpha_{m,b}$	0.0166	0.0132	0.0053	0.0009	0.00001
11	S	S	$\beta_{r(max)}$	0.4248	0.3312	0.1800	0.0660	—
(Case 24)			$\beta_{b,t}$	1.4406	0.2778	0.0606	0.0090	—
			α_m	0.0060	0.0029	0.0008	0.0001	—
12	C	S	$\beta_{b,r}$	1.4754	0.5634	0.2358	0.07542	—
(Case 25)			α_m	0.0040	0.0014	0.0004	0.00004	—
13	S	C	$\beta_{a,r}$	0.3804	0.2772	0.1572	0.0612	0.0072
(Case 26)			$\beta_{b,t}$	0.7356	0.1326	0.0288	0.0042	—
			α_m	0.0025	0.0012	0.0003	—	—
14	C	C	$\beta_{a,r}$	0.3240	0.2082	0.1122	0.0420	—
(Case 27)			$\beta_{b,r}$	0.8598	0.3420	0.1482	0.0486	—
			α_m	0.0018	0.0006	0.0002	—	—

The numerical values for Cases 7–14 are for a value of Poisson's ratio, $\nu = 0.3$. The case numbers in parentheses refer to Table 2-13 for Cases 7–14

Section 4.2. In this table the absolute values of σ and d are given, with the value of σ always chosen at the plate surface, and related to the unit bending moment by

$$\sigma = 6M/t^2 \text{ lbf/in.}^2 \tag{4-25}$$

where $M = M_r$ = unit radial bending moment (in.-lbf per inch of circumference) or
$M = M_t$ = unit tangential bending moment (in.-lbf per inch of radius)
t = plate thickness (in.)

In Cases 1–6 the magnitudes and locations of the maximum values of σ and d are provided, as indicated by the subscript, m. In Cases 7–14, the maximum values of σ are sometimes not known, and the locations of the maximum values of d are sometimes not known. In these cases, 7–14, additional subscripts are used to indicate location (a, b) or direction of stress (r, t) as defined in the table. The "Description" column of the table provides the case number of the applicable frequency formula from Table 2-13, Chapter 2, for use in finding G_{pk}. The constants α and β in Cases 7–14 were calculated using a Poisson's ratio of $\nu = 0.3$.

4.6 STRESSES AND DISPLACEMENTS IN SIMPLE STRUCTURES

Some of the stress and displacement information contained in Section 4.5 can be applied to those structures in Chapter 3 for which frequency formulas are provided in conjunction with referenced material from Chapter 2. The fre-

TABLE 4-8 Estimation of Peak Stresses and Displacements for Simple Structures[a]

Frequency	Stress and Displacement
Beams	
Composite beams: Sections 3.1.1, 3.1.2, and 2.4.1.	Section 4.5.1 with Tables 4-1–4-5
Stepped beams: Sections 3.2 and 2.4.1.	
Housings (treated as beams): Sections 3.6.1, 3.6.2, 3.6.3, and 2.4.1; Section 3.6.4, Fig. 3.10*a*, Example 1, and Fig. 3.10*d*, Example 1, with Section 2.4.1.	
Plates	
Housings (treated as plates): Section 3.6.4: Fig. 3.10*a*, Examples 2 and 3; Fig. 3.10*b*; Fig. 3.10*c*; Fig. 3.10*d*, Example 2; with Section 2.6.1.	Section 4.5.2 with Table 4-6.
Stiffened plates: Sections 3.5 and 2.7.	

[a] Use the frequency formulas from Chapter 3 to estimate G_{pk} as described in Section 4.2. Multiply G_{pk} times the 1-g stress and displacement data from Section 4.5 to find peak stresses and displacements.

quency formulas allow the determination of the fundamental natural frequency that can be used to determine peak acceleration G_{pk}, as described in Section 4.2. The value of G_{pk} can then be used with the 1-g stress and displacement results of Section 4.5 to obtain peak stress and displacement. Table 4-8 outlines the sections from Chapters 2 and 3 that can be used in conjunction with the material in Section 4.5.

REFERENCES

1 R. E. Doyle, *The Shock and Vibration Digest*, **19** (6), 3–10 (1987).

2 R. J. Roark and W. C. Young, *Formulas for Stress and Strain*, 5th ed., New York: McGraw-Hill, 1975.

3 D. S. Steinberg, *Institute of Environmental Sciences Proceedings*, March 25–26, 1982, Los Angeles, CA, pp. 13–15.

4 R. T. Fenner, *Engineering Elasticity*, New York: Halsted Press, John Wiley, 1986, pp. 8–12.

5 R. T. Fenner, in Ref. 4, pp. 235 and 256.

6 R. J. Roark and W. C. Young, in Ref. 2, pp. 96–101, 186, and 188.

7 R. J. Roark and W. C. Young, in Ref. 2, pp. 147–152.

8 R. J. Roark and W. C. Young, in Ref. 2, pp. 172–180.

9 R. J. Roark and W. C. Young, in Ref. 2, pp. 290–296.

10 R. J. Roark and W. C. Young, in Ref. 2, pp. 386–404.

11 R. J. Roark and W. C. Young, in Ref. 2, pp. 332–373.

5

SHOCK

This chapter describes the response of a structure to a shock environment, and the attenuation of the shock due to transmission through structures.

5.1 METHODS OF REPRESENTING SHOCKS

The shocks of interest here are pyrotechnic, those due to the detonation of ordnance devices such as separation nuts. These shocks result in a rapid transfer of energy through the structure and can cause significant relative structural displacements and material stresses. Fatigue is usually not of importance in this type of shock environment, and the limiting factors are material strength and the relative dynamic displacement of parts of the structure, which may impact and damage each other.

There are several ways of specifying a shock environment, including the acceleration–time history, the pulse shock, the Fourier integral spectrum, the acceleration response spectrum, and the normalized four-coordinate response spectrum. The acceleration–time history is the actual acceleration versus time that would be recorded by an ideal accelerometer (Fig. 5.1). The pulse shock is a simple acceleration-versus-time pulse (Fig. 5.2) that may be approximated by a laboratory impact shock tester, which usually does not accurately represent the actual shock environment over the entire frequency range.

The Fourier integral spectrum results from the application of the Fourier series to a nonperiodic transient function, yielding a frequency amplitude $A(\omega)$ and a phase-angle value $\theta(\omega)$ at each frequency $F = \omega/2\pi$ Hz, where ω is the angular frequency (rad/sec). As an example, the rectangular pulse of

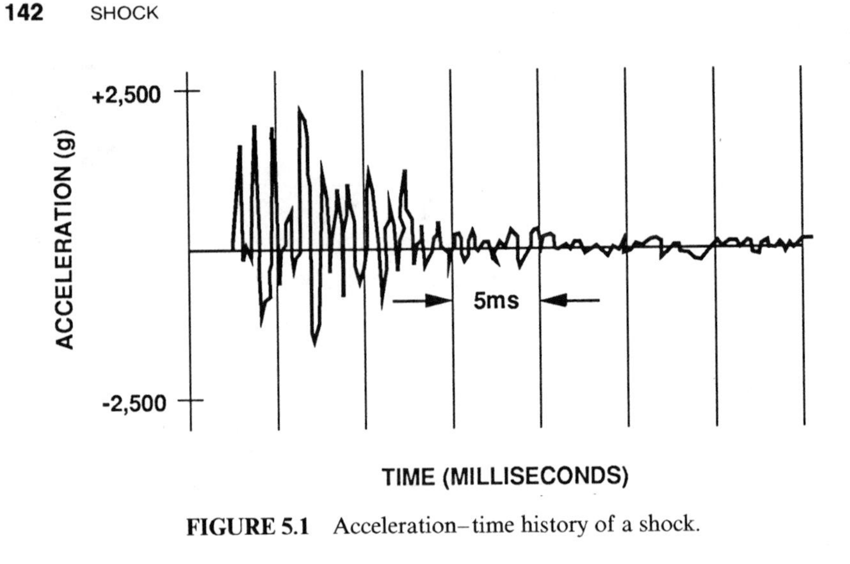

FIGURE 5.1 Acceleration–time history of a shock.

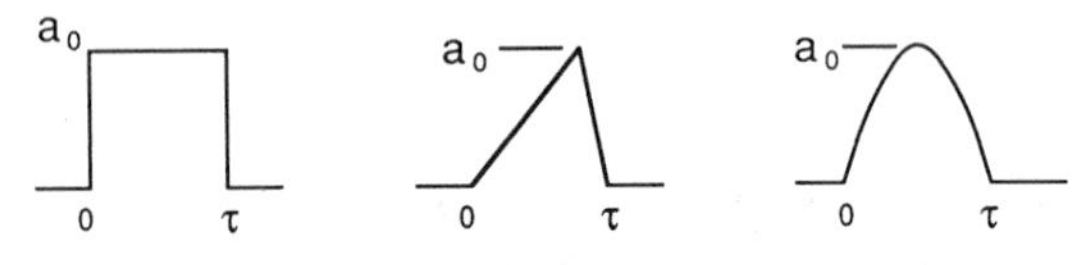

FIGURE 5.2 Simple acceleration-versus-time pulses.

Fig. 5.3 will have the values of $|A(\omega)|$ and $\theta(\omega)$ shown in Fig. 5.4. For the rectangular pulse of duration τ and acceleration a_0, the acceleration at any time is given by

$$a(t) = \frac{1}{2\pi}\int_{-\infty}^{+\infty} A(\omega)e^{i\omega t}\,d\omega \tag{5-1}$$

where

$$|A(\omega)| = (2a_0/\omega)\sin(\omega\tau/2) \tag{5-2}$$

and

$$\theta(\omega) = \omega\tau/2 \tag{5-3}$$

The acceleration response spectrum is the acceleration that would be experienced by a single-degree-of-freedom oscillator of natural frequency F subjected to the shock, plotted over the entire frequency range of interest for

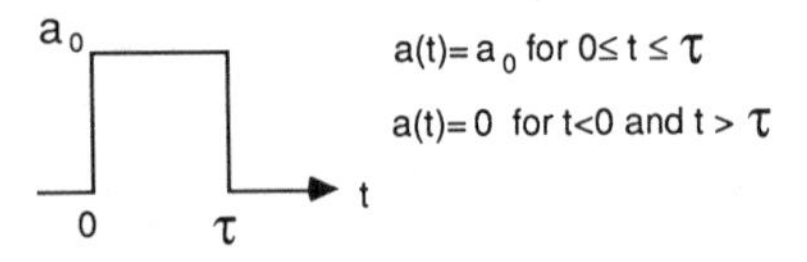

FIGURE 5.3 Rectangular acceleration pulse.

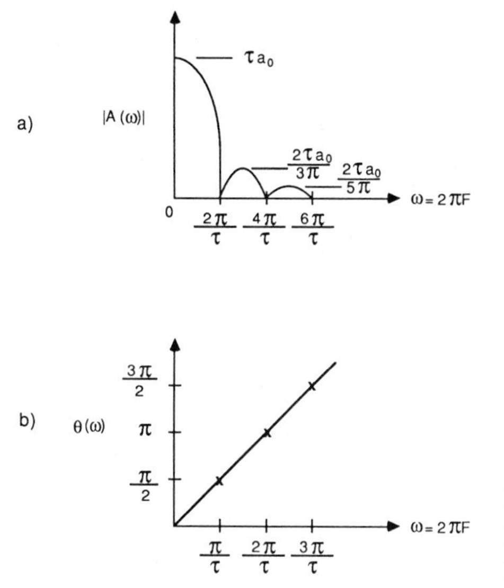

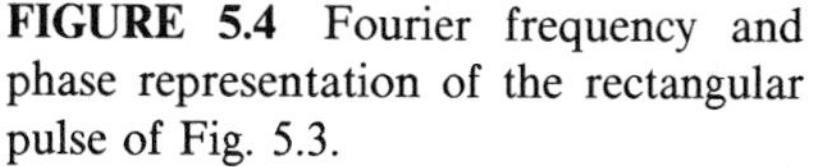
FIGURE 5.4 Fourier frequency and phase representation of the rectangular pulse of Fig. 5.3.

an infinite number of such oscillators, each at a different frequency in the range (Fig. 5.5). The deflection, velocity, and acceleration response of a single-degree-of-freedom system may be referenced to the mounting surface (relative response) or to an inertial reference frame (absolute response). These two responses are related and are identical after the shock input is terminated. Because of their significance in impact damage due to relative dynamic displacement in parts of the structure, and their direct relation to structural strain and stress, the relative responses are commonly used in analysis (Fig. 5.6).

The normalized four coordinate response spectrum provides a way of representing relative displacement, velocity and acceleration responses as a function of frequency, in a nondimensional form.[1] The relative response parameters are normalized with respect to $a_0\tau$, which is the velocity of a rectangular pulse having the same peak acceleration a_0 and duration τ as the shock input being analyzed. The normalized relative parameters are:

$$\text{Dimensionless frequency} = F\tau, \quad \text{where } F \text{ is frequency (Hz)}$$

$$\text{Dimensionless deflection} = D = d/a_0\tau^2,$$

where d is the relative deflection response (in.)

$$\text{Dimensionless velocity} = V = v/a_0\tau,$$

where v is the pseudovelocity response (in./sec)

$$\text{Dimensionless acceleration} = A = a/a_0,$$

where a is the equivalent static acceleration (in./sec^2)

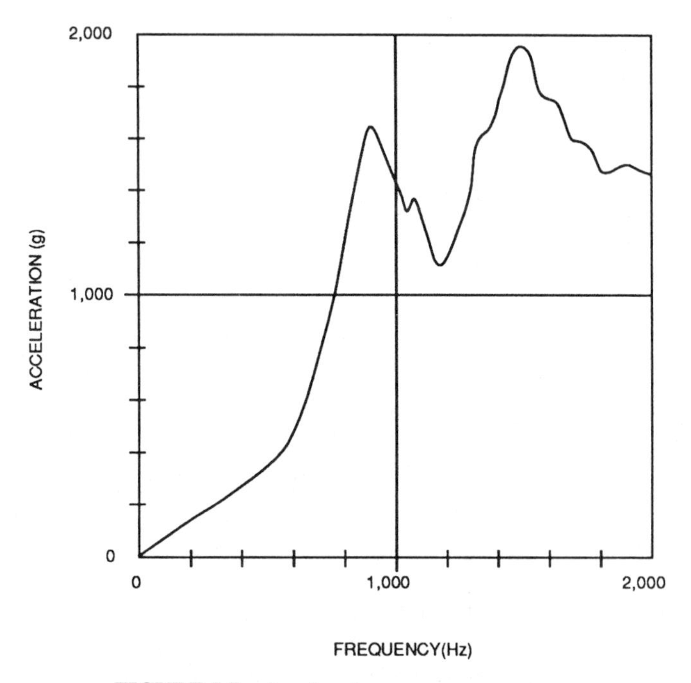

FIGURE 5.5 Acceleration response spectrum.

The relative deflection response is equal to the absolute deflection response minus the mounting surface displacement from the equilibrium position referenced to inertial space. The maximum pseudovelocity response $v_{\max}$ is found from

$$v_{\max} = 2\pi F d_{\max} \tag{5-4}$$

where F is the natural frequency of the structure and $d_{\max}$ is the maximum relative deflection response. The pseudovelocity response is generally in close agreement with the true relative velocity response, but is lower at low frequencies and higher at high frequencies than the true relative velocity response. The equivalent static acceleration[2,3] is that acceleration which, if steadily applied, would produce the same value of $d_{\max}$ as the shock. For an undamped system the equivalent static acceleration is identical to the absolute response acceleration. With damping, the equivalent static acceleration will be less than the absolute response acceleration at low frequencies, but will tend to approach the value of the absolute response acceleration at high frequencies.

Damping may also be included in the normalized four-coordinate response spectrum, as shown in Fig. 5.7. In this figure, c is the damping, in units of

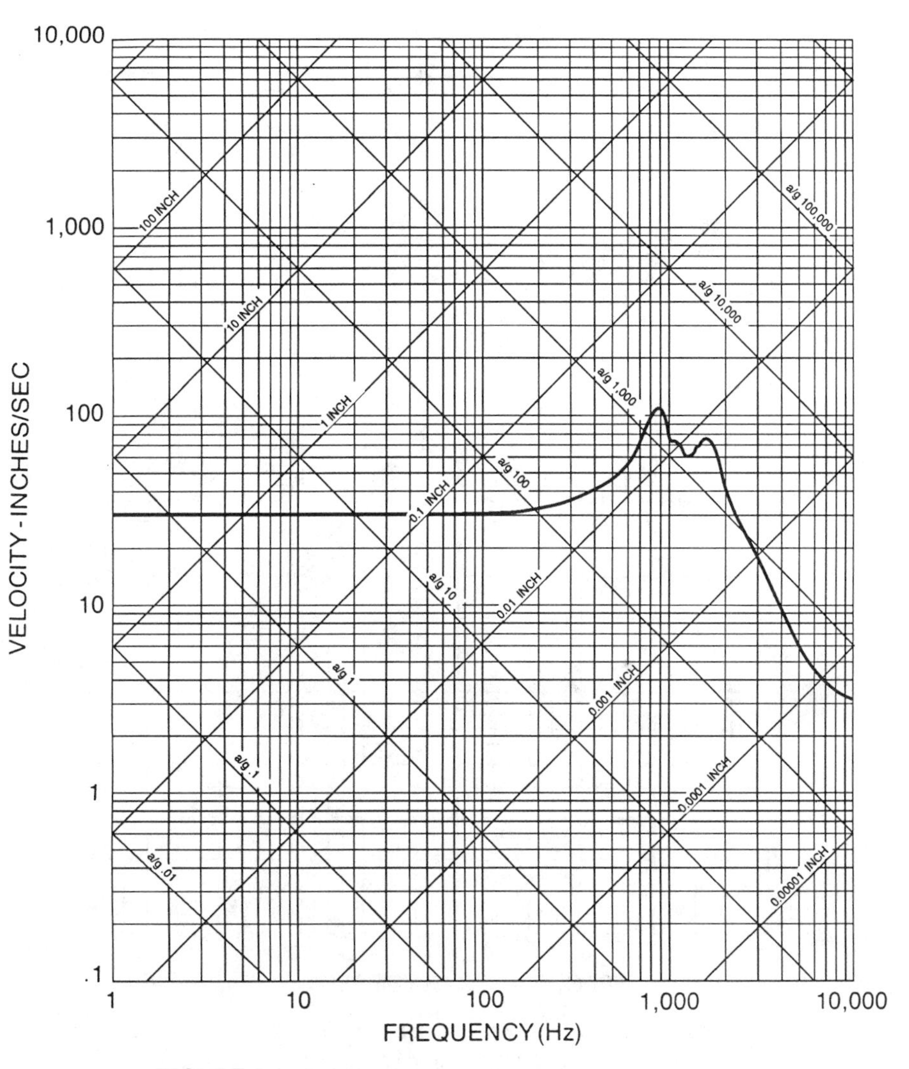

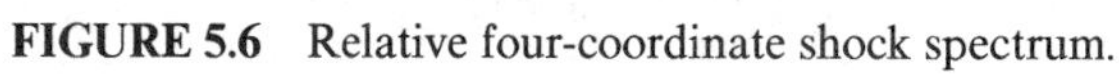
FIGURE 5.6 Relative four-coordinate shock spectrum.

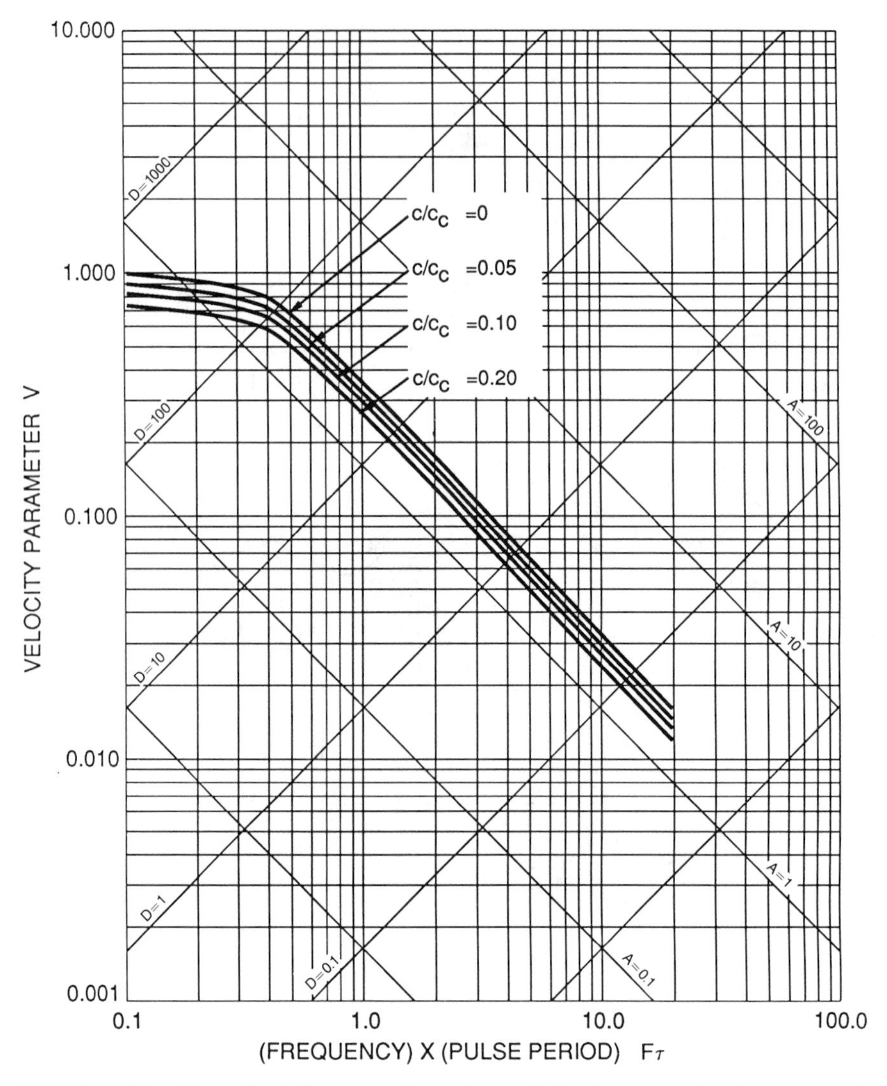

FIGURE 5.7 Normalized four-coordinate damped shock spectra for the rectangular acceleration pulse of Fig. 5.3.

lbf · sec/in.; when multiplied by the velocity, v in./sec, it yields the damping force cv lbf. The critical damping, c_c, is the smallest magnitude of damping at which oscillatory motion does not occur. For $c = 0$, the structural displacements will oscillate freely at their natural frequency and at constant amplitude after cessation of the shock. For $0 < c < c_c$, the oscillations will decrease in amplitude with time until all of the energy is dissipated by damping. For $c_c \leq c$, there is no oscillatory motion, only a monotonic decay of displacement amplitude with time until all of the energy is dissipated by damping. The damping ratio of a structure is defined as

$$R_c = c/c_c \tag{5-5}$$

and this is the parameter identified in Fig. 5.7. R_c is related to the transmissibility Q as shown in Eqs. (4.6) and (4.7).

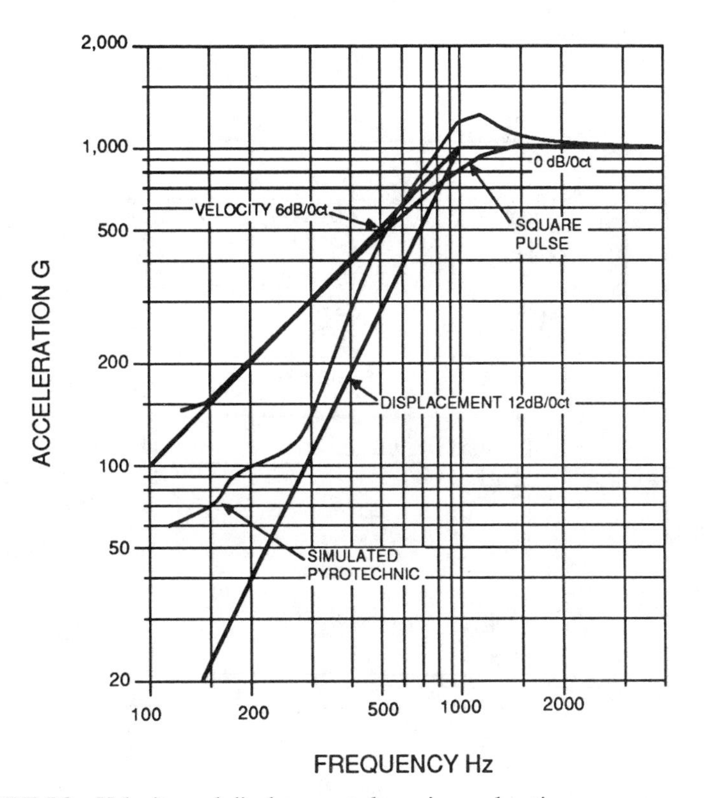

FIGURE 5.8 Velocity and displacement slopes in acceleration response spectra.

5.2 NORMALIZED, RELATIVE SHOCK RESPONSE SPECTRA

This section presents a number of normalized four-coordinate shock response spectra, which provide relative displacement, velocity, and acceleration responses as a function of frequency for specified, idealized shock pulses. The shock pulses are presented in acceleration–time history form. Some of these acceleration-versus-time shock pulses can be approximated by laboratory drop-test machines or shaker tables. If a specified acceleration response shock spectrum is to be simulated in the laboratory, the response spectra tabulated here can be useful in selecting a shock pulse form that produces an acceleration response spectrum matching the specification, at least over that portion of the frequency range which is of greatest importance.

A standard drop shock machine imparts the shock to the test specimen by the sudden change in velocity when the following table impacts the base. This change in velocity will yield a 6-dB-per-octave decrease in energy with decreasing frequency below the frequency value at which the acceleration is a

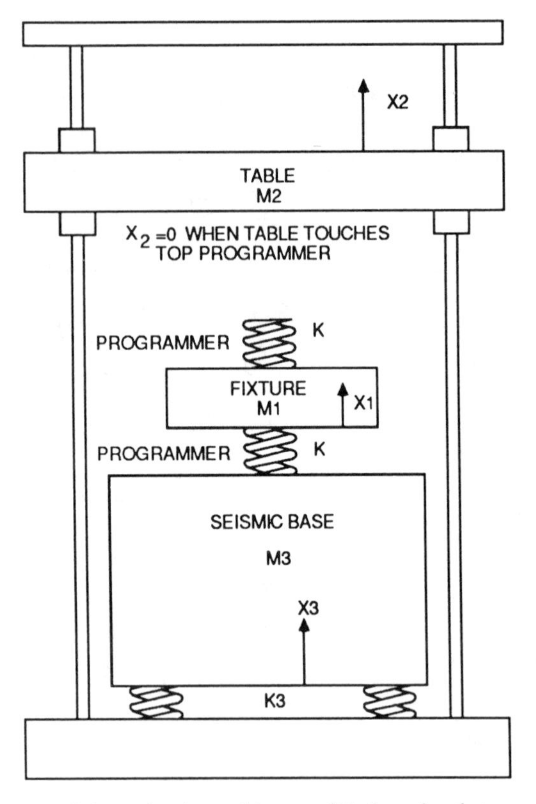

FIGURE 5.9 Standard drop shock machine modified to simulate a pyrotechnic shock by the bounded impact method.

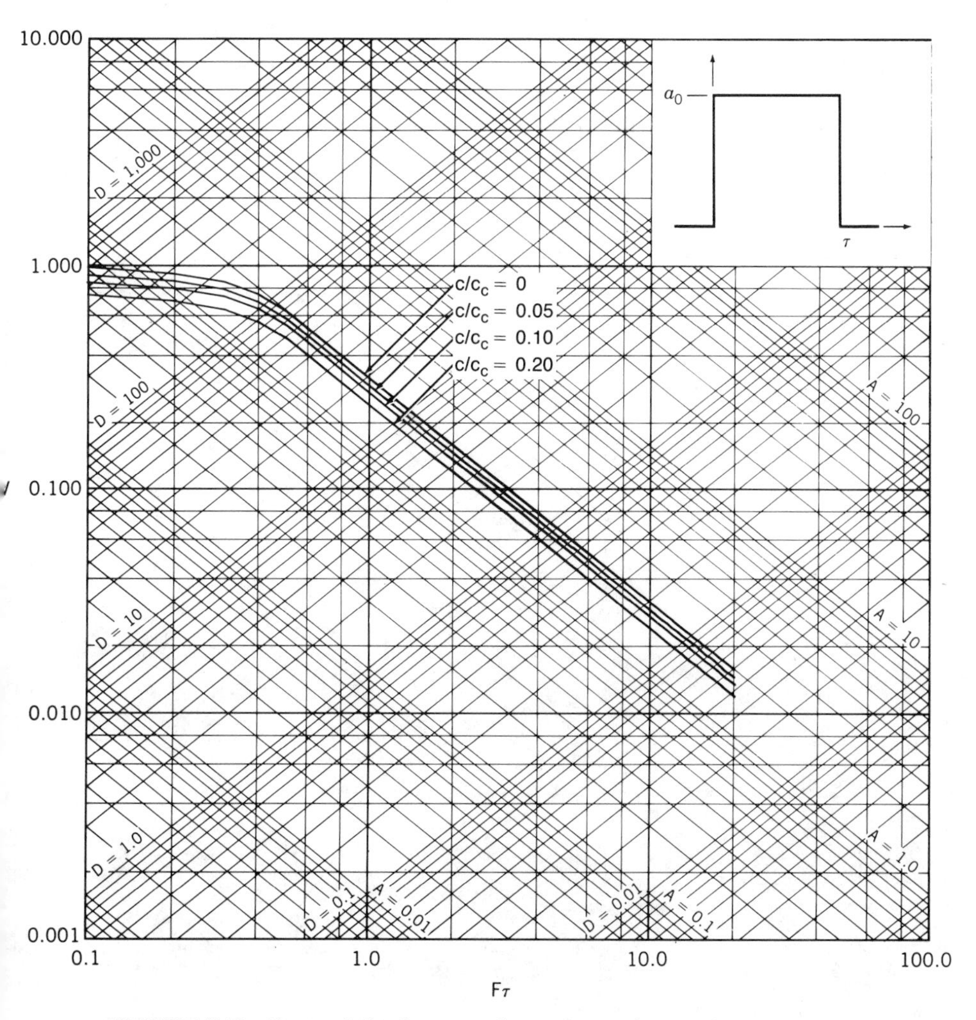

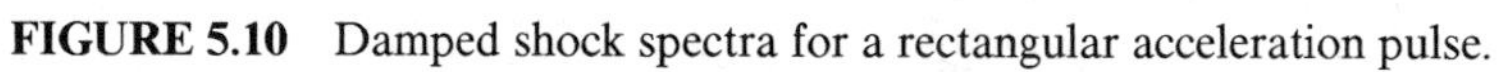
FIGURE 5.10 Damped shock spectra for a rectangular acceleration pulse.

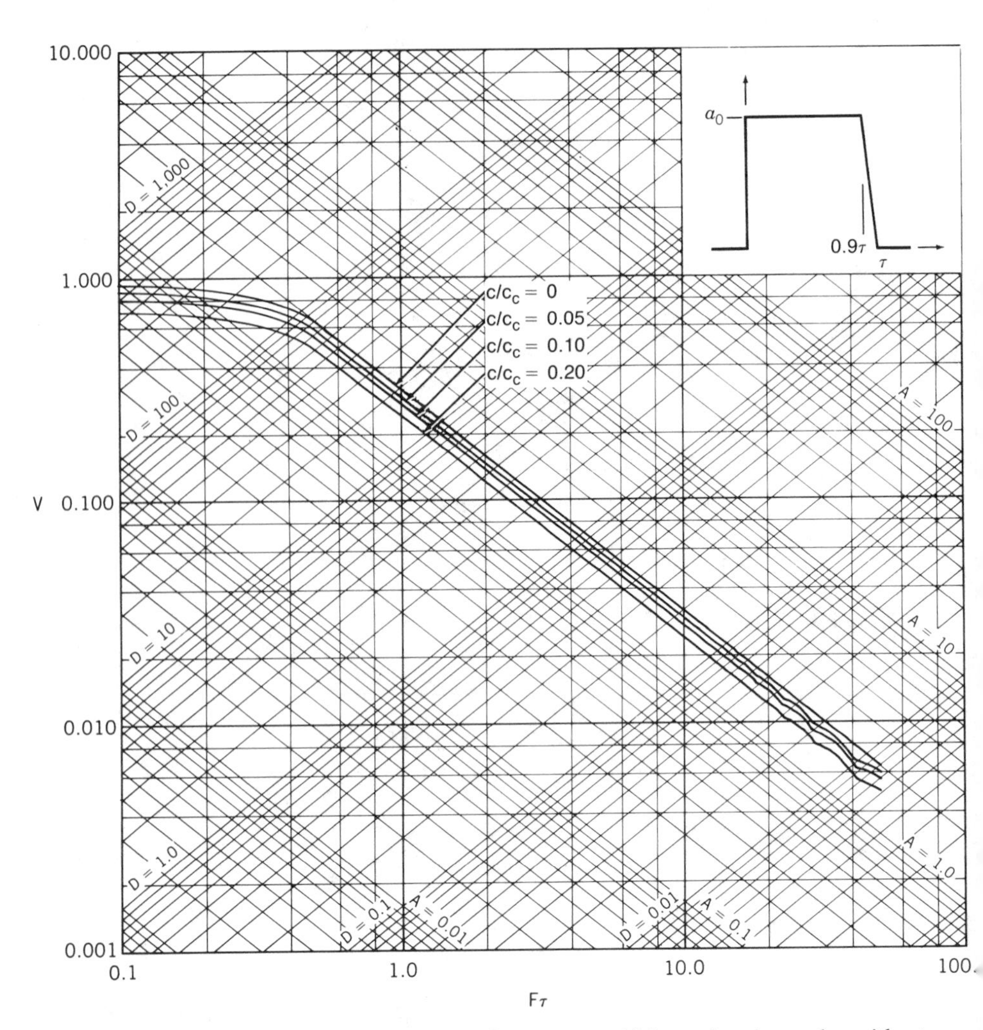

FIGURE 5.11 Damped shock spectra for a trapezoidal acceleration pulse with step rise and constant-slope decay; decay time = 0.1τ.

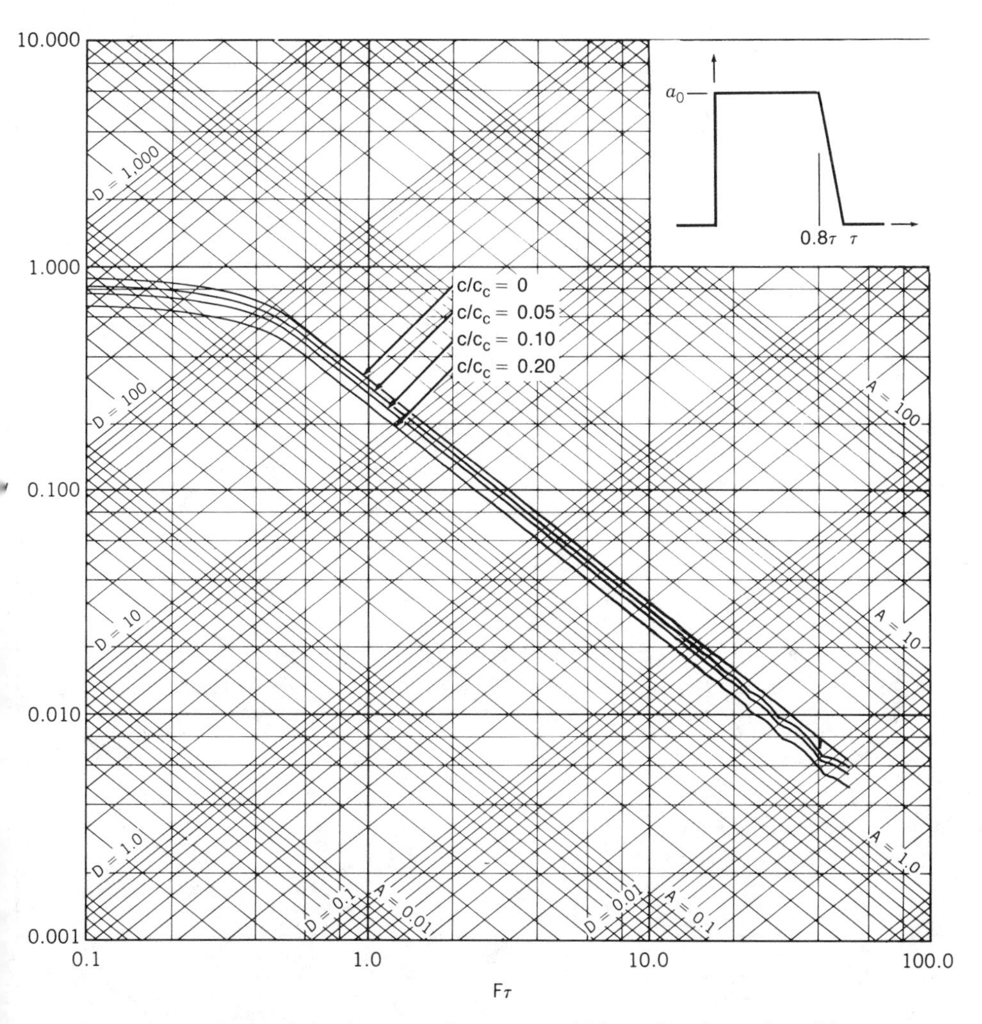

FIGURE 5.12 Damped shock spectra for a trapezoidal acceleration pulse with a step rise and constant-slope decay; decay time = 0.2τ.

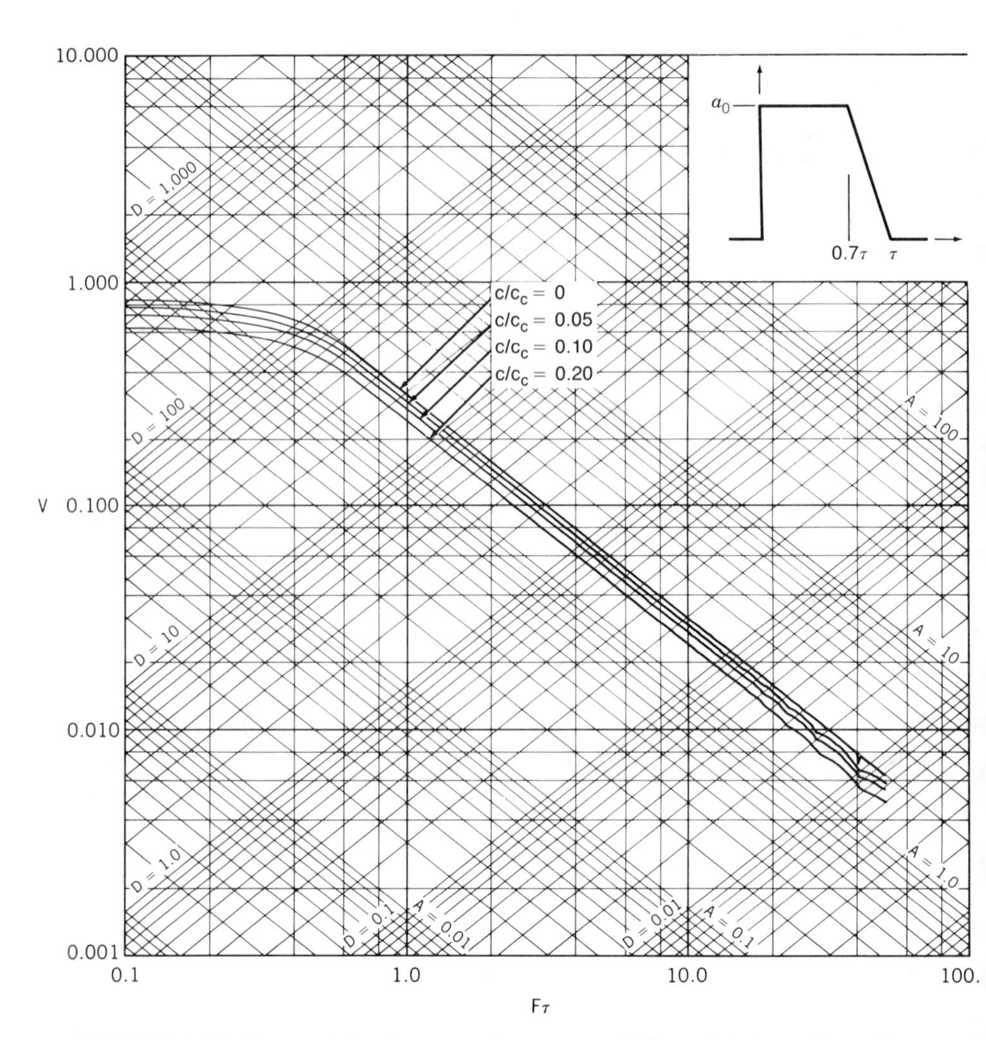

FIGURE 5.13 Damped shock spectra for a trapezoidal acceleration pulse with step rise and constant-slope decay; decay time = 0.3τ.

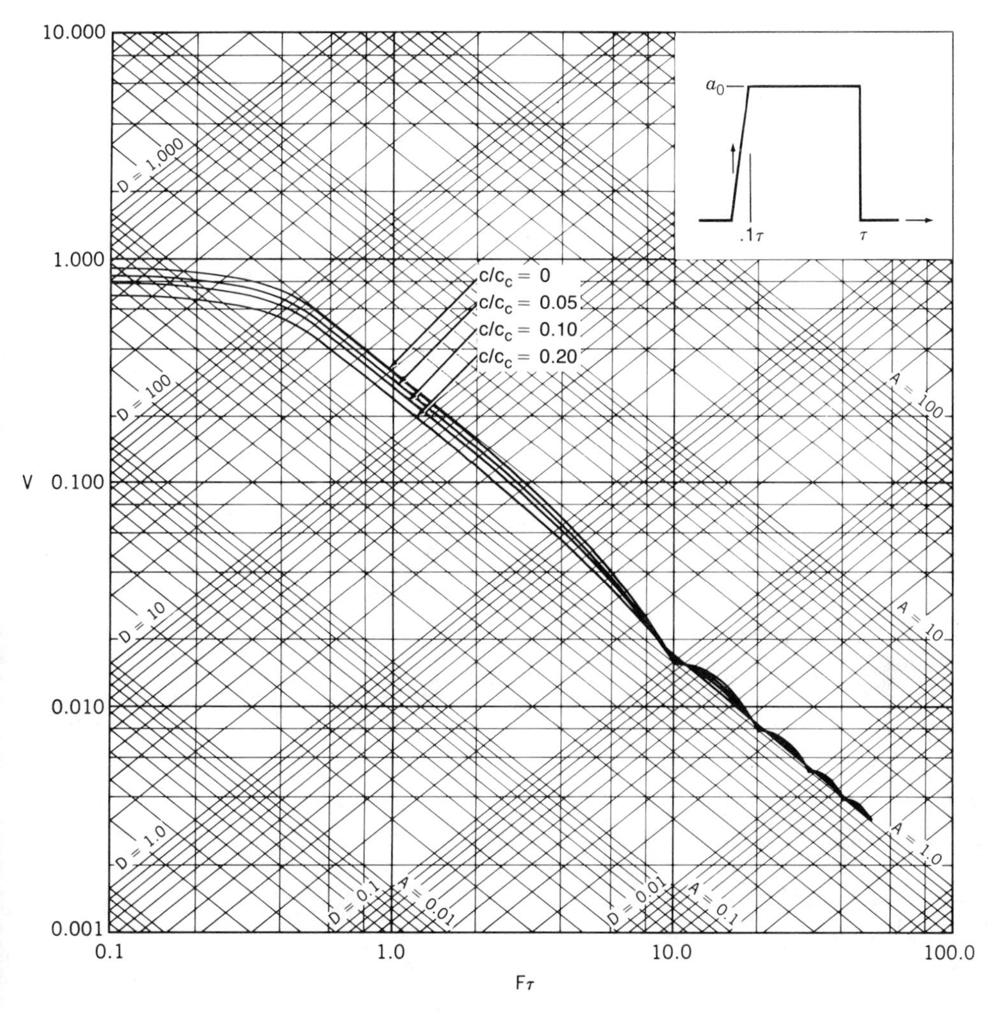

FIGURE 5.14 Damped shock spectra for a trapezoidal acceleration pulse with a constant-slope rise and vertical decay; rise time = 0.1τ.

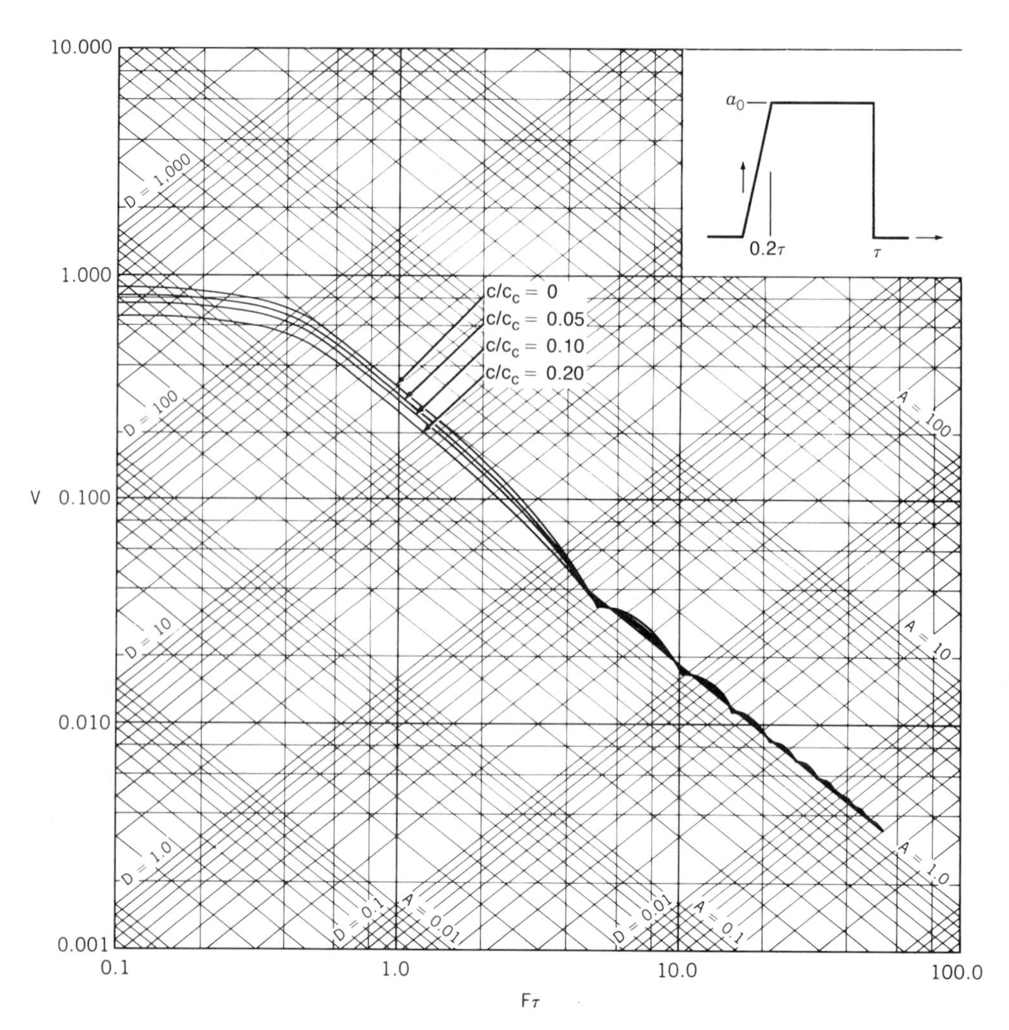

FIGURE 5.15 Damped shock spectra for a trapezoidal acceleration pulse with a constant-slope rise and vertical decay; rise time = 0.2τ.

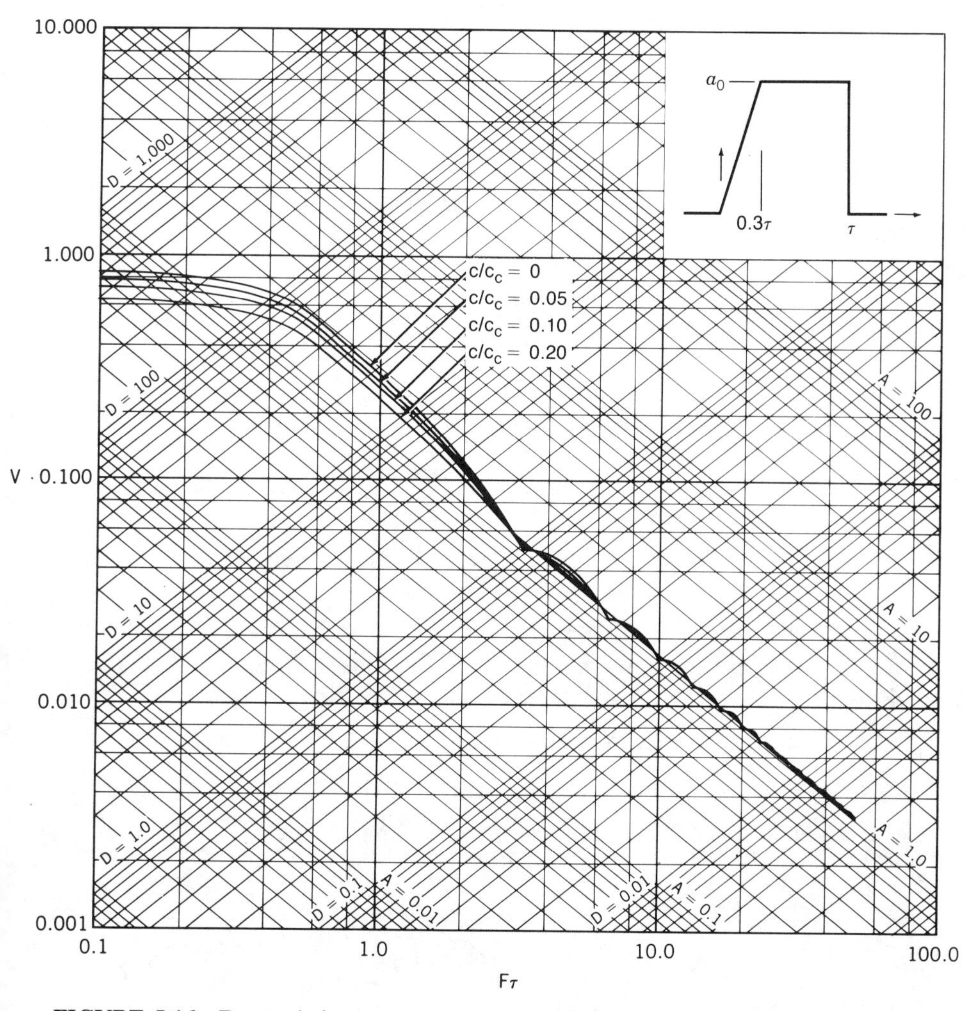

FIGURE 5.16 Damped shock spectra for a trapezoidal acceleration pulse with constant-slope rise and vertical decay; rise time = 0.3τ.

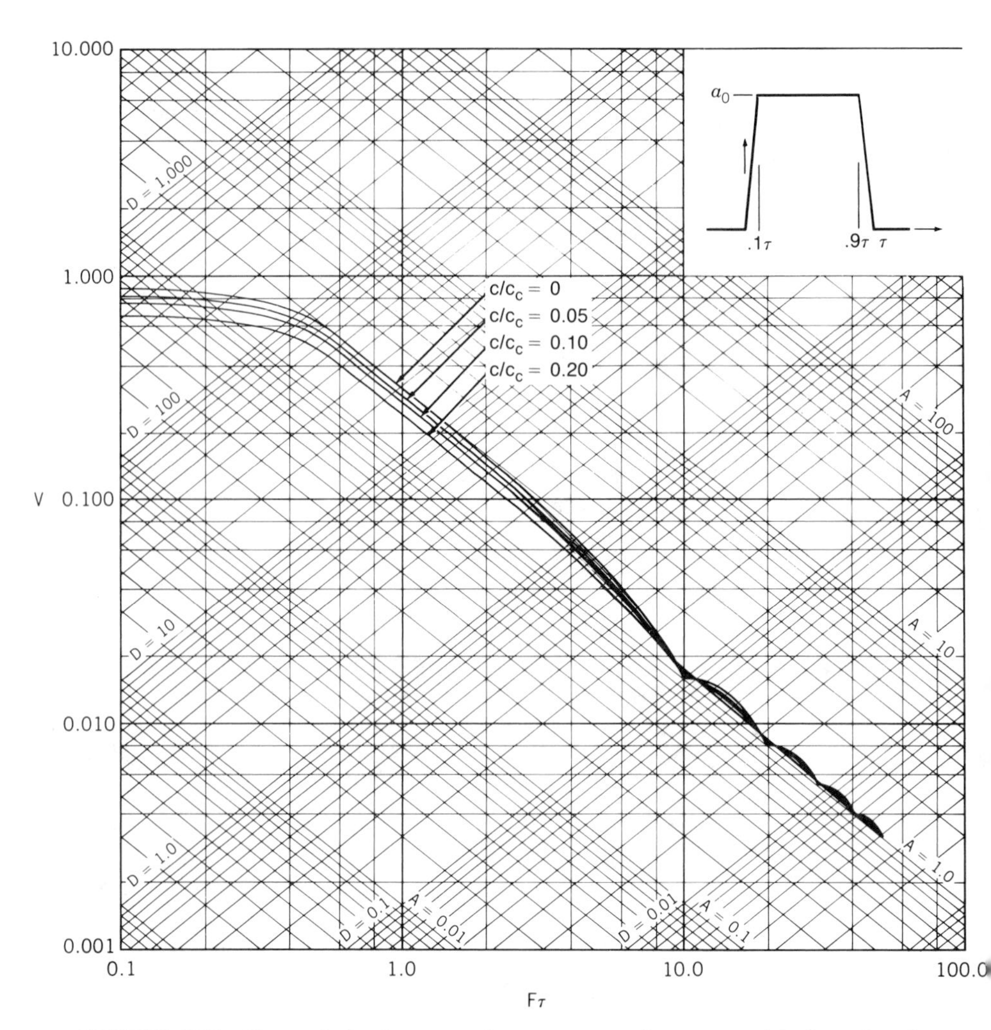

FIGURE 5.17 Damped shock spectra for a constant-slope symmetrical acceleration pulse; rise time = 0.1τ, decay time = 0.1τ.

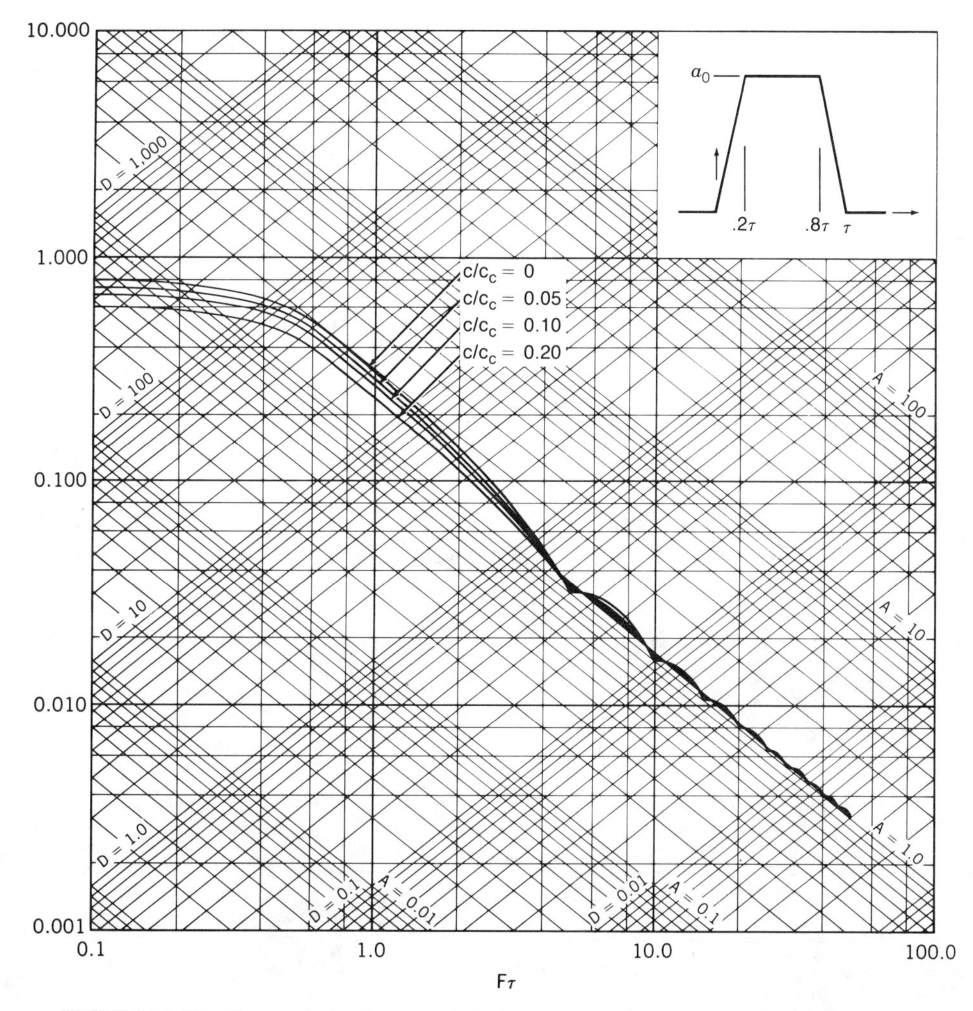

FIGURE 5.18 Damped shock spectra for a constant-slope symmetrical acceleration pulse; rise time = 0.2τ, decay time = 0.2τ.

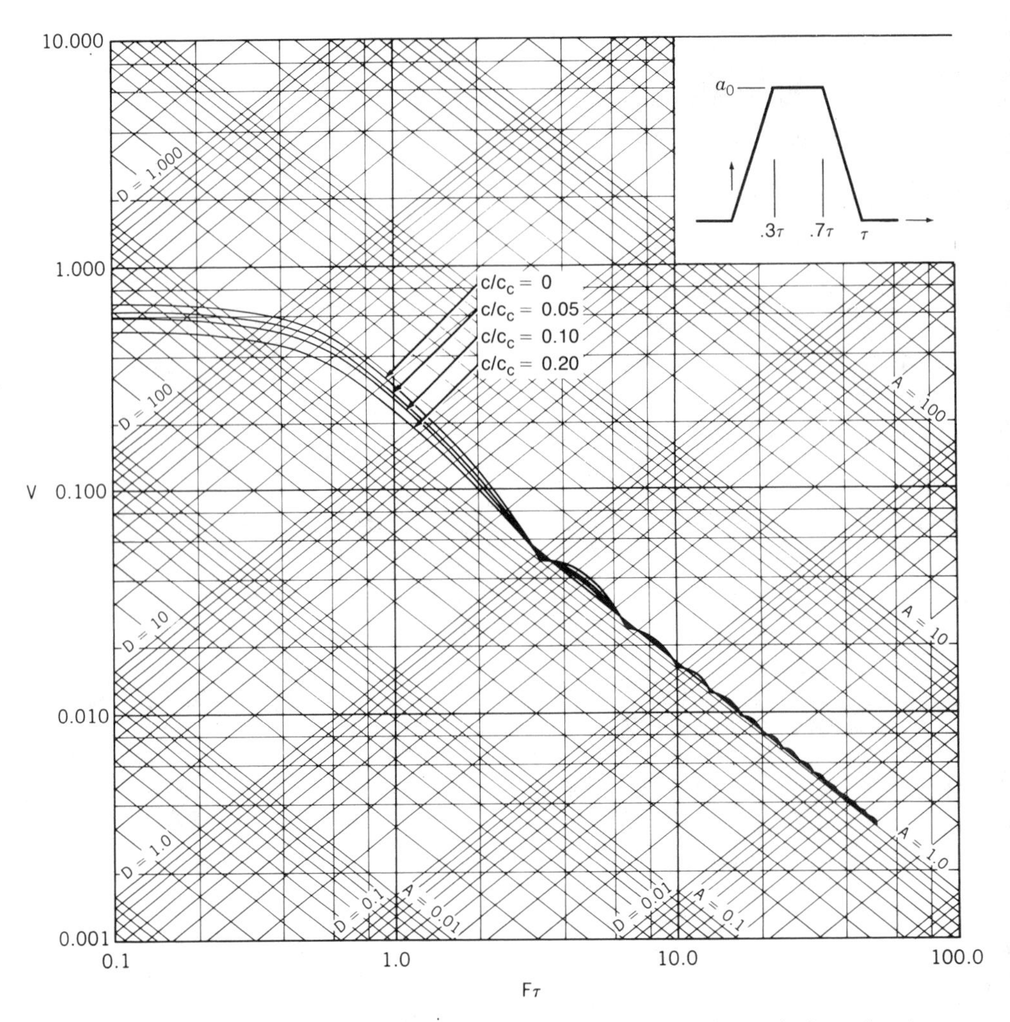

FIGURE 5.19 Damped shock spectra for a constant-slope symmetrical acceleration pulse; rise time = 0.3τ, decay time = 0.3τ.

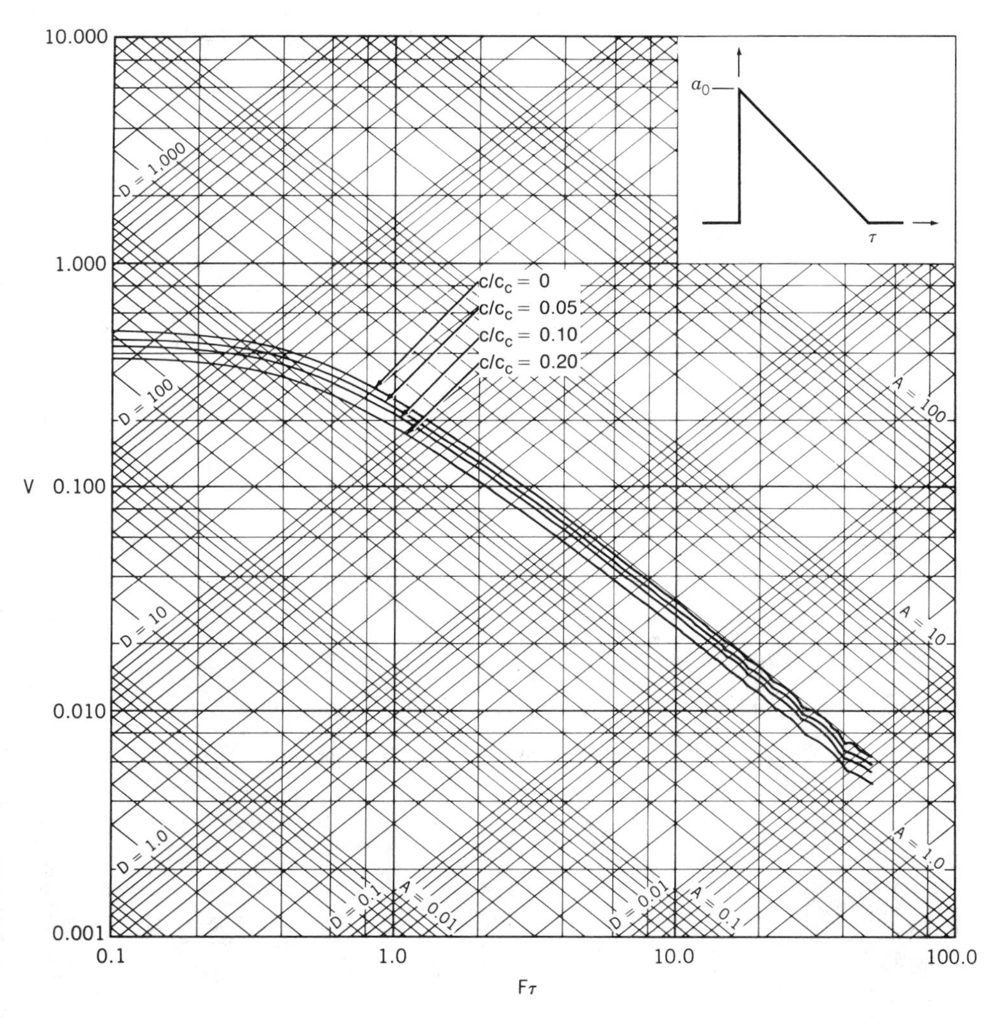

FIGURE 5.20 Damped shock spectra for a triangular acceleration pulse with step rise.

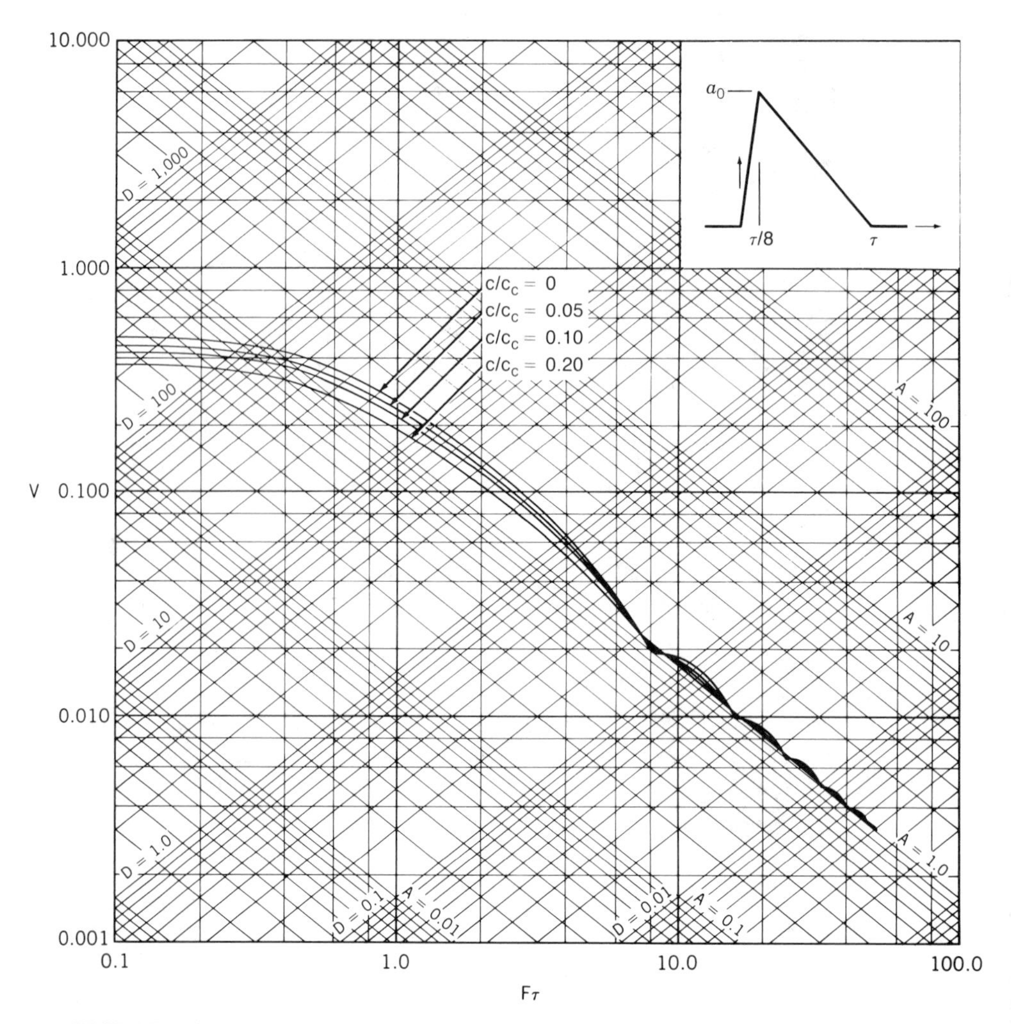

FIGURE 5.21 Damped shock spectra for a triangular acceleration pulse with rise time = $\tau/8$.

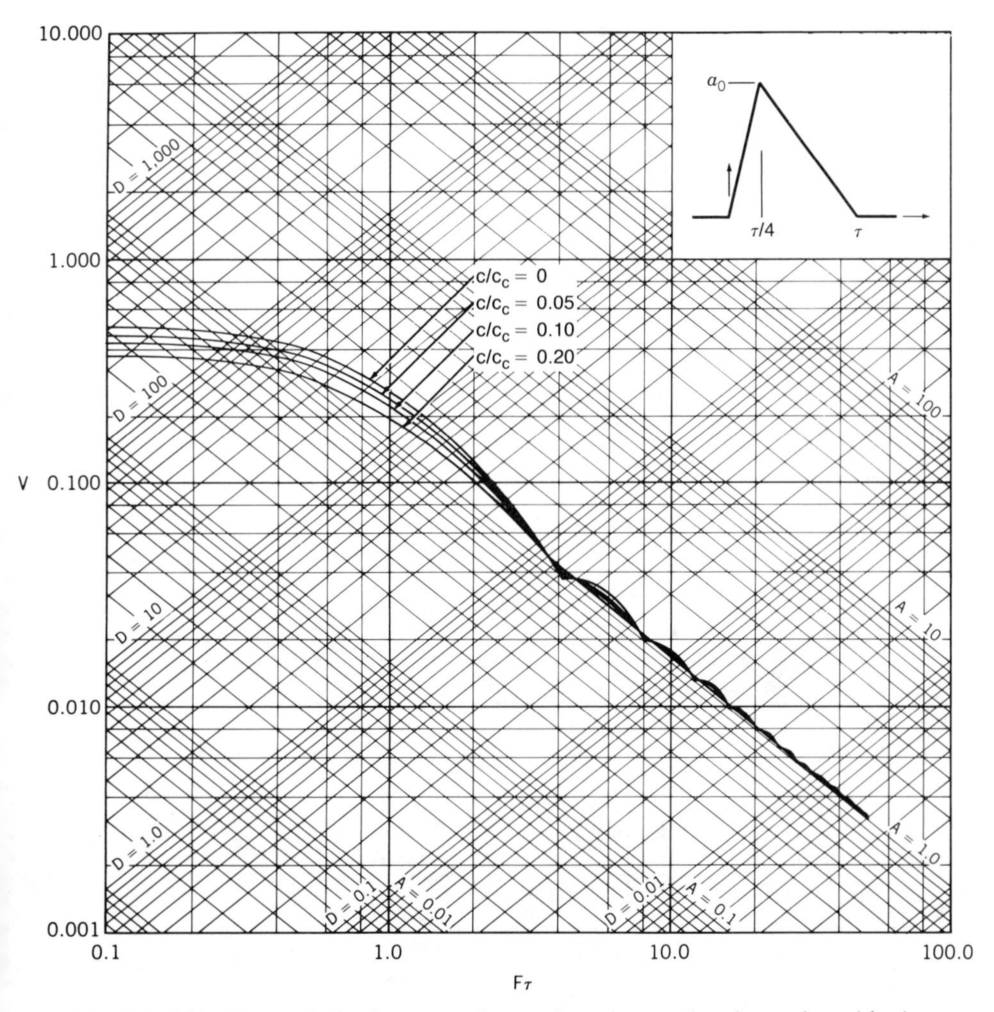

FIGURE 5.22 Damped shock spectra for a triangular acceleration pulse with rise time = $\tau/4$.

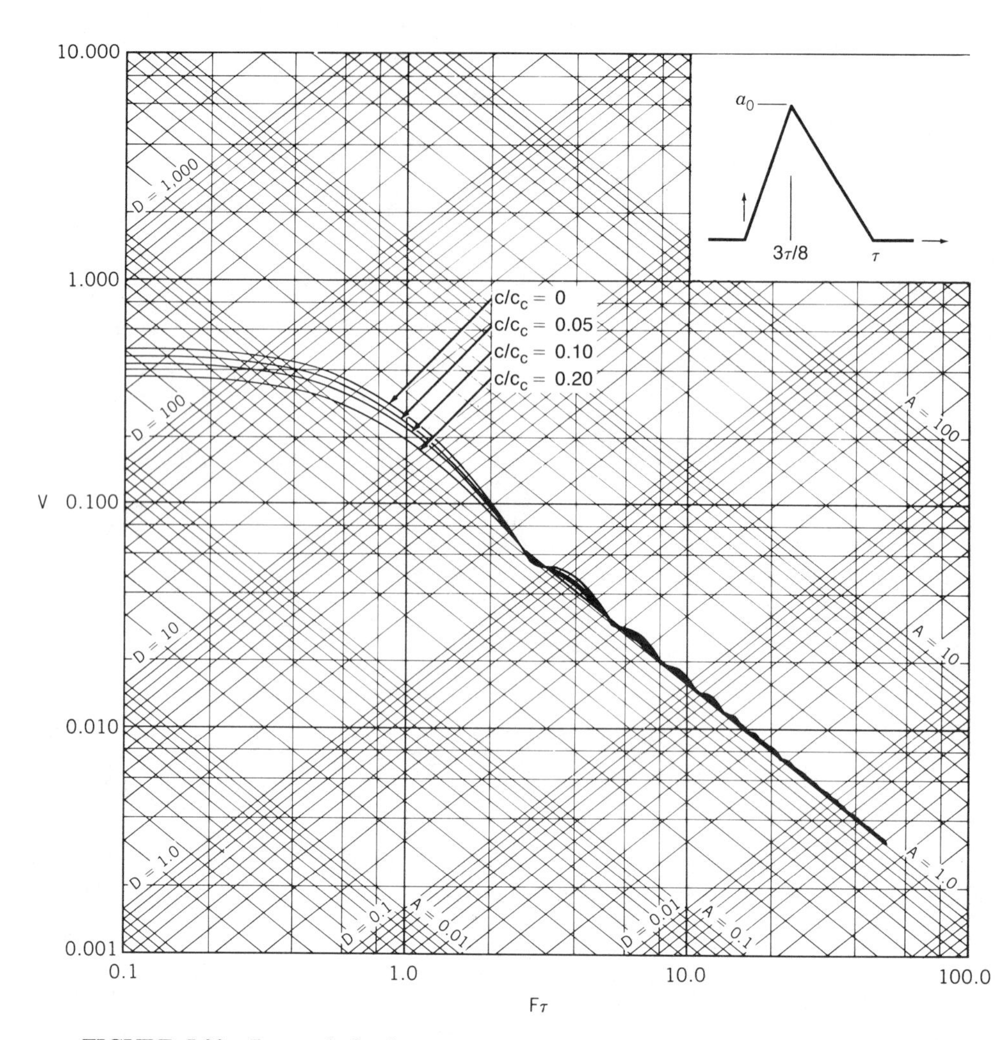

FIGURE 5.23 Damped shock spectra for a triangular acceleration pulse with rise time = $3\tau/8$.

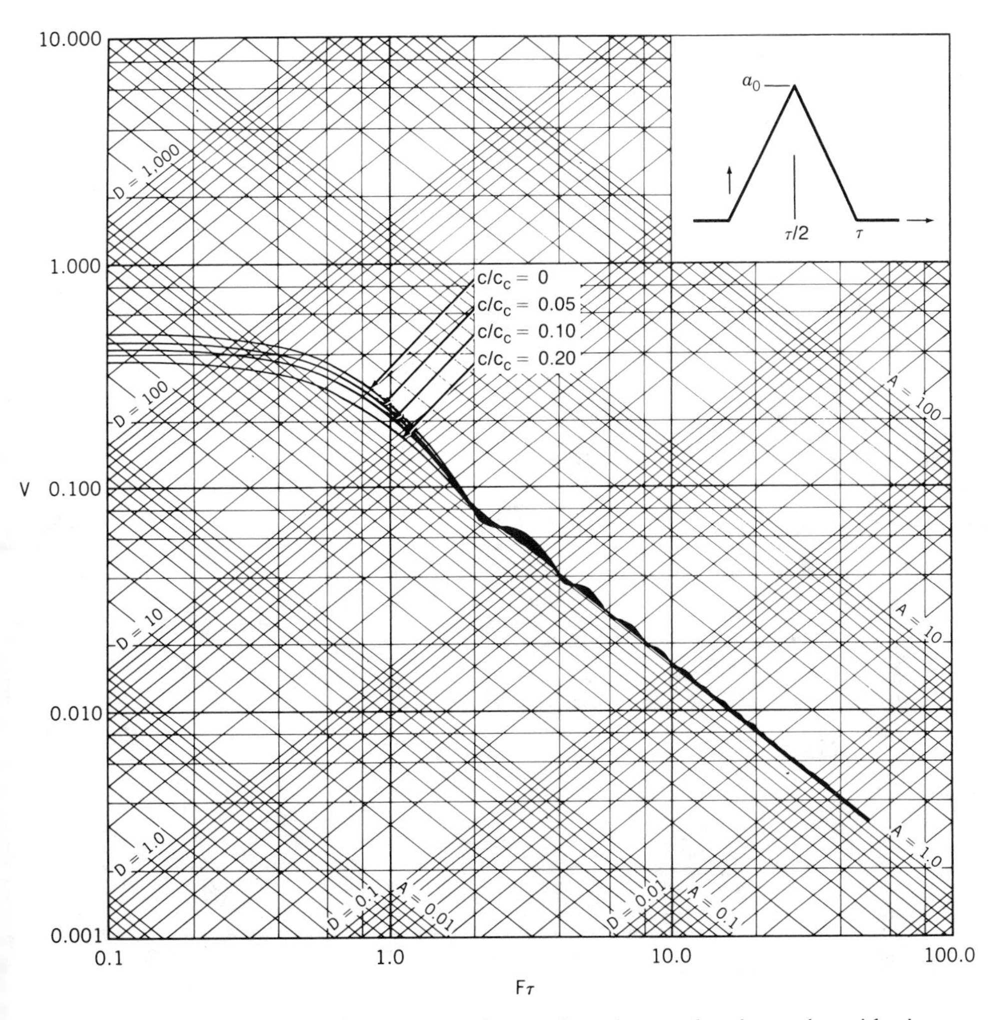

FIGURE 5.24 Damped shock spectra for a triangular acceleration pulse with rise time = $\tau/2$.

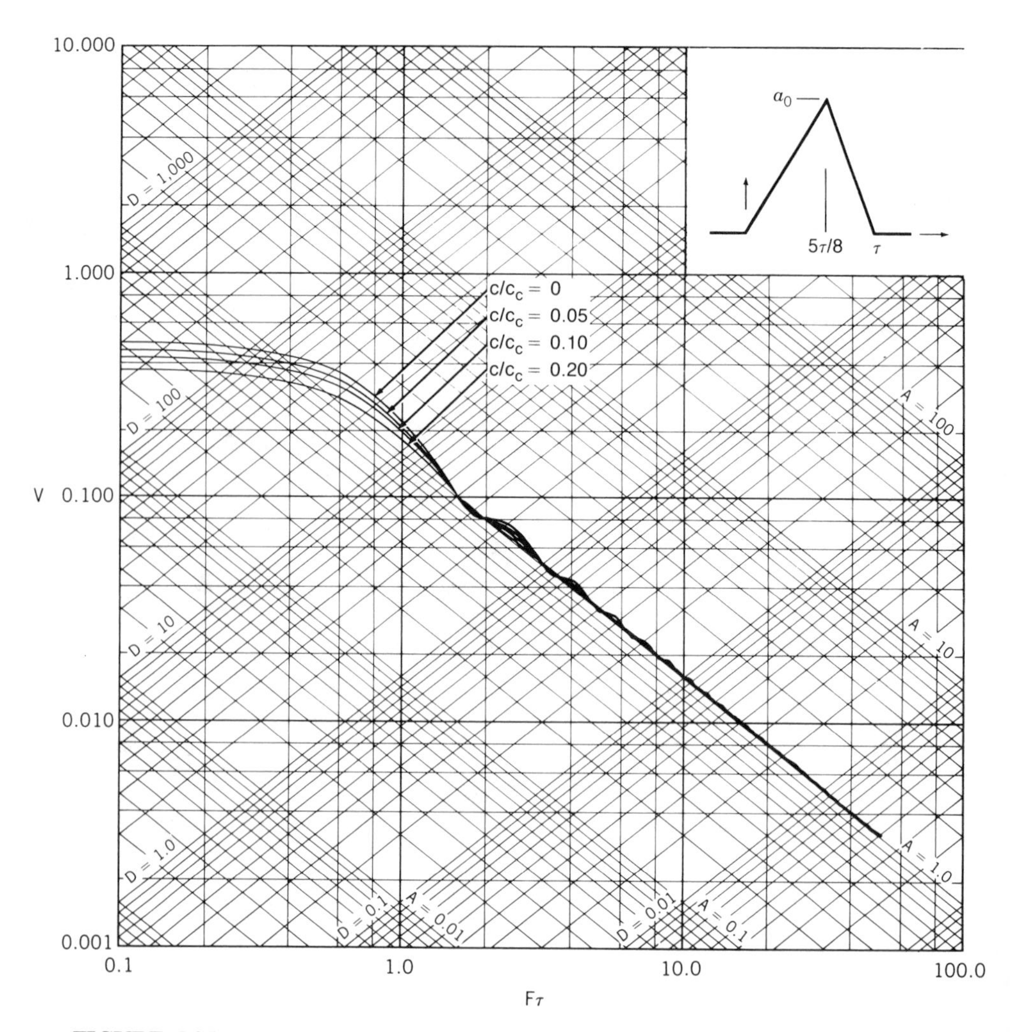

FIGURE 5.25 Damped shock spectra for a triangular acceleration pulse with rise time = $5\tau/8$.

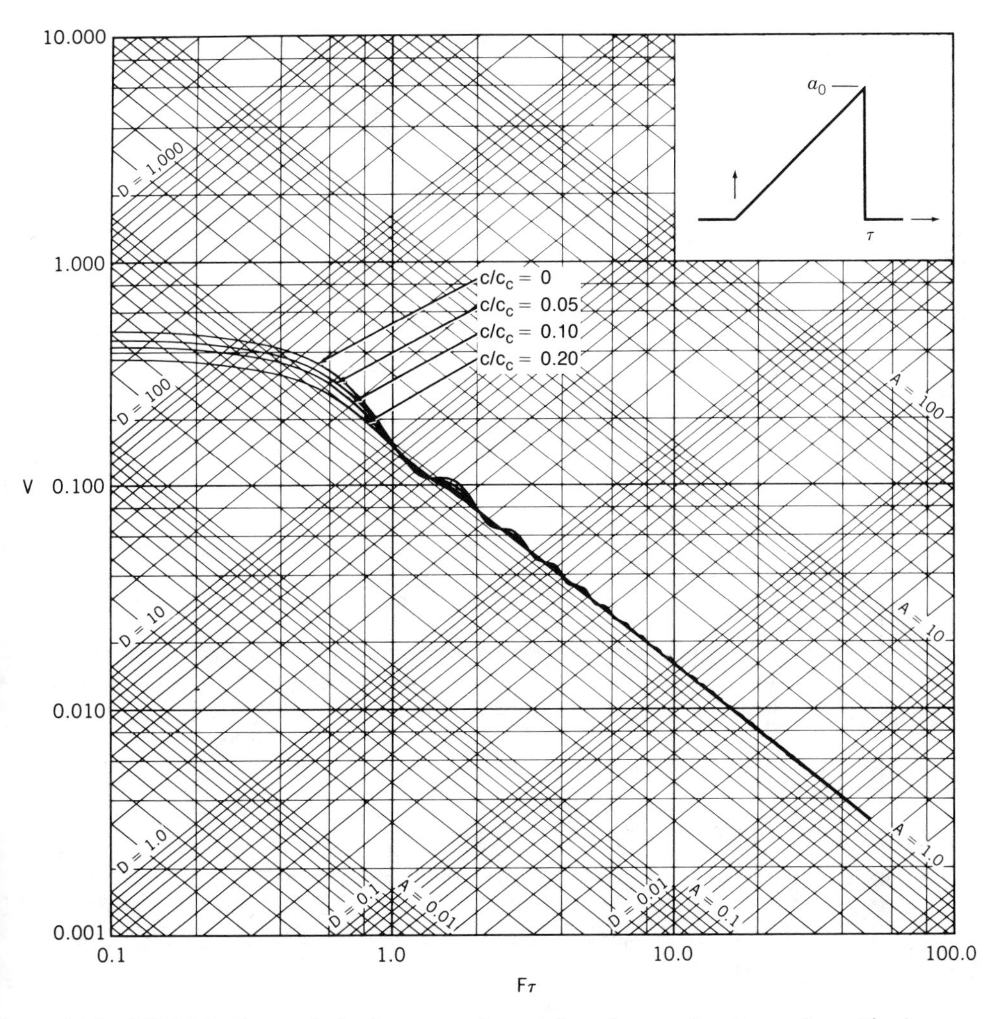

FIGURE 5.26 Damped shock spectra for a triangular acceleration pulse with rise time = τ.

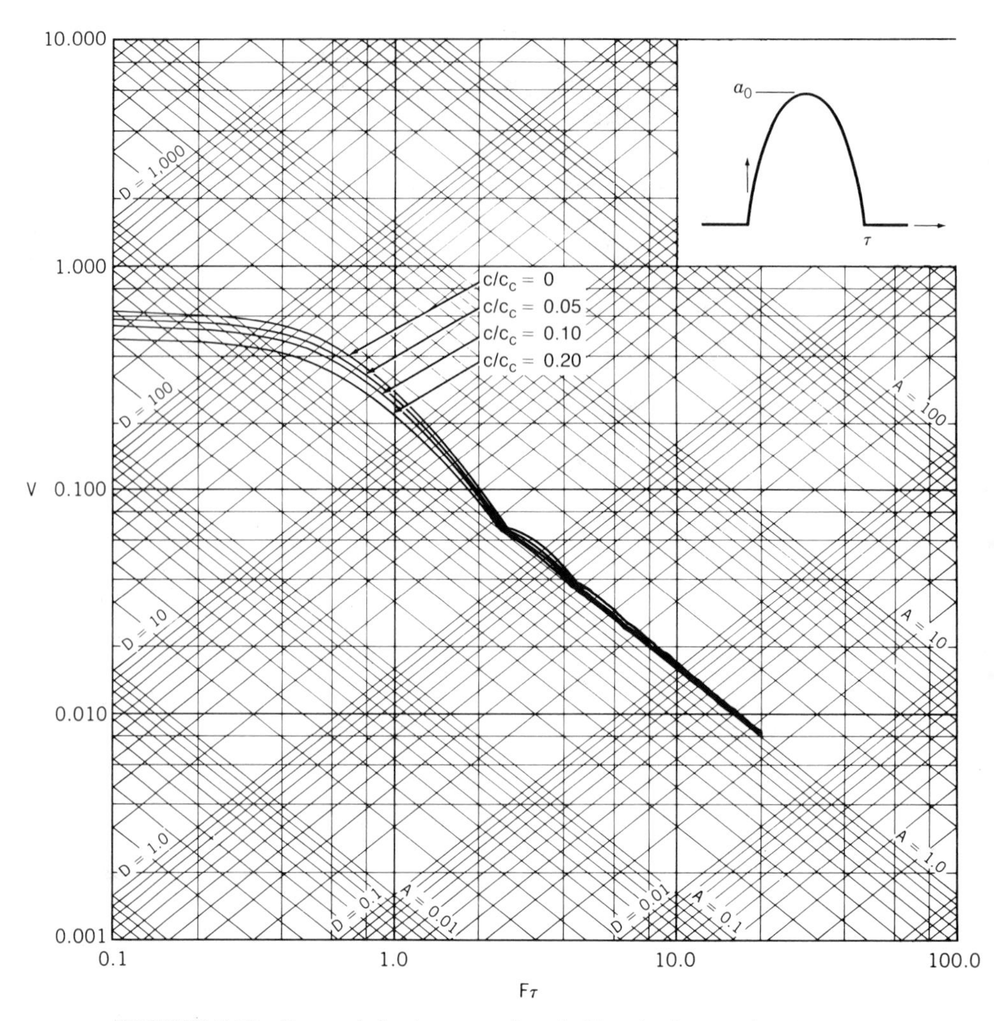

FIGURE 5.27 Damped shock spectra for a half-cycle sine acceleration pulse.

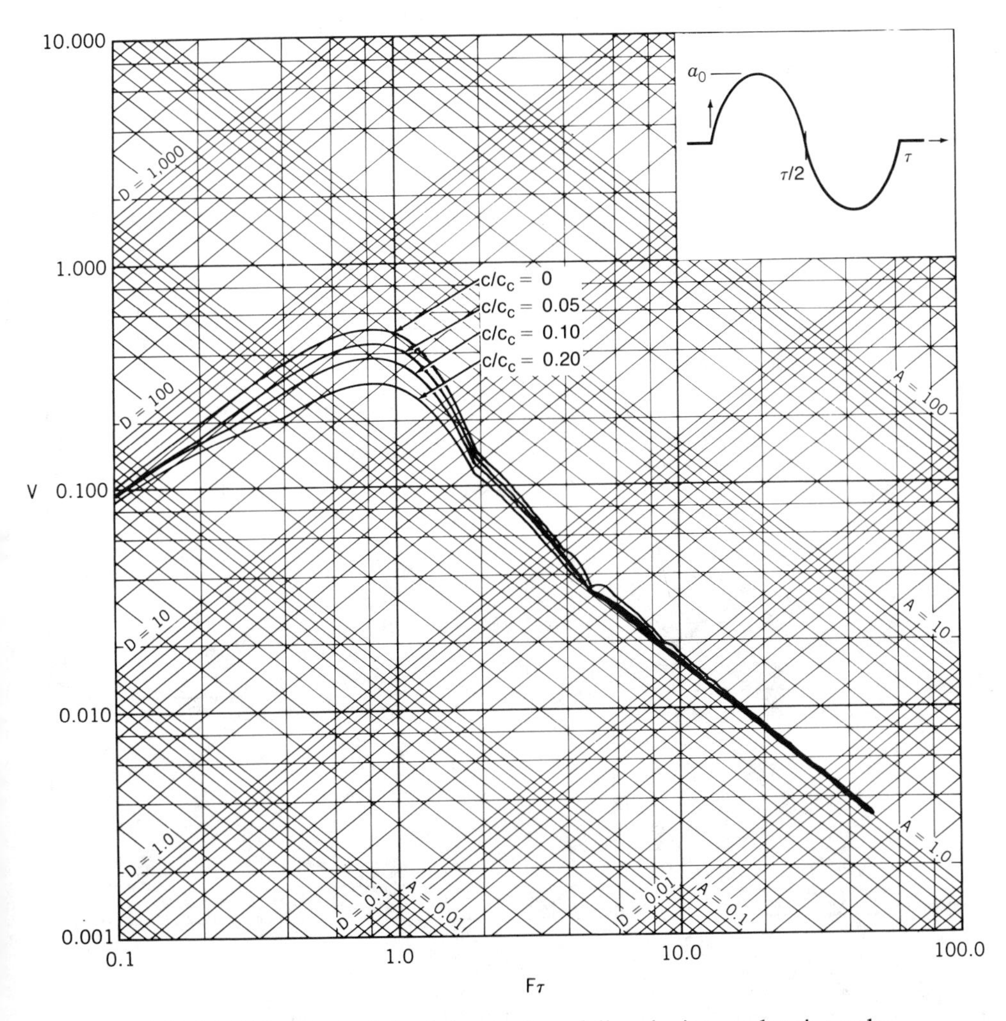

FIGURE 5.28 Damped shock spectra for a full-cycle sine acceleration pulse.

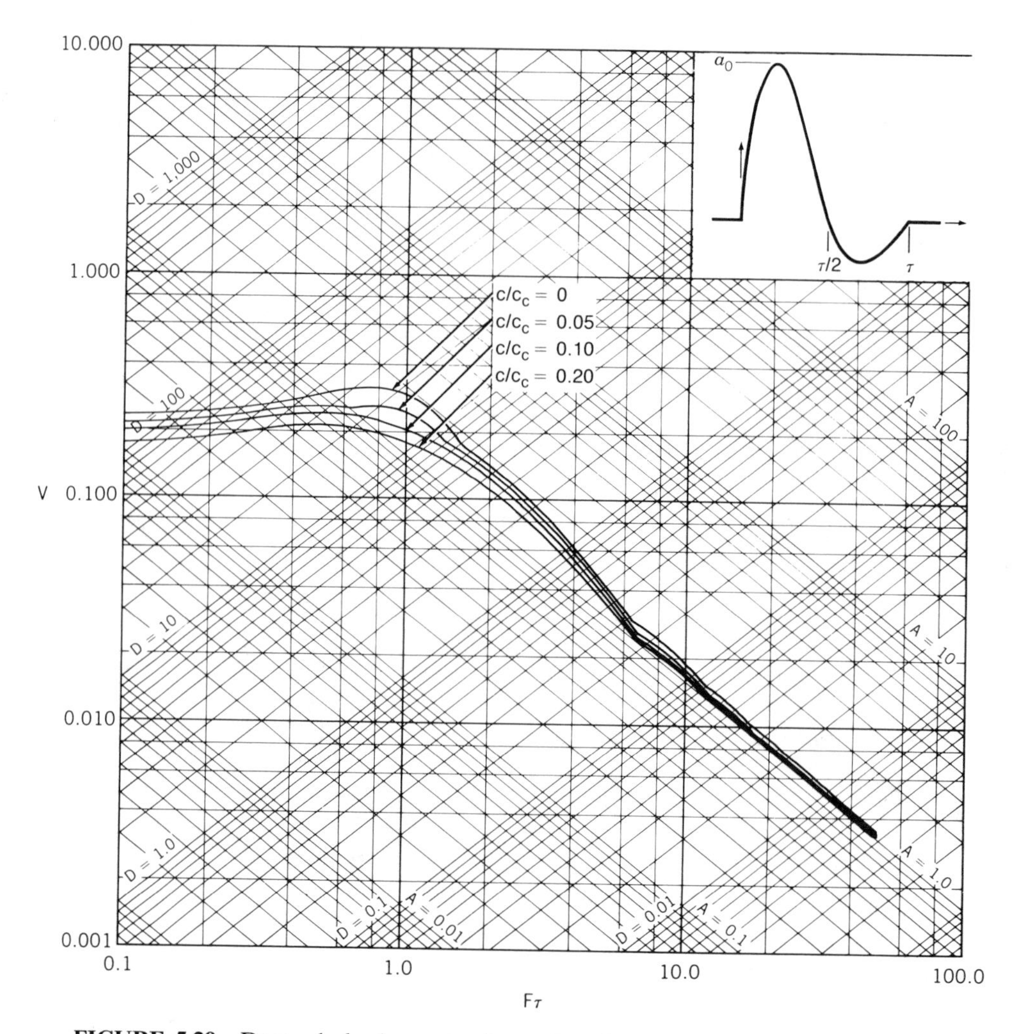

FIGURE 5.29 Damped shock spectra for a decaying sinusoidal acceleration pulse with one cycle and amplitude ratio = $\frac{1}{16}$.

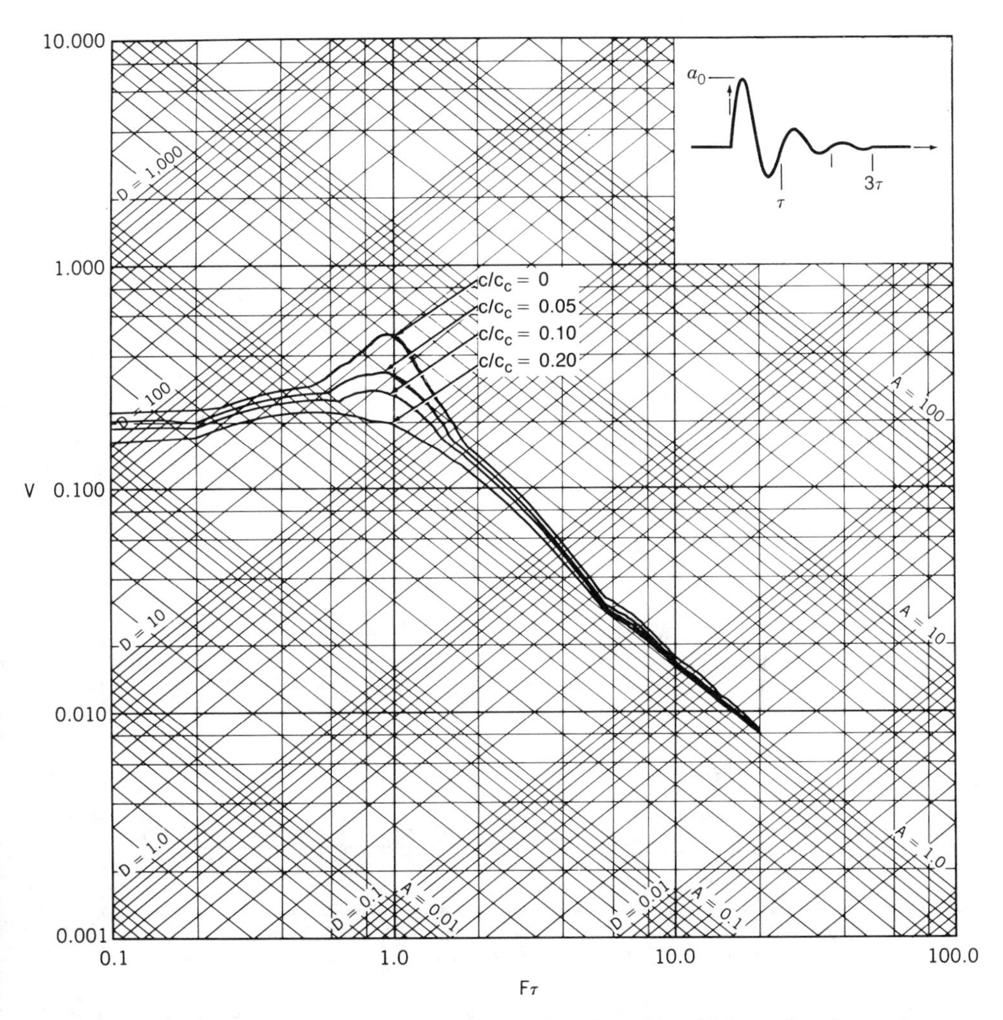

FIGURE 5.30 Damped shock spectra for a decaying sinusoidal acceleration pulse with three cycles and amplitude ratio = $\frac{1}{4}$.

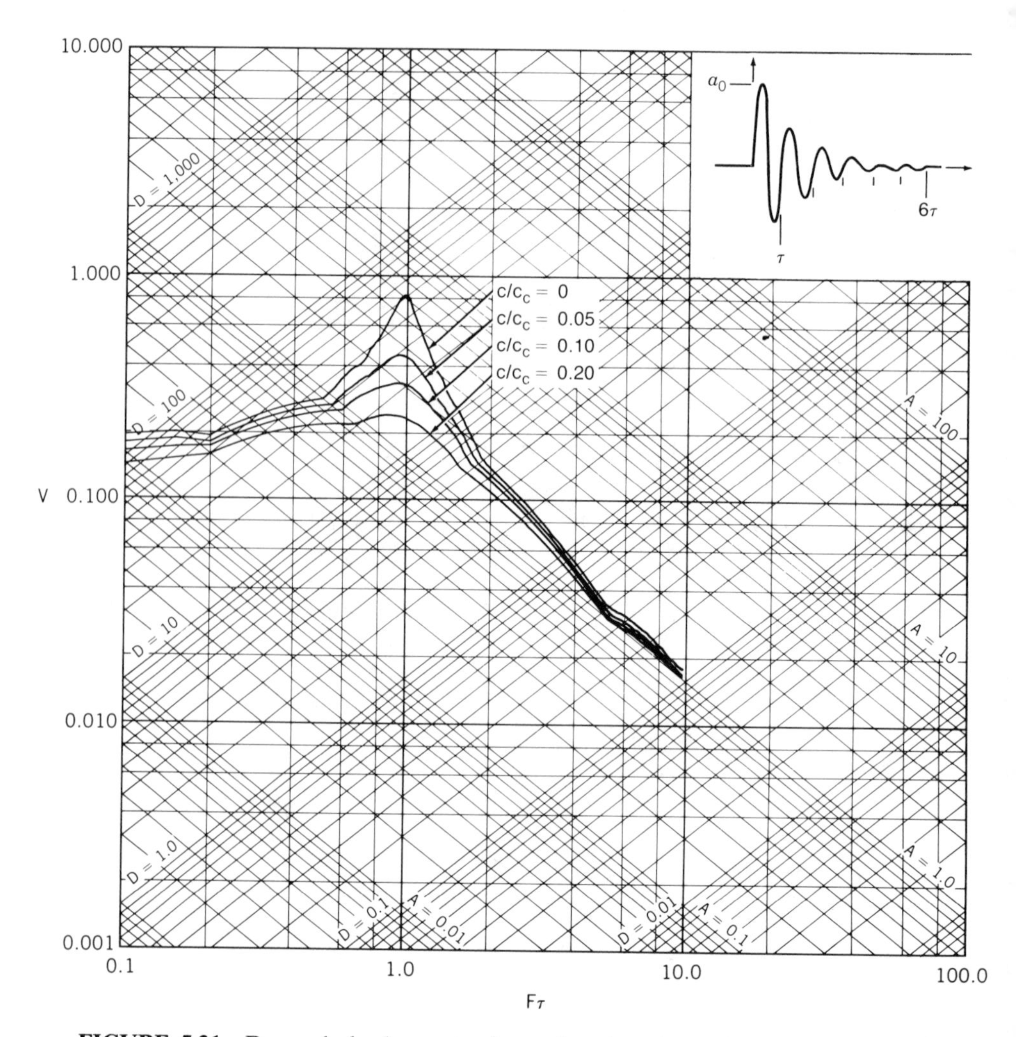

FIGURE 5.31 Damped shock spectra for a decaying sinusoidal acceleration pulse with six cycles and amplitude ratio = $\frac{1}{2}$.

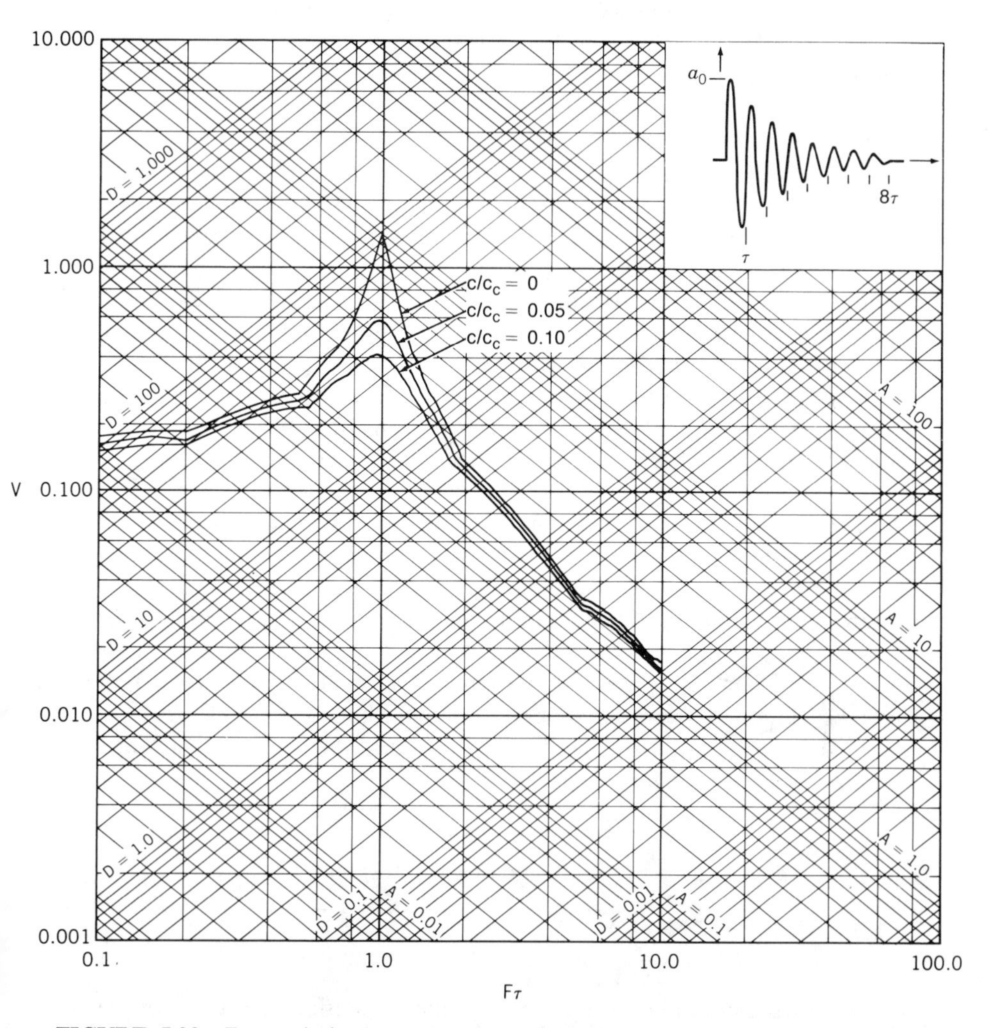

FIGURE 5.32 Damped shock spectra for a decaying sinusoidal acceleration pulse with eight cycles and amplitude ratio = $1/\sqrt{2}$.

maximum. This corresponds to a 3-db-per-octave decrease in acceleration with decreasing frequency; however, the established convention is to use the energy slope value on the acceleration-versus-frequency spectrum. See the "Velocity" line in Fig. 5.8. Most pyrotechnic shocks are due to the use of explosive separation devices during satellite or missile staging, which impart no significant relative velocity but which allow the stages to be moved in different directions. The absence of a residual velocity produces a 12-dB-per-octave decrease in energy with decreasing frequency (a 6-dB-per-octave decrease in acceleration) at frequencies below the acceleration peak. See the "Displacement" line in Fig. 5.8. The standard drop shock machine may be modified to simulate a pyrotechnic shock, as illustrated in Fig. 5.9, adopted from Fandrich.[4,5] This test configuration is called the "bounded impact" method. The 12-dB-per-octave decreasing energy slope at low frequencies may be achieved with mass ratios M_2/M_1 of 1.62, 5.72, 11.73, 19.74, or 29.74. Increasing the stiffness K of the programmers increases the peak frequency of the acceleration response spectrum, while increasing the drop height X_2 increases the acceleration amplitude. Other types of shocks with low-frequency slopes larger than 6 dB per octave may be simulated by changing the mass ratio M_2/M_1 and allowing the programmers to have unequal stiffness K_1 and K_2.

Figures 5.10–5.32 present the normalized four-coordinate shock spectra for the acceleration–time history pulses shown in the inset drawing of each figure, and for several values of the damping ratio R_c. The original drawings for Figures 5.10 through 5.32 were kindly provided by Maurice Gertel of Kinetic Systems Inc., Boston, with permission granted through the courtesy of NASA, George C. Marshall Space Flight Center. The relative response parameters may be obtained from each figure through the use of Eqs. (5-6), where a_0 = peak acceleration (in./sec^2) and τ = pulse duration (sec) as shown in the acceleration–time history pulse:

$$\begin{aligned} F &= F\tau/\tau = \text{frequency (Hz)} \\ d &= Da_0\tau^2 = \text{relative deflection response (in.)} \\ v &= Va_0\tau = \text{pseudovelocity response (in./sec)} \\ a &= Aa_0 = \text{equivalent static acceleration (in./sec}^2\text{)} \end{aligned} \qquad (5\text{-}6)$$

where the dimensionless frequency $F\tau$, the dimensionless deflection D, the dimensionless velocity V, and the dimensionless acceleration A are read off of the figure.

5.3 PYROTECHNIC SHOCK SOURCES

As discussed in Section 5.2, the low-frequency part of the pyrotechnic acceleration response spectrum (below the frequency at which the acceleration response peaks) usually has a 12-dB-per-octave decrease in energy (a 6-dB-

TABLE 5-1 Peak Acceleration Response and Frequency for Pyrotechnic Devices[a]

Type of Device	Peak Acceleration Response (g)	Peak Frequency (Hz)	t (msec)[b]
Linear, 0.17-in. joint, 50 grains/ft	45,000	6,000	3
Linear, 0.19-in. joint, 30 grains/ft	16,000	2,600	3
Linear, 0.19-in. joint, 10 grains/ft	11,000	2,800	3
Linear, 0.09-in. joint, 10 grains/ft	8,000	3,000	3
Pin puller	5,000–7,000	3,000–10,000	5–15
Pin pusher	800	500	50
Separation nuts, 5/16–1/2 in.	15,000	10,000	3
Bolt, pin, cable cutters	4,000–6,000	4,000	—
Ordnance valves, 3/4 in.	3,000	4,000	—

[a]Adapted from Refs. 6 and 7.
[b]This column is the typical time history effective duration.

per-octave decrease in acceleration response) with decreasing frequency. Table 5-1 presents typical values of peak acceleration response and frequency, and the time history effective duration, for various types of pyrotechnic devices, usually measured 4–5 in. from the pyrotechnic source.[6,7]

5.4 SHOCK ATTENUATION

As the shock travels from the pyrotechnic source through the structure it dissipates energy in the structural materials, and some of the shock energy is reflected at structural joints and at discontinuities in material properties and geometric cross sections. The high-frequency energy is dissipated more rapidly than the low-frequency energy, so that as the shock travels away from the source it decreases in amplitude and flattens out, with the peak acceleration decreasing faster than the low-frequency slope, and the low-frequency slope becoming less steep.

Figures 5.33–5.39 show the attenuation of the pyrotechnic shock peak acceleration with distance from the shock source. The abscissa is the distance in inches from the pyrotechnic source, and the ordinate is the remaining percentage of the peak acceleration, normalized to 100% at a distance of 5 in. from the source.

Testing of the shock reduction due to reflection of energy at mechanically fastened structural joints[8] indicates approximately 50% reduction in shock

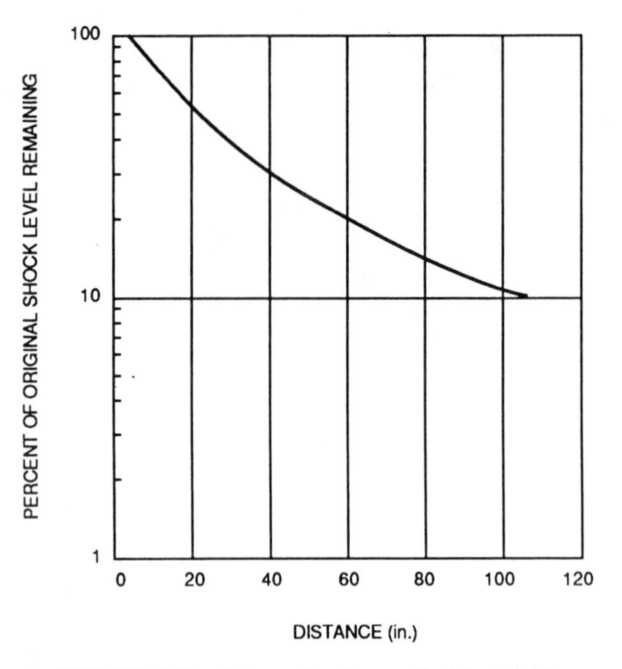

FIGURE 5.33 Attenuation for a cylindrical shell.

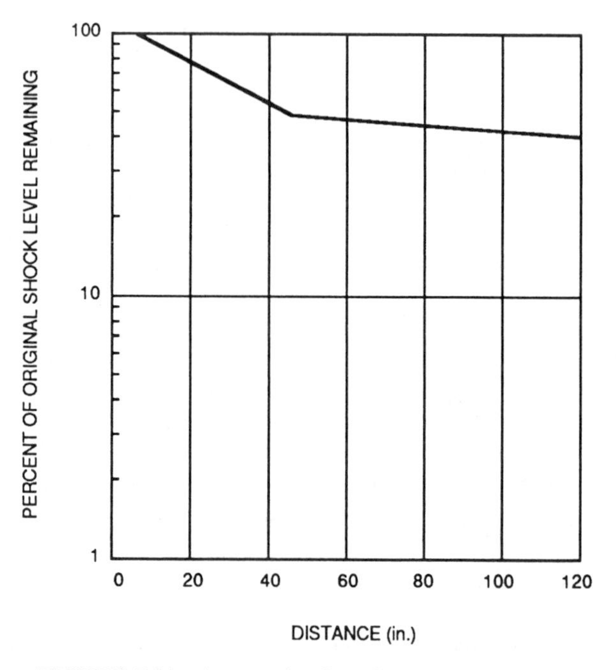

FIGURE 5.34 Attenuation for a longeron or stringer.

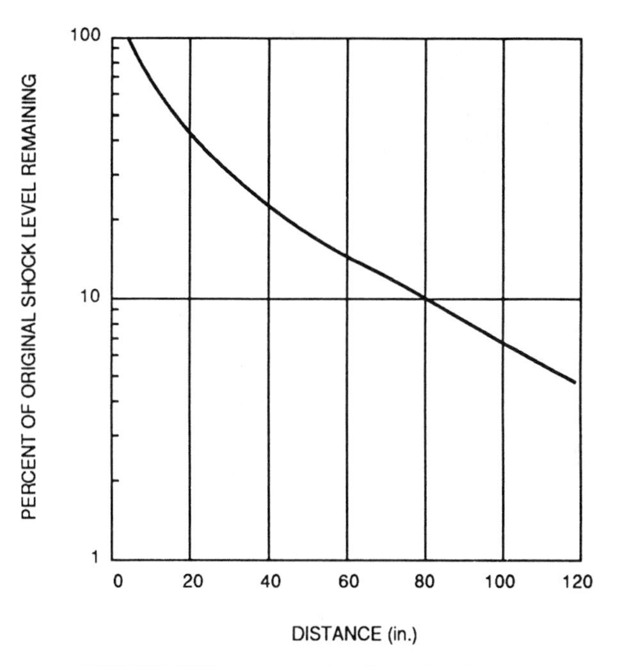

FIGURE 5.35 Attenuation for a ring frame.

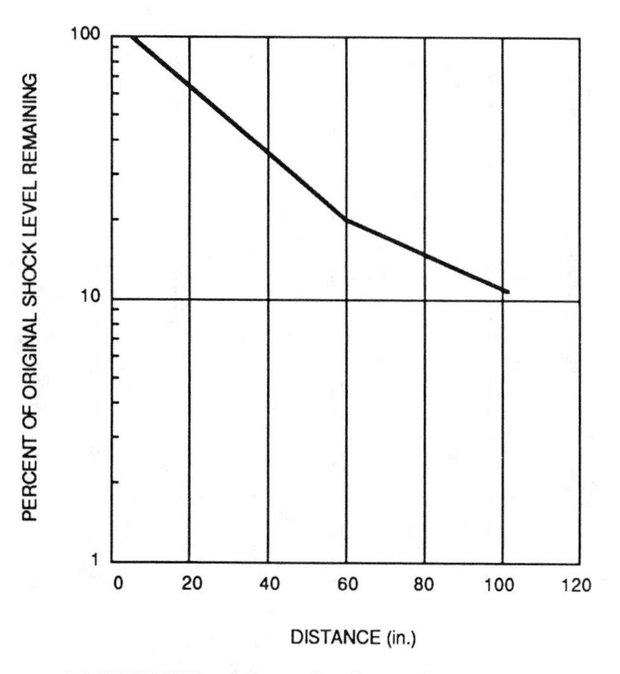

FIGURE 5.36 Attenuation for a primary truss.

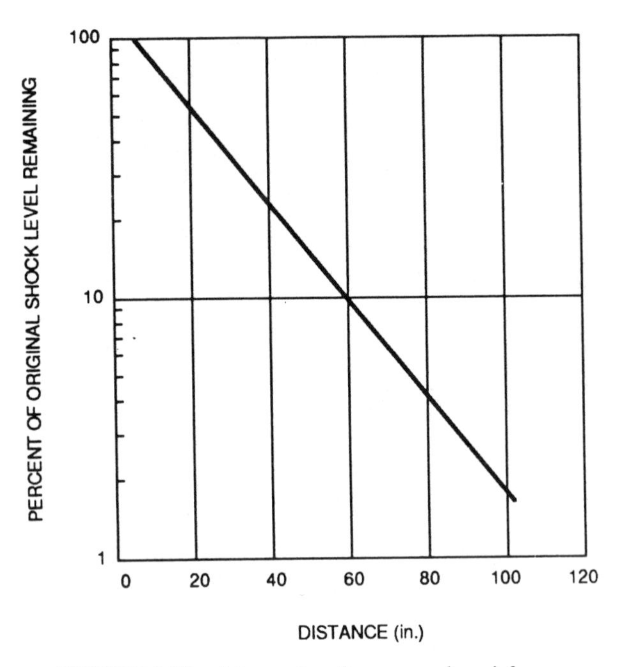

FIGURE 5.37 Attenuation for a complex airframe.

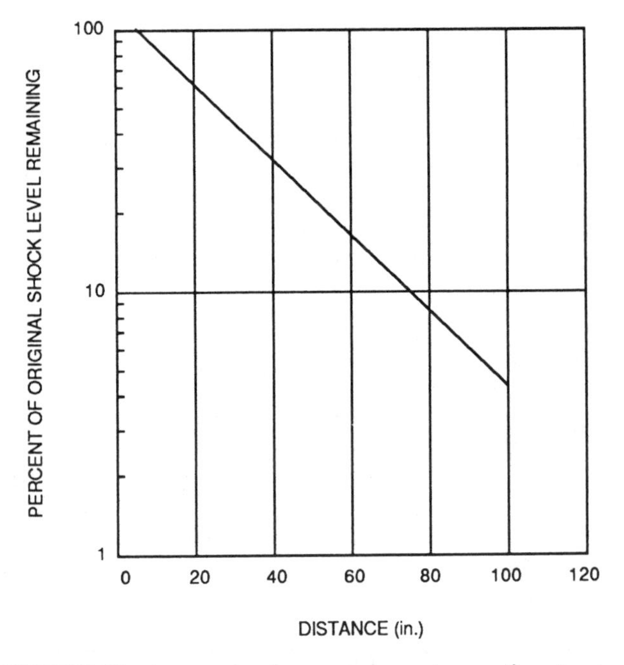

FIGURE 5.38 Attenuation for an equipment-mounting structure.

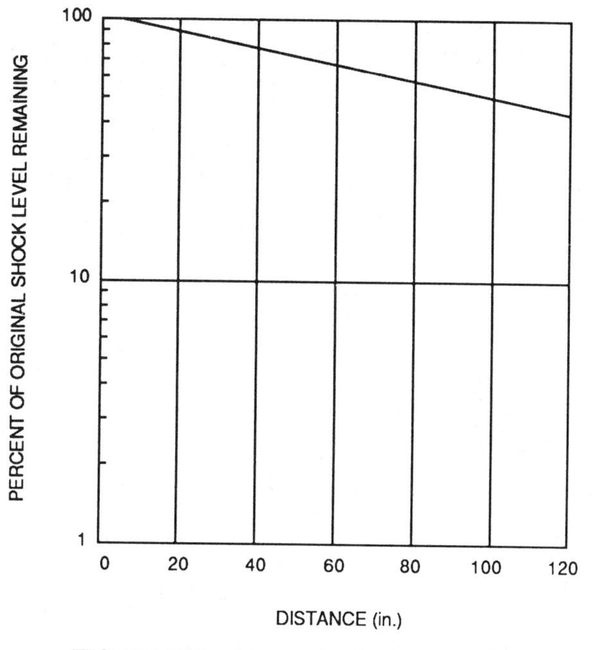

FIGURE 5.39 Attenuation for honeycomb.

spectrum acceleration responses at all frequencies. A heat-shrink rubber sleeve around the bolts fastening the joint significantly reduces the acceleration response at most frequencies, while inserts of lead or alternate layers of steel and magnesium reduce accelerations only in the upper frequency range (above 1500 Hz). When multiple mechanically fastened structural joints are present between the shock source and the location of interest, a rule of thumb is to assume a peak acceleration attenuation of 40% (60% remaining) for each joint up to a maximum of three joints.[7]

5.5 PEAK DISPLACEMENT AND STRESS DUE TO SHOCK

If the acceleration response due to a pyrotechnic shock source, attenuated through a structure, and acting on a subsystem is labeled G_s, then the maximum displacement d_s may be estimated by

$$d_s(G_s/g)\,d_0 \text{ in.} \tag{5-7}$$

or

$$d_s = 9.8G_s/F_0^2 \text{ in.} \tag{5-8}$$

where g = acceleration of gravity at surface of earth = 386 in./sec^2
d_0 = maximum static displacement due to a 1-g acceleration (in.)
F_0 = natural frequency of the subsystem (Hz)

Equations (5-7) and (5-8) are analogous to Equations (4-9) and (4-10). Similar to Eq. (5-7), the maximum stress due to the acceleration response G_s is given by

$$\sigma_s = (G_s/g)\sigma_0 \text{ lbf/in.}^2 \tag{5-9}$$

where σ_0 is the maximum stress due to a 1-g acceleration (lbf/in.^{2}). Values of d_0 and σ_0 may be found as discussed in Sections 4.5 and 4.6.

A rule of thumb[9] based on extensive testing of large dual in-line-pin (DIP) parts mounted on printed wire boards (PWB) specifies the maximum allowable displacement due to shock, to avoid failure of solder joints and lead wires, as

$$d_s(\text{max allowable}) = 0.018a/L^2 \text{ in.} \tag{5-10}$$

where a is the length of PWB parallel to the DIP (in.) and L is the length of the DIP (in.).

Equation (5-10) applies to the case where the DIP is located where the PWB undergoes its maximum deflection, for example, in the center of a rectangular PWB. If located in regions of less PWB deflection, the allowable value of d_s may be increased proportionately.

REFERENCES

1 M. Gertel and R. Holland, *Development of a Technique for Determination of Component Shock Specifications*, NASA Report No. 607-4-1, Contract No. NAS-8-11090, December 1964.

2 I. Vigness, *Elementary Considerations of Shock Spectra*, The Shock and Vibration Bulletin, Bulletin 34, Part III, Washington, DC: Shock and Vibration Information Center, Naval Research Laboratory.

3 D. D. Smallwood, *The Shock Response Spectrum at Low Frequencies*, The Shock and Vibration Bulletin, Bulletin 56, Part I, August, 1986, Washington, DC: Shock and Vibration Information Center, Naval Research Laboratory.

4 R. T. Fandrich, Jr., *Pyrotechnic Shock Testing on a Standard Drop Machine*, Institute of Environmental Sciences, 1974 Proceedings, April 28–May 1, 1974, Washington, DC, pp. 269–273.

5 R. T. Fandrich, Jr., *Bounded Impact: A Repeatable Method For Pyrotechnic Shock Simulation*, The Shock and Vibration Bulletin, Bulletin 46, Part II, August 1976, Washington, DC: Shock and Vibration Information Center, Naval Research Laboratory, pp. 101–108.

6 M. B. McGrath, *A Discussion of Pyrotechnic Shock Criteria*, The Shock and Vibration Bulletin, Bulletin 41, Part 5, December 1970, Washington, DC: Shock and Vibration Information Center, Naval Research Laboratory, pp. 1–7.
7 S. Barrett, *The Development of Pyro Shock Test Requirements for Viking Lander Capsule Components*, Institute of Environmental Sciences, 1975 Proceedings, April 14–16, 1975, Anaheim, CA, Volume II, pp. 5–10.
8 S. Barrett and W. J. Kacena, *Methods of Attenuating Pyrotechnic Shock*, The Shock and Vibration Bulletin, Bulletin 42, Part 4, January 1972, Washington, DC: Shock and Vibration Information Center, Naval Research Laboratory, pp. 21–32.
9 D. S. Steinberg (private communication).

6

ISOLATION

The vibration and shock acceleration environments usually result in the most severe stresses and displacements experienced by a structure. It is sometimes necessary to reduce the force and displacement levels to which components or structures are exposed in order to avoid performance degradation or failure. This may be done by designing the component or structure so that its fundamental natural frequency of vibration is at least one octave removed from the frequency of any sinusoidal vibration environment to which it may be exposed. The structure's fundamental natural frequency will also determine the acceleration level due to the random vibration power spectra density (PSD) (Chapter 4) and the shock acceleration response (Chapter 5). It may be necessary to introduce an additional structure between the component and its mounting surface to reduce the transmitted forces and displacements over a frequency range where damage could occur. This technique of reducing the force and displacement levels is called isolation, and if an additional structure is used, it is called an isolator.

6.1 SINUSOIDAL VIBRATION ENVIRONMENT

6.1.1 Single-Degree-of-Freedom (DOF) Systems

The response of a component or structure to a sinusoidally varying displacement of its mounting surface is described by a dimensionless ratio called the transmissibility Q. In Fig. 6.1a, the one-degree-of-freedom (DOF) structure is represented by a mass (top box) and a stiffness/damping element (middle

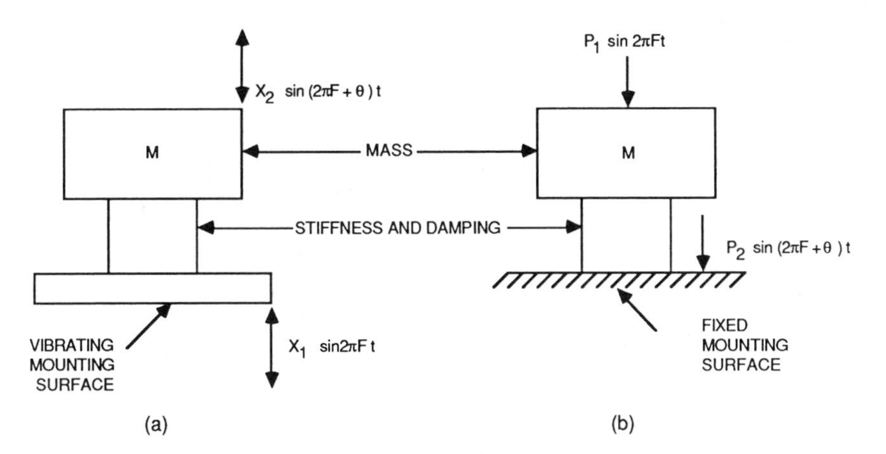

FIGURE 6.1 One-degree-of-freedom system with rubberlike stiffness/damping element.

box) that connects the mass to the mounting surface vibrating at frequency F. The structural mass is out of phase with the mounting surface by an angle θ. The transmissibility is the ratio of the maximum displacement of the structure to the maximum displacement of the mounting surface:

$$Q = X_2/X_1 \tag{6-1}$$

In Fig. 6.1*b*, a sinusoidally varying input force P_1 is applied to the structural mass and transmitted through the stiffness/damping element to exert a force P_2 on the fixed mounting surface. In this case, the transmissibility is the ratio of the maximum force on the mounting surface to the maximum force on the structural mass:

$$Q = P_2/P_1 \tag{6-2}$$

It may be shown by reciprocity that the magnitude of the displacement ratio is the same as the magnitude of the force ratio:

$$Q = X_2/X_1 = P_2/P_1 \tag{6-3}$$

The stiffness and damping model commonly used to represent mechanical structures that do not contain rubberlike damping materials is a parallel spring and dashpot combination (Fig. 6.2). In the model the dashpot has viscous damping, where the damping force is directly proportional to the velocity of the displacement of the vibrating structure. In this case, the transmissibility is

$$Q = \left[1 + (2R_F R_c)^2\right]^{1/2} \Big/ \left[(1 - R_F^2)^2 + (2R_F R_c)^2\right]^{1/2} \tag{6-4}$$

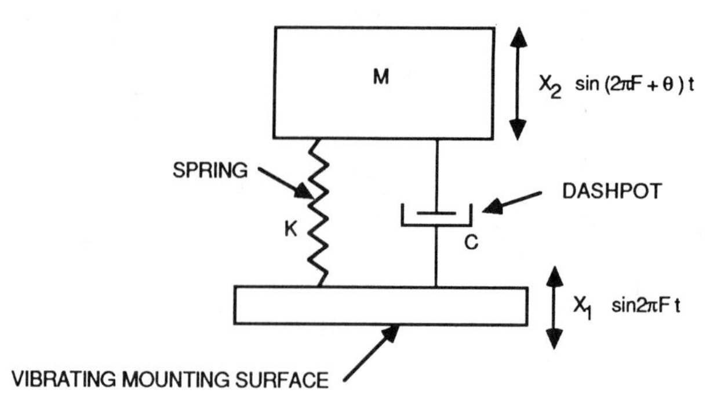

FIGURE 6.2 One-degree-of-freedom system with parallel spring and dashpot combination.

where $R_F = F_v/F_0$
F_v = frequency of vibration of the mounting surface (Hz)
F_0 = fundamental natural frequency of the structure (Hz)
$R_c = C/C_c$
C = damping coefficient of dashpot (lbf · sec/in.)
C_c = critical value of C (lbf · sec/in.)

and

$$F_0 = (K/M)^{1/2}/2\pi \text{ Hz} \tag{6-5}$$

$$C_c = 2(KM)^{1/2} \text{ lbf} \cdot \text{sec/in.} \tag{6-6}$$

where K is the spring rate (lbf/in.) and M is the mass of structure (lbm). See the discussion following Eq. (4-6), Chapter 4, for an explanation of C_c.

The phase angle is

$$\theta = \tan^{-1}\{-(2R_F^3R_c)/[1 - R_F^2 + 4R_F^2R_c^2]\} \tag{6-7}$$

When rubberlike materials are used to connect a structure to a mounting surface, Fig. 6.1*a* may be used to model the system, with the stiffness/damping element (middle box) representing the rubberlike material.[1] This rubberlike material will have a shear modulus G and a damping factor δ defined by

$$\delta = \tan\phi = (1/\pi)(\text{logarithmic decrement}) \tag{6-8}$$

where ϕ is the angle by which strain lags in phase behind stress and the logarithmic decrement is the natural logarithm of the ratio of successive maximum amplitudes of the damped vibration response to a transient input. The right-hand side of the equality in Eq. (6-8) is only valid for $\delta < 0.3$. For many commonly used rubberlike materials G and δ are both independent of frequency, in which case the transmissibility is

$$Q = (1 + \delta^2)^{1/2} \Big/ \left[\left(1 - R_F^2\right)^2 + \delta^2\right]^{1/2} \tag{6-9}$$

and the phase angle is

$$\theta = \tan^{-1}\left[-\delta R_F^2/\left(1 - R_F^2 + \delta^2\right)\right] \tag{6-10}$$

where $R_F = F_v/F_0$, F_v is the frequency of vibration of the mounting surface (Hz), and

$$F_0 = (1/2\pi)(kG/M)^{1/2}\ \text{Hz} \tag{6-11}$$

The constant k in Eq. (6-11) has the units of length, and the product kG is the stiffness (lbf/in.). For the rubberlike materials considered here, with stiffness/damping elements of uniform cross-sectional area A and length l

$$k = (3A/l)(1 + bS^2) \tag{6-12}$$

where S = shape factor = (area of one loaded surface)/(total force free area)
= 0.25 for a cube; = $D/4l$ for a cylinder
b = 2 for stiffness/damping elements that are circular, square, or moderately rectangular in cross section.

When the rubberlike element is used directly in shear,

$$k = A/l \tag{6-13}$$

Equation (6-4) is plotted in Fig. 6.3 and Eq. (6-9) is plotted in Fig. 6.4. When there is no damping ($R_c = 0$ and $\delta = 0$), the phase angle $\theta = 0$ and Eqs. (6-4) and (6-9) both reduce to

$$Q = 1/\left(1 - R_F^2\right) \tag{6-14}$$

which becomes degenerate ($Q = \infty$) at $R_F = 1$. When $R_c > 0$ and $\delta > 0$ and $R_F = 1$, the resonance condition, Q reduces to

$$Q = \left(1 + 4R_c^2\right)^{1/2}/2R_c \tag{6-15}$$

and

$$Q = (1 + \delta^2)^{1/2}/\delta \tag{6-16}$$

for Eqs. (6-4) and (6-9), respectively. For large damping, $R_c \gg 1$ and $\delta \gg 1$, $Q \simeq 1$, and $\theta \simeq 0$, even at resonance ($R_F = 1$).

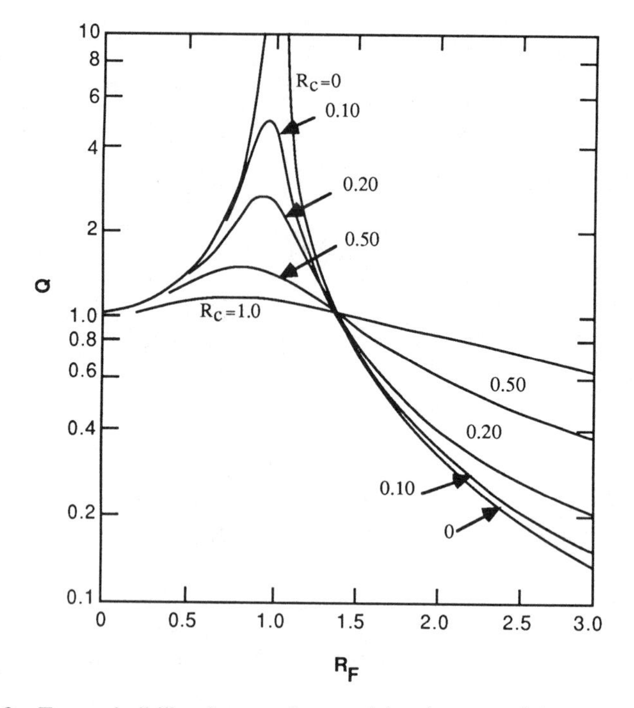

FIGURE 6.3 Transmissibility for one-degree-of-freedom parallel spring and dashpot combination.

For high input frequencies F_v, $R_F \gg 1$,

$$Q \approx 1/F_v \tag{6-17}$$

and

$$Q \approx 1/F_v^2 \tag{6-18}$$

for Eqs. (6-4) and (6-9), respectively. Equations (6-17) and (6-18) show that Q decreases faster with increasing input frequency for the rubberlike stiffness/damping element than for a parallel spring and dashpot combination.

6.1.2 Two-Degree-of-Freedom (DOF) Systems

The two DOF systems with parallel spring and dashpot combinations in series (Fig. 6.5) have a transmissibility[1,2]

$$Q = \left(R_N^2 + I_N^2\right)^{1/2} / \left(R_D^2 + I_D^2\right)^{1/2} \tag{6-19}$$

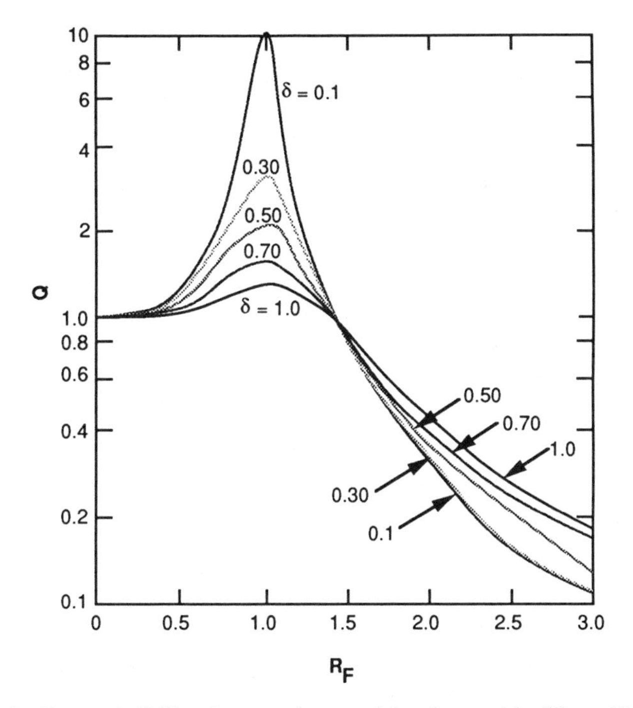

FIGURE 6.4 Transmissibility for one-degree-of-freedom rubberlike stiffness/damping element.

and a phase angle

$$\theta = \tan^{-1}[(R_D I_N - R_N I_D)/(R_N R_D + I_N I_D)] \tag{6-20}$$

where $R_N = (1 - 4R_F^2 R_{c,1} R_{c,2})$

$I_N = 2R_F(R_{c,1} + R_{c,2})$

$R_D = [\alpha\beta R_F^4/(1+\alpha)^2 - (1+\alpha+\beta)R_F^2/(1+\alpha) + (1 - 4R_F^2 R_{c,1} R_{c,2})]$

$I_D = [-2(R_{c,1} + \beta R_{c,1} + \alpha R_{c,2})R_F^3/(1-\alpha) + 2R_F(R_{c,1} + R_{c,2})]$

$R_F = 2\pi F_v[(K_1 + K_2)M_1/K_1K_2]^{1/2}$

F_v = frequency of vibration of the mounting surface (Hz)

$R_{c,1} = C_1/C_c$

$R_{c,2} = C_2/C_c$

$\alpha = K_2/K_1$

$\beta = M_2/M_1$

and C_1 and C_2 are the dashpot damping coefficients; K_1 and K_2 are the spring rates; and M_1 and M_2 are the masses, all shown in Fig. 6.5.

When rubberlike materials are used for the stiffness/damping elements in the two-DOF series system (Fig. 6.6), with both G and δ independent of

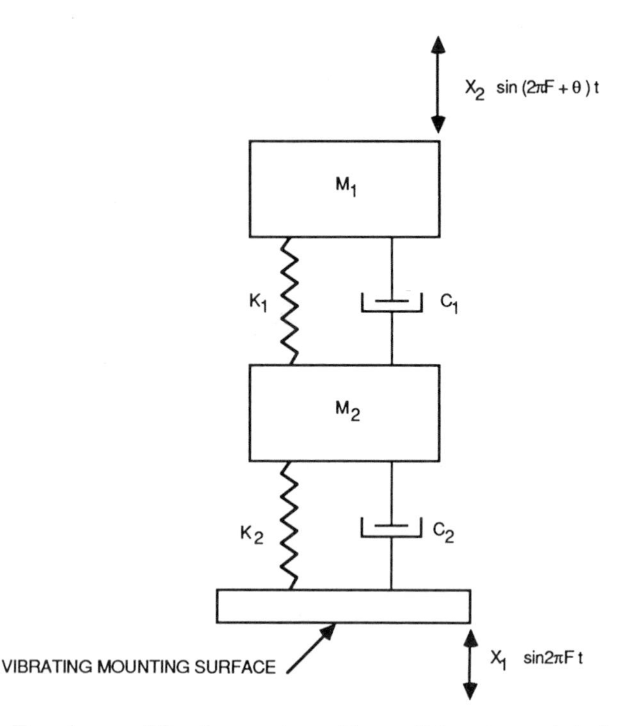

FIGURE 6.5 Two-degree-of-freedom system with parallel spring and dashpot combination in series.

frequency as in the one-DOF case, the transmissibility and phase equations have the same form as Eqs. (6-19) and (6-20) but with different parameters:

$$Q = \left(R_N^2 + I_N^2\right)^{1/2} / \left(R_D^2 + I_D^2\right)^{1/2} \tag{6-21}$$

$$\theta = \tan^{-1}[(R_D I_N - R_N I_D)/(R_N R_D + I_N I_D)] \tag{6-22}$$

where $R_N = 1 - \delta_1\delta_2$
$I_N = \delta_1 + \delta_2$
$R_D = [\alpha\beta R_F^4/(1+\alpha)^2 - (1+\alpha+\beta)R_F^2/(1+\alpha) + (1-\delta_1\delta_2)]$
$I_D = [-(\delta_1 + \beta\delta_1 + \alpha\delta_2)R_F^2/(1+\alpha) + (\delta_1 + \delta_2)]$
$R_F = 2\pi F_v[(k_1G_1 + k_2G_2)M_1/(k_1G_1k_2G_2)]^{1/2}$

M_1, M_2, α, β, and F_v are defined following Eq. (6-20), k_1 and k_2 are defined following Eq. (6-11), and G_1, G_2, δ_1, and δ_2 are the shear moduli and damping factors for the stiffness/damping elements shown in Fig. 6.6.

Equations (6-19) and (6-21) show that for large F_v, Q decreases faster with increasing input frequency for the rubberlike stiffness/damping elements than for the parallel spring and dashpot combinations. Comparison of the one- and

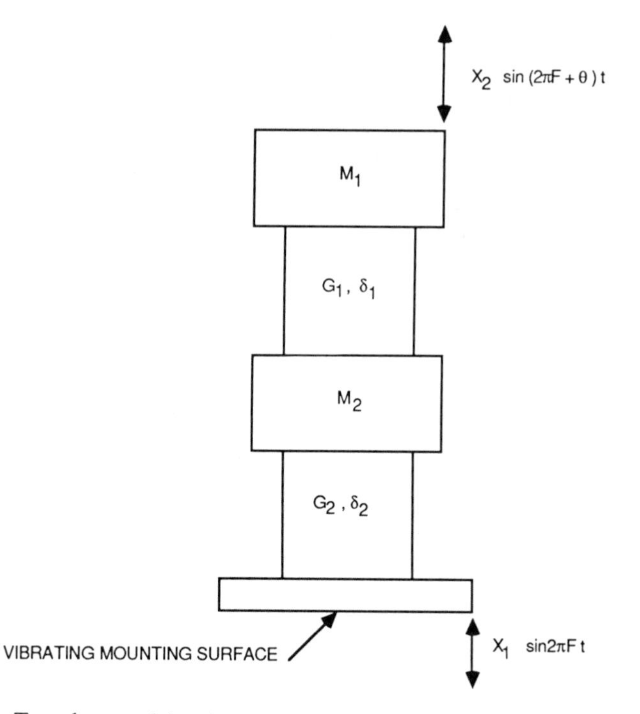

FIGURE 6.6 Two-degree-of-freedom systems with rubberlike stiffness/damping elements in series.

two-DOF systems discussed previously shows that at high input frequencies where $R_F \gg 1$, the transmissibilities decrease with increasing F_v as follows:

$$\begin{aligned}
\text{One DOF, parallel spring and dashpot:} \quad & Q \approx (F_v)^{-1} \\
\text{One DOF, rubberlike stiffness/damper:} \quad & Q \approx (F_v)^{-2} \\
\text{Two DOF in series, parallel spring and dashpot:} \quad & Q \approx (F_v)^{-2} \\
\text{Two DOF in series, rubberlike stiffness/damper:} \quad & Q \approx (F_v)^{-4}
\end{aligned} \tag{6-23}$$

The two DOF systems whose transmissibilities are given by Eq. (6-19) for the parallel spring and dashpot and Eq. (6-21) for the rubberlike stiffness/damper each have a secondary resonance frequency F_2 as well as a primary resonance frequency F_1. To be an effective isolator at higher frequencies it is desirable to have the secondary resonance occur at the lowest possible frequency or to make the ratio F_2/F_1 a minimum. This condition occurs for the rubberlike stiffness/damper system when

$$\alpha = 1 + \beta \tag{6-24}$$

when δ_1 and δ_2 are small in Eq. (6-21). When Eq. (6-24) is observed, the frequency ratio is

$$F_2/F_1 = \left[1 + (1 + \beta)^{1/2}\right] \Big/ \beta^{1/2} \tag{6-25}$$

Equation (6-24) is generally a good choice for α even when the δ values are not small in Eq. (6-21). When each stage of the two-DOF rubberlike stiffness/damping system of Eq. (6-21) has identical moduli and damping factors,

$$G_1 = G_2 = G \quad \text{and} \quad \delta_1 = \delta_2 = \delta \tag{6-26}$$

The use of Eqs. (6-24) and (6-26) in Eq. (6-21) yields

$$Q = (1 + \delta^2) \Big/ \left[\left(\lambda\gamma R_F^4 - 2\lambda R_F^2 + 1 - \delta^2 \right)^2 + 4\delta^2 \left(1 - \lambda R_F^2\right)^2 \right]^{1/2} \tag{6-27}$$

where $\lambda = (1 + \beta)/(2 + \beta)$ and $\gamma = \beta/(2 + \beta)$.

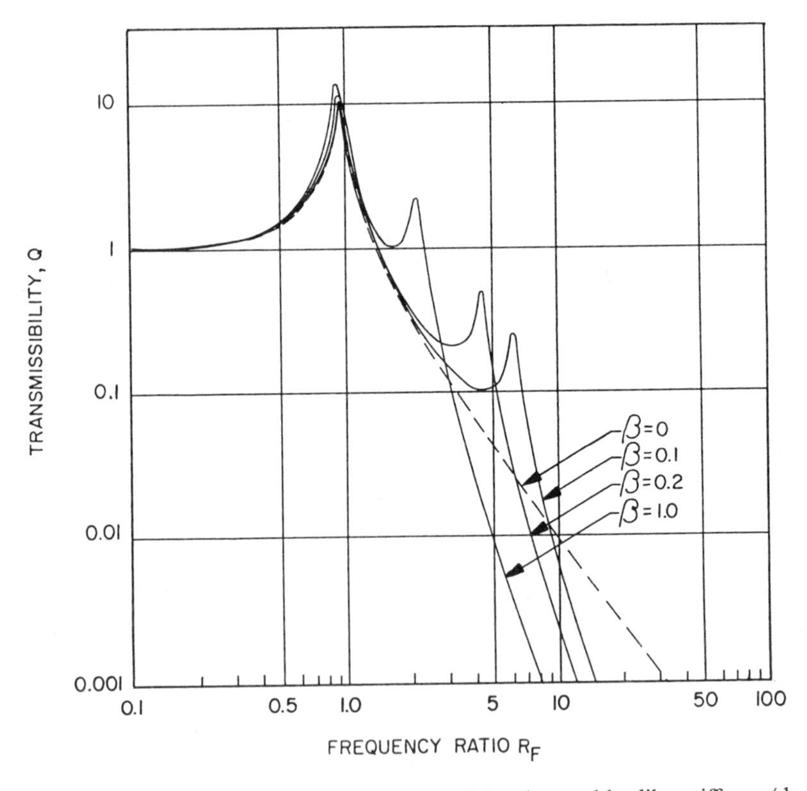

FIGURE 6.7 Transmissibility for two-degree-of-freedom rubberlike stiffness/damping elements in series.

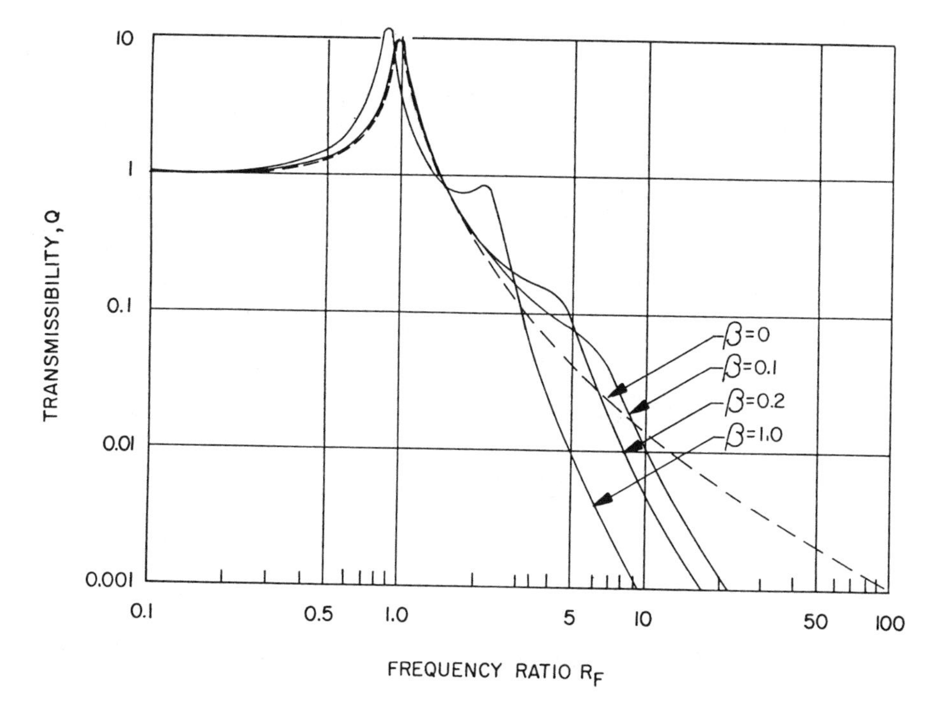

FIGURE 6.8 Transmissibility for two-degree-of-freedom parallel spring and dashpot combinations in series.

It is often assumed that each stage of the two-DOF parallel spring and dashpot combination of Eq. (6-19) has identical spring rates and damping ratios,

$$K_1 = K_2 = K \quad \text{and} \quad R_{c,1} = R_{c,2} = R_c \tag{6-28}$$

Use of Eq. (6-28) in Eq. (6-19) yields

$$Q = \left(1 + 4R_F^2 R_c^2\right) \Big/ \Big\{\left[(\beta/4)R_F^4 - \left(1 + \beta/2 + 4R_c^2\right)R_F^2 + 1\right]^2 + 4R_c^2 R_F^2 \left[2 - (1 + \beta/2)R_F^2\right]^2\Big\}^{1/2} \tag{6-29}$$

When M_2 is set equal to zero, then $\beta = 0$, $\lambda = \frac{1}{2}$, and $\gamma = 0$. In this case, Eqs. (6-27) and (6-29) reduce to the one-DOF equations (6-9) and (6-4), respectively.

Equation (6-27) is plotted in Fig. 6.7 for a damping factor of $\delta = 0.1$ and mass ratios of $\beta = 0.1$, 0.2 and 1.0. Also shown in this figure is the one-DOF system, $\beta = 0$. Equation (6-29) is plotted in Fig. 6.8 for $R_c = 0.05$ and values of $\beta = 0.1$, 0.2, and 1.0, with the one-DOF system, $\beta = 0$.

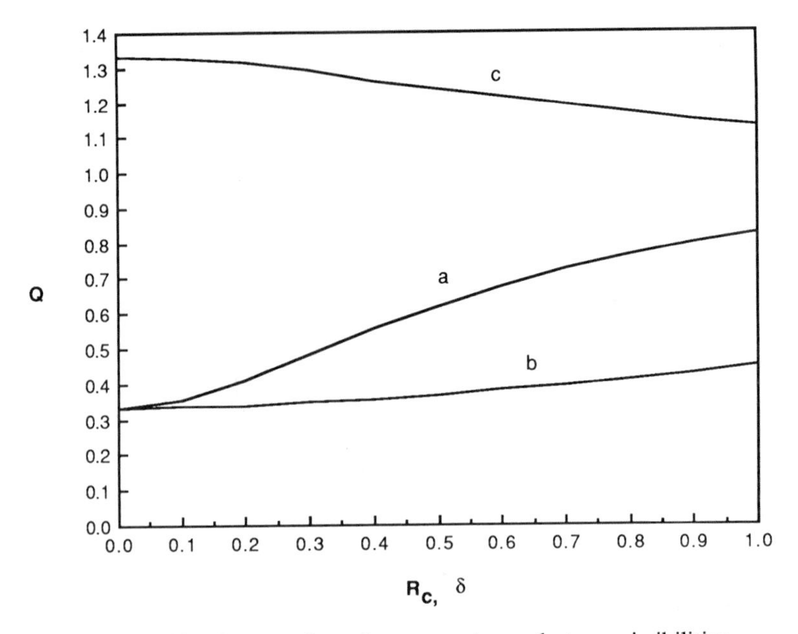

FIGURE 6.9 Octave rule and reverse octave rule transmissibilities.

In many cases the component under consideration is part of a complex structure that may be modeled as a system with several DOF. For example, a housing with a resonant frequency (fundamental natural frequency) F_1 may contain a printed wire board (PWB) with a resonant frequency F_2 on which is mounted a component with resonant frequency F_3. The displacements and forces due to a vibration environment can be greatly amplified if two or more of these resonant frequencies have similar values, resulting in a multiplication of the peak transmissibilities. This may be avoided by separating successive frequencies by at least an octave.[3] The octave rule states that the resonant frequency of a substructure should be at least twice that of its support; for the preceding example,

$$F(\text{component}) \geq 2F(\text{PWB}) \geq 2F(\text{housing}) \tag{6-30}$$

The reverse octave rule states that the resonant frequency of a substructure should be no more than one-half that of its support:

$$F(\text{component}) \leq \tfrac{1}{2}F(\text{PWB}) \leq \tfrac{1}{2}F(\text{housing}) \tag{6-31}$$

Figure 6.9, calculated from Eqs. (6-4) and (6-9), illustrates the effectiveness of these rules of thumb. For the reverse octave rule, with $R_F = 2$, the Q values are less than unity, ranging from $Q = 0.333$ for no damping to $Q = 0.825$ for $R_c = 1.0$ (parallel spring and dashpot, line a) to $Q = 0.447$ for $\delta = 1.0$

(rubberlike stiffness/damping, line b). For the octave rule, with $R_F = 0.5$, the Q values are greater than unity, ranging from 1.333 for no damping to 1.131 for $R_c = \delta = 1.0$, line c. Although Q values slightly greater than unity occur with the octave rule, the large Q values at resonance are avoided.

6.2 STIFFNESS FORMULAS FOR SOME RUBBERLIKE MATERIAL MOUNTINGS

There are several isolator mounting configurations utilizing rubberlike materials for which the stiffness, or spring constant K, has been evaluated and expressed in terms of a formula containing geometric terms and material properties.[4] Some of these cases are presented in the following subsections. Calculated stiffnesses are generally within $\pm 15\%$ of actual stiffness when strains are limited to 10–20%. The stiffness is usually specific to a given direction of motion and may differ by an order of magnitude for a different direction of motion.

6.2.1 Compression Block

When a rectangular block of rubber, Fig. 6.10, is in compression and its loaded surfaces are prevented from slipping by mechanical connection to the mounting structure, the spring constant is

$$K = E_c A/tR \text{ lbf/in.} \tag{6-32}$$

where $E_c = E_0(1 + 2kS^2)$ (psi)
E_0 = Young's modulus (psi)
k = a numerical factor, function of E_0 or G, from Table 6-1
S = shape factor = (loaded surface area)/(force free area) = $LB/2t(L + B)$
L = length of block (in.)
b = breadth of block (in.)

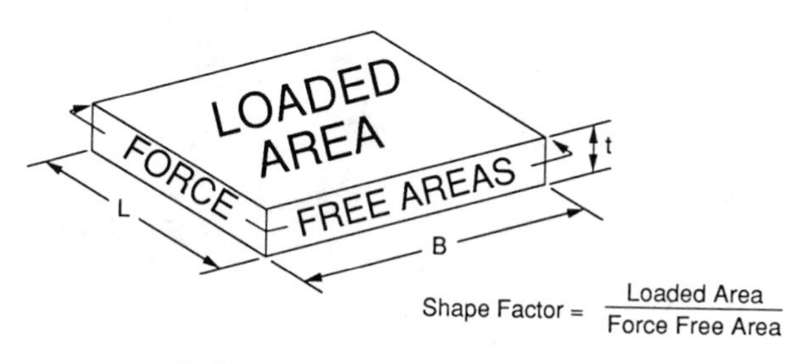

FIGURE 6.10 Compression block.

TABLE 6-1 Natural Rubber Properties versus Hardness

Shore A Hardness	Young's Modulus E_0 (psi)	Shear Modulus G (psi)	Bulk Modulus B (psi)	Numerical Factor k
30	130	43	142,000	0.93
35	168	53	142,000	0.89
40	213	64	142,000	0.85
45	256	76	142,000	0.80
50	310	90	146,000	0.73
55	460	115	154,000	0.64
60	630	150	163,000	0.57
65	830	195	171,000	0.54
70	1040	245	180,000	0.53
75	1340	317	189,000	0.52

Source: Ref. 3.

t = thickness of block (in.)
A = cross sectional area of block, LB
$R = 1 + E_c/B$
B = bulk modulus of compression (psi)

Table 6-1 presents values of E_0, G (shear modulus), B, and k for natural rubber, referenced to hardness.

6.2.2 Compression Strip

When a long strip is compressed normal to its length, Fig. 6.11, the spring constant per inch length of strip is

$$K_0 = 4bE_0(1 + kS^2)/3tR \text{ lbf/in.} \tag{6-33}$$

where b is the width of strip (in.), $S = b/2t$, and the other parameters are defined following Eq. (6-32).

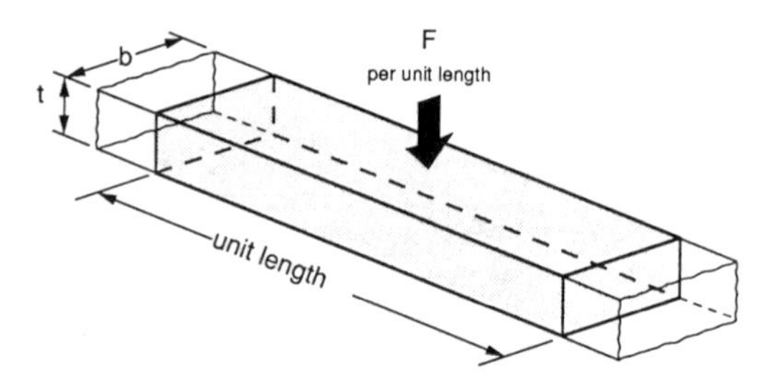

FIGURE 6.11 Compression strip.

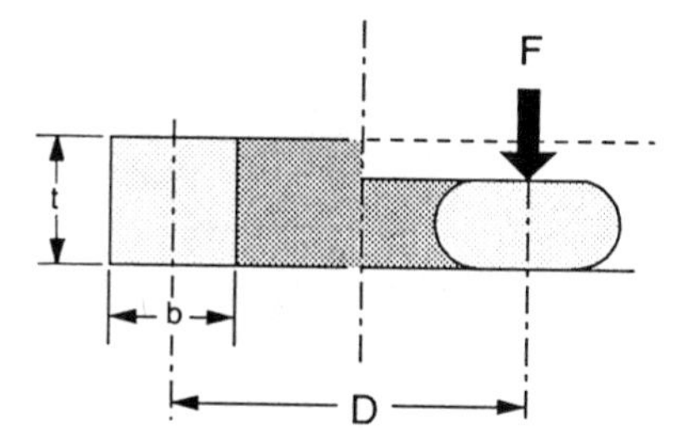

FIGURE 6.12 Rectangular section rubber ring.

6.2.3 Solid Rubber Ring in Compression

Rectangular Cross Section

When a solid rubber ring of rectangular cross section is compressed parallel to the ring axis, Fig. 6.12, the spring constant is

$$K = \tfrac{4}{3}E_0\pi D(b/t)(1 + kb^2/4t^2)\ \text{lbf/in.} \tag{6-34}$$

where D is the mean diameter of ring (in.), b is the radial width of section (in.), t is the thickness of section (in.), and the other parameters are defined following Eq. (6-32).

Circular Cross Section

When a solid rubber ring of circular cross section is compressed parallel to the ring axis, Fig. 6.13, the spring constant is

$$K = 3.95E_0D(\delta/d)^{1/2}\ \text{lbf/in.} \tag{6-35}$$

where δ is the deflection due to compression (in.), d is the diameter of circular cross section (in.), and other parameters are previously defined.

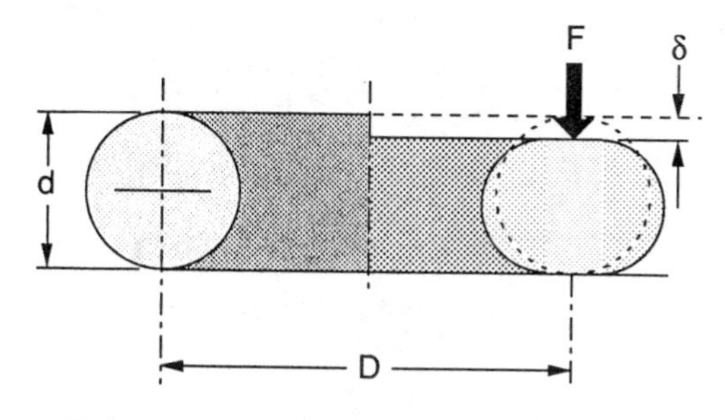

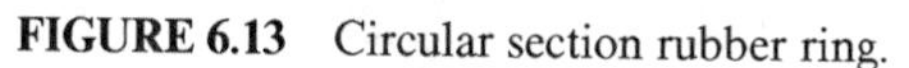

FIGURE 6.13 Circular section rubber ring.

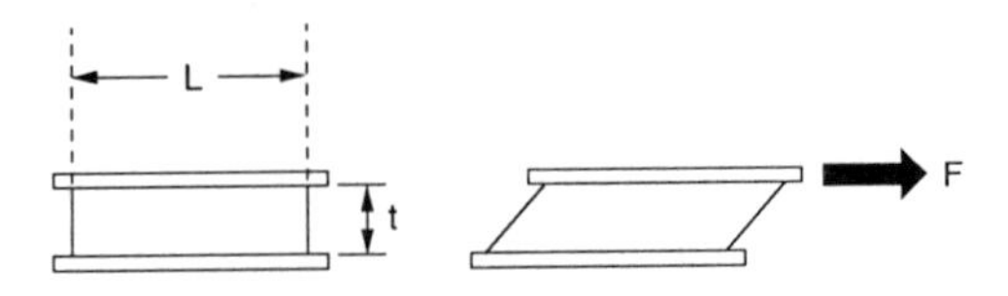

FIGURE 6.14 Shear block.

6.2.4 Shear Block

When a solid rubber block is in shear, Fig. 6.14, its spring constant is

$$K = GA/t \text{ lbf/in.} \tag{6-36}$$

where G is the shear modulus (psi), A is the cross-sectional area of block parallel to the direction of shear (in.^{2}), and t is the thickness of block normal to the direction of shear (in.).

6.2.5 Torsion Disk

An annulus of rubber bonded to two end plates normal to the axis of the annulus, Fig. 6.15, will have a torsional stiffness K_θ given by

$$K_\theta = T/\theta = \pi G\left(R_1^4 - R_0^4\right)/2t + \pi G\theta^2\left(R_1^6 - R_0^6\right)/9t^3 \text{ lbf} \cdot \text{in./rad} \tag{6-37}$$

where T = torque (lbf · in.)
G = shear modulus, psi
R_0 = inner radius of annulus (in.)
R_1 = outer radius of annulus (in.)
t = thickness of disk (in.)
θ = angular rotational displacement around axis of annulus, from equilibrium position (rad)
$\phi = R\theta/t$ = shear strain (rad)

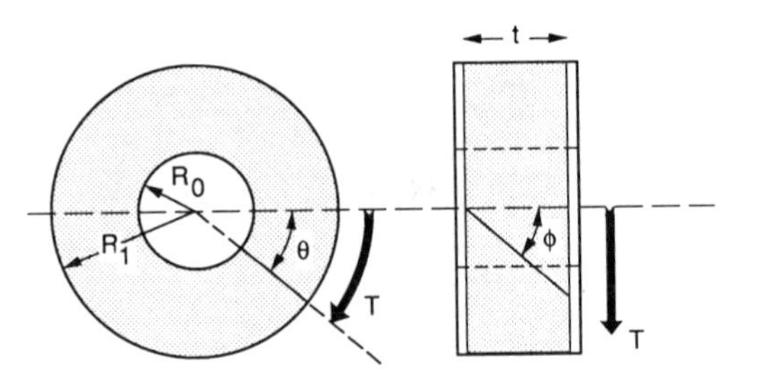

FIGURE 6.15 Torsion disk.

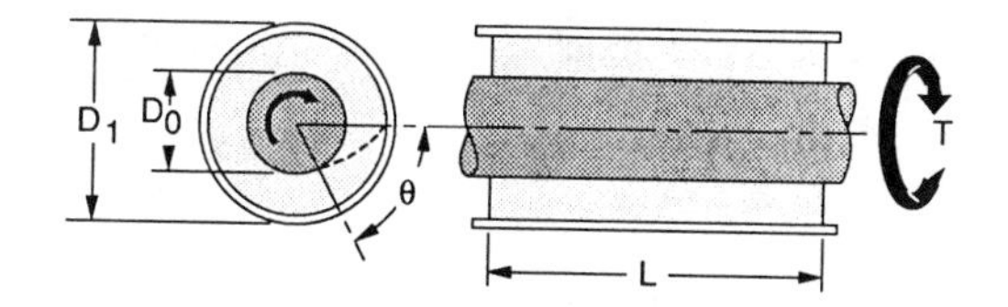

FIGURE 6.16a Bush mounting in torsion.

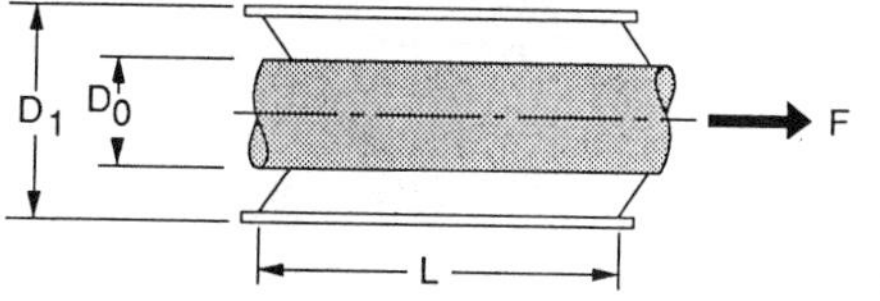

FIGURE 6.16b Bush mounting in axial deformation.

For $\phi < 0.5$, the error in omitting the second term in Eq. (6-37) is less than 6%.

6.2.6 Bush Mounting

A bush mounting will have a torsional stiffness, an axial stiffness and a radial stiffness, as shown in Fig. 6.16.

Torsional Stiffness

The torsional stiffness K_θ, Fig. 6.16*a*, is given by

$$K_\theta = T/\theta = \pi GL/\left(1/D_0^2 - 1/D_1^2\right) \text{ lbf} \cdot \text{in./rad} \tag{6-38}$$

where T = torque (lbf · in.)
θ = angular rotational displacement around axis of bush mounting, from equilibrium position (rad)
G = shear modulus (psi)
L = length of bush mounting (in.)
D_0 = inner diameter of rubber cylinder (in.)
D_1 = outer diameter of rubber cylinder (in.)

Axial Stiffness

The axial stiffness K_a, Fig. 6.16*b*, is given by

$$K_a = \left[2.73GL/\log_{10}(D_1/D_0)\right]/\left[1 + \alpha(D_1/L)^2\right] \text{ lbf/in.} \tag{6-39}$$

where α is found in Table 6-2.

TABLE 6-2 Values of α and β for Axial and Radial Bush Stiffness

Diameter Ratio D_1/D_0	Axial Stiffness Constant, α, for Eq. (6-39)	Radial Stiffness Constants for Eq. (6-40)	
		Long Bushes β_L	Short Bushes β_S
1.05	0.0001	320,000	322
1.10	0.0005	43,700	165
1.25	0.0025	3,400	70
1.50	0.0068	602	38
1.75	0.0111	212	27
2	0.0148	135	21
3	0.0244	42	12.4
4	0.0289	25	9.4
5	0.0309	18.3	7.9
7	0.0321	12.7	6.3
10	0.0315	9.5	5.2
20	0.0282	6.3	3.8
100	0.0204	3.4	2.4
1000	0.0135	2.1	1.5

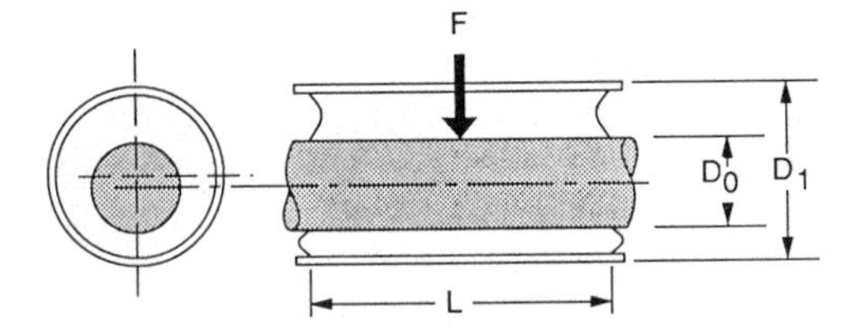

FIGURE 6.16c Bush mounting in radial deformation.

Radial Stiffness

The radial stiffness K_r, Fig. 6.16c, is given by

$$K_r = \beta LG \text{ lbf/in.} \tag{6-40}$$

where β is found in Table 6-2 for long bushes β_L and short bushes β_S. For bushes of intermediate length, the values of β will be between the β_L and β_S values.

6.3 RANDOM-VIBRATION ENVIRONMENT

In a random-vibration environment, the transmissibility values from Section 6.1 may be used in Eq. (4-8) of Chapter 4 to obtain estimates of the peak acceleration response at frequencies of interest in one- and two-DOF systems, using the specified random-vibration PSD. In this application, the transmissibility (Q) values obtained for the specific mechanical configuration are used in place of the estimates described in Section 4.2.2.

6.4 SHOCK ENVIRONMENT

For a shock acceleration environment due to an acceleration–time pulse of duration τ sec and peak acceleration a_0, the shock frequency is

$$F_s = 1/\tau \text{ Hz} \tag{6-41}$$

and the ratio of the acceleration response (the equivalent static acceleration of Section 5.1) a to a_0 is given by

$$A = a/a_0 = \text{function}(F_0/F_s, R_c) \tag{6-42}$$

where F_0 is the fundamental natural frequency of the structure, Hz. Values of A may be found from Figs. 5-10–5-32 in Section 5.2. Examination of these figures shows that there is a region, $F_0/F_s = F_0\tau < 0.2–0.5$, where A is less than unity, and that A continues to decrease with decreasing $F_0\tau$. Given a specified shock environment in terms of a_0, τ, and shock pulse time history, the acceleration response $a = Aa_0$ may be estimated, from the appropriate above referenced figure, as a function of F_0 and R_c.

6.5 EXAMPLE OF ISOLATOR ANALYSIS AND DESIGN

This section provides an example of an isolator analysis and design[5] utilizing rubberlike isolator mounting stiffness formulas provided in Sections 6.2.3 and 6.2.6. As an example, a shock isolator was required to protect a satellite accelerometer against pyrotechnic shock. The isolator was also required to yield an acceptable response to the random-vibration environment, maintain angular alignment within 0.25 deg of arc, and meet restrictive weight and volume requirements. The specified shock and vibration environments are presented in Tables 6-3 and 6-4.

The accelerometer is tested by the manufacturer to meet all performance specifications after being subjected to a 250g peak, 11 msec, half-sine shock along three mutually perpendicular axes. This quality assurance screening test may be compared to the pyrotechnic shock environment of Table 6-3 to which

TABLE 6-3 Pyrotechnic Shock Response Spectrum[a]

Frequency (Hz)	Shock Response Acceleration (g)
100	48
100–3,400	+7.0 dB/octave
3,400–6,000	2,828
6,000–10,000	−3.4 dB/octave
10,000	2,121

[a] The shock is imposed in both directions along each of three mutually perpendicular axes for a total of six shocks.

TABLE 6-4 Random-Vibration Power Density Spectrum[a]

Frequency (Hz)	Power Spectral Density (PSD)(g^2/Hz)
20	0.2
20–60	+10 dB/octave
60–200	0.8
200–400	−4 dB/octave
400	0.32
400–2,000	−9 dB/octave
2,000	0.0026

[a]The random vibration is imposed for 2 min along each of three mutually perpendicular axes for a total of 6 min.

the accelerometer will be exposed in flight. Based on the manufacturer's screening procedure, it was decided to require that the shock isolator fundamental natural frequency F_0 correspond to a response acceleration of no more than $250g$ as determined from Table 6-3. This results in an upper limit on F_0, which is determined as follows. From Table 6-3, $G = 48g$ at $F = 100$ Hz, and the energy spectrum increases at 7 dB per octave, which means that the acceleration spectrum increases at 3.5 dB per octave (see Section 5.2). The dB increase in going from $48g$ to $250g$ is $10\log(250/48) = 7.1670$ dB, which corresponds to $7.1670/3.5 = 2.0477$ octaves, or a factor of $2^{2.0477} = 4.134$ times in frequency, resulting in $F_0 \leq 4.134 \times 100$ Hz $= 413$ Hz.

It is also imperative that the random-vibration environment not result in accelerations of more than $250g$. In order for the maximum value of the 3σ peak acceleration G_{pk} to equal the maximum allowable value of $250g$, the shock isolator transmissibility would have to be

$$Q = (G_{pk})^2/9(\pi/2)(\text{PSD})F = (250)^2/(\pi/2)(0.8)200 = 27.6 \quad (6\text{-}43)$$

where Eq. (4-8) was used at a frequency of 200 Hz, which, based on the PSD values from Table 6-4, yields the maximum value of G_{pk}. This value of Q is much greater than the actual isolator Q values that were measured as $Q \simeq 6g$ and indicates that the actual G_{pk} will be much less than the allowable $250g$ maximum.

The lower limit on F_0 is determined by the maximum allowable displacement in the moving parts of the isolator, found from Eq. (4-10):

$$D = 9.8G/(F_0)^2 \text{ Hz} \quad (6\text{-}44)$$

where $G = G_{pk}$ in units of $g = 386$ in./sec^2 and D is the maximum isolator displacement (in.). Consideration of several design concepts led to the selection of commercially available O-rings as spring elements for the isolator. Considering the size limitations, it was decided that the maximum practical values for O-ring compression (moving part displacement) was $D = 0.030$ in. In the random-vibration environment, Eqs. (4-8) and (4-10) lead to

$$F_0 = 11.073(\text{PSD} \cdot Q)^{1/3}/D^{2/3} = 194 \text{ Hz} \quad (6\text{-}45)$$

TABLE 6-5 Calculated Accelerations and Displacements

	Pyrotechnic Shock		Random Vibration 3σ Values[a]	
F_0 (Hz)	Acceleration (g)	Displacement (in.)	Acceleration (g)	Displacement (in.)
250	139	0.022	112	0.018
300	172	0.019	109	0.012
350	206	0.017	106	0.009

[a] Values based on an assumed transmissibility of $Q = 6$.

where a value of $Q = 6$ was used (based on prior experience) and PSD = 0.8 from Table 6.4. In the pyrotechnic shock environment, using Eq. (4-10), $G/F_0^2 = 9.8D = 0.294$ in. By successive approximations, it is found that at $F_0 = 171$ Hz, $G = [\log(171/100)/\log 2]$ octaves $\times$ 3.5 dB/octave = 2.709 dB above the $48g$ value at 100 Hz, so $G = 48 \times 10^{0.2709} = 89.57g$ and $D = 9.8 \times 89.57/(171)^2 = 0.0300$ in.

The median value of F_0 between the upper limit of 413 Hz imposed by acceleration in the shock environment and the lower limit of 194 Hz imposed by displacement in the vibration environment is 303.5 Hz. A design goal of $F_0 = 300$ Hz was chosen, with a desired range of $F_0 = 250$–350 Hz to allow for uncertainties in spring constants due to material variability and analytical estimates. Table 6-5 shows the calculated accelerations and displacements due to the shock and random-vibration environments. At $F_0 = 250$ Hz, the 0.022-in. displacement is 73% of the design clearance of 0.030 in. At $F_0 = 350$ Hz, the acceleration of $206g$ is 82% of the design limit of $250g$. In both cases it is the pyrotechnic shock which provides the more severe environment. See Table 6-5.

6.5.1 Shock Isolator Design

The O-ring shock isolator design is shown in Fig. 6.17. The base plate, outside support, and center support are 6061-T6 aluminum; the mounting plate, top support, and screws are 303 stainless steel. The O-rings are commercially available and are silicone. The Z axis is parallel to the accelerometer sensing axis and normal to the base plate. The X axis is normal to the Z axis and also represents the third (Y) axis in analysis and test. The larger, upper and lower O-rings are the springs for the Z-axis motion; the smaller, center O-ring is the spring for the X-axis motion. All three O-rings have the same thickness of 0.103 in. The shock isolator design allows a 0.030-in. clearance between metal parts separated by O-rings, that is, the O-rings must be compressed 0.030 in. from their equilibrium preload position before the metal parts touch. The total weight of the shock isolator is 0.952 lb, and the weight supported by the O-rings is 0.577 lb. The center of gravity (CG) of the 0.577 lb is at the center of the three O-ring configuration.

The base plate and the outside support have fundamental natural frequencies much larger than the 300 Hz shock isolator design goal and are effectively isolated from the accelerometer by the shock isolator.

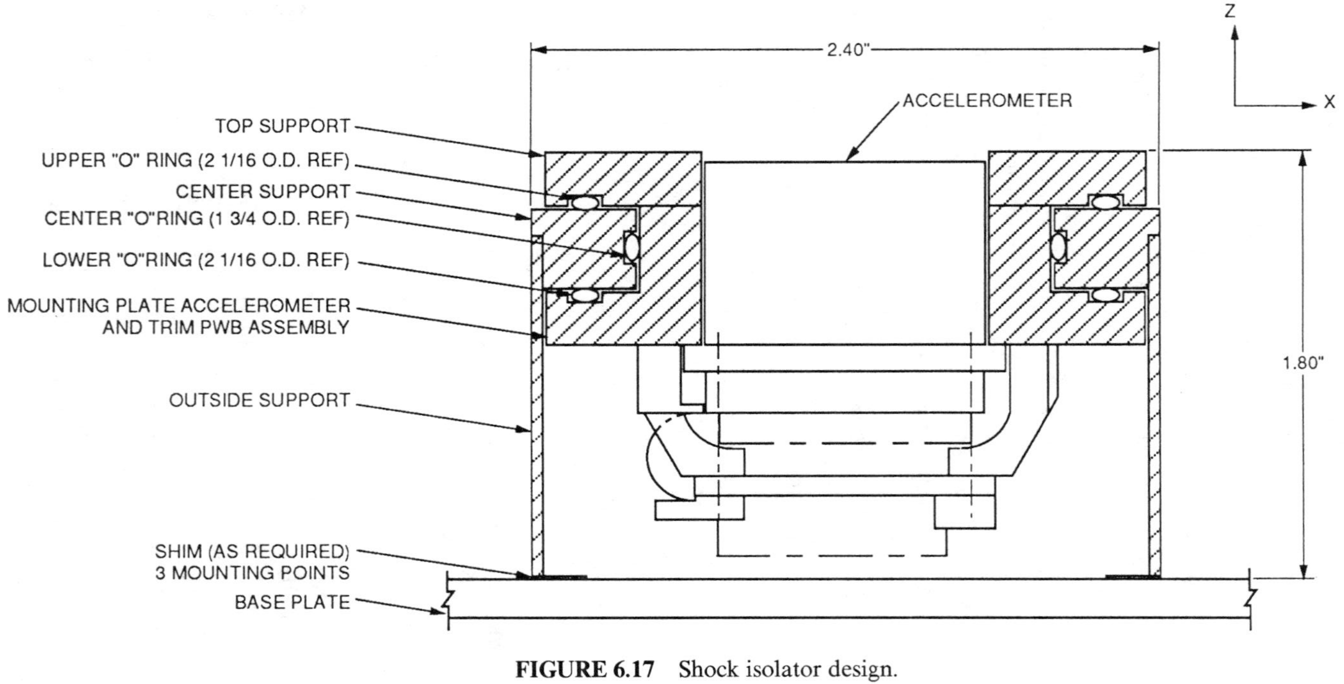

FIGURE 6.17 Shock isolator design.

6.5.2 Shock Isolator Analysis

The O-rings support a total mass of $m = 0.577$ lb. With a design goal of $F_0 = 300$ Hz for the shock isolator fundamental natural frequency, the required spring stiffness K may be found from the expression

$$F_0 = (1/2\pi)(Kg/m)^{1/2} \tag{6-46}$$

Using the preceding values of F_0 and m in Eq. (6-46) yields

$$K = 5{,}310 \text{ lb/in.} \tag{6-47}$$

the spring constant required for motion in both the Z and X directions.

The silicone O-ring material is typical of many rubberlike materials and has nonlinear stress–strain characteristics, with the static modulus of elasticity increasing with strain and also dependent on the shape factor (geometric configuration) of the material being strained. This results in the static spring stiffness K increasing with strain. The O-rings are preloaded by compressing them to a designed fraction of their unloaded thickness, in order to accommodate the displacements shown in Table 6-5. The spring stiffness calculated for this preloaded condition is used as the mean value of K and is designed to be equal to the value of 5,310 lb/in. obtained in Eq. (6-47).

Z-Direction Analysis

In the Z direction the two O-rings act as parallel springs between the movable mass m and the fixed center support (see Fig. 6.17). The static spring constant for a single O-ring along an axis normal to the plane of the O-ring is given by (Section 6.2.3)

$$K = 3.94ED(\delta/d)^{1/2} \text{ lb/in.} \tag{6-48}$$

where E = Young's modulus (lbf/in.^{2})
D = mean diameter of O-ring (in.)
d = thickness of O-ring (in.)
δ = preload compression of O-ring (in.)

This formula provides results in close agreement with experiment up to values of $\delta/d = 0.15$, and yields lower K values than experiment for $\delta/d > 0.15$.

The O-rings chosen for the Z-direction springs have values of $D = 1.97$ in. and $d = 0.103$ in., and the shock isolator is designed to preload and compress them by $\delta = 0.025$ in. Since the two O-rings are in parallel, each one must produce half the stiffness specified by Eq. (6-47) or 2,655 lb/in. Substituting these values into Eq. (6-48) and solving for the Young's modulus yields a value of $E = 693$ psi. The O-ring material is characterized by its hardness as measured on the Shore Durometer A Scale, which is related to Young's modulus E (see Table 6-1). A value of $E = 693$ psi corresponds to a durometer of 62 ± 2. As stated above, however, the actual K value will be greater than calculated for the $\delta/d = 0.24$ design value, and the 62 durometer value is an overestimate. Also, the correlation between hardness and Young's modulus is only reliable under static conditions. In rubberlike materials such as silicone, the dynamic modulus of elasticity is complex and is a function of frequency and displacement amplitude from the equilibrium position, as well as other factors. Fundamental natural frequencies calculated from static measurements of E may be in error by a factor of 2 or more, either too large or too small, compared to dynamic measured values of F_0. Since the dynamic complex values of E versus frequency are not known for most materials used in commercially available O-rings, the best that can be done is to use the static value of E as a starting point and modify the choice of durometer based on the experimental results.

From Eq. (6-48) it is seen that the spring constant is nonlinear and proportional to the square root of the compression δ. However, when two O-rings are used in parallel as shown in Fig. 6.17, and the supported mass is dynamically displaced a distance γ from the equilibrium position where both O-rings have the same value of δ due to the designed preload, then one O-ring has a larger value of $\delta' = \delta + \gamma$ and the other O-ring has a corresponding smaller value of $\delta' = \delta - \gamma$. The addition of the resulting spring constants in parallel yields

$$\begin{aligned} K &= K(\delta + \gamma) + K(\delta - \gamma) \\ &= \left(3.95ED/d^{1/2}\right)\left[(\delta + \gamma)^{1/2} + (\delta - \gamma)^{1/2}\right] \\ &= 3.95ED(\delta/d)^{1/2}(2)\left[1 - \tfrac{1}{8}(\gamma/\delta)^2\right] \end{aligned} \quad (6\text{-}49)$$

where γ is the dynamic displacement from equilibrium position.

From Eqs. (6-48) and (6-49) it may be shown that a dynamic displacement by γ in. from a preload position results in much less nonlinearity for the two O-rings in parallel, compared to a single O-ring:

$$\begin{aligned} (K' - K)/K &= \tfrac{1}{2}(\gamma/\delta) - \tfrac{1}{8}(\gamma/\delta)^2 \text{ for one O-ring} \\ &= -\tfrac{1}{8}(\gamma/\delta)^2 \text{ for two O-rings in parallel} \end{aligned} \quad (6\text{-}50)$$

where K' is the spring constant when the supported mass is dynamically displaced a distance γ from the preload compression δ, and K is the spring constant at preload compression δ.

For $\gamma/\delta = 0.5$, a typical dynamic operating value, Eq. (6-50) will yield a value of $(K' - K)/K = +0.219$ for one O-ring versus -0.031 for two O-rings in parallel, or a 22% increase in K versus a 3% decrease in K. The two O-ring design will introduce less nonlinearity than would a single O-ring.

X-Direction Analysis

In the X direction one O-ring acts as a spring between the movable mass m and the fixed center support (see Fig. 6.17). The static spring constant for an O-ring along an axis lying in the plane of the O-ring is given by (Section 6.2.6)

$$K = \beta LG \quad (6\text{-}51)$$

with

$$\beta = \frac{80\pi(A^2 + B^2)}{25(A^2 + B^2)\ln(A/B) - 9(A^2 - B^2)} \tag{6-52}$$

where G = shear modulus (lbf/in.^{2})
L = thickness of compressed O-ring in the direction normal to the plane of the O-ring (in.)
A = outside diameter of compressed O-ring in the O-ring plane (in.)
B = inside diameter of compressed O-ring in the O-ring plane (in.)

The single O-ring chosen for the X-direction spring has a mean diameter of 1.652 in. and a thickness of 0.103 in. It is placed in the center support groove and compressed by the narrow O-ring groove to conform to values of A = 1.709 in. and B = 1.552 in. This compression results in a value of L = 0.106 in., based on a constant volume O-ring cross section. Using these values in Eqs. (6-51) and (6-52) results in $\beta = 163$ and

$$K = 17.3G \text{ lb/in.} \tag{6-53}$$

Using in Eq. (6-53) the value of K = 5,310 lb/in. from Eq. (6-47), the shear modulus is found to be

$$G = 307 \text{ psi} \tag{6-54}$$

for the single O-ring, corresponding to a durometer of 74 ± 2.

As noted previously, values of F_0 calculated from static measurements of G may be in error by a factor of 2 or more in either direction, and the previously found durometer value for the X-axis O-ring should be used only as a starting point to be modified based on experimental results.

The isolator shown in Fig. 6.17 was constructed and sine sweep, random-vibration, and pyrotechnic shock tests were conducted with O-rings of different durometers. It was found that a durometer of 40 for the two Z-axis O-rings and a durometer of 70 for the X-axis O-ring resulted in the isolator meeting all of the performance requirements.

REFERENCES

1 J. C. Snowdon, *Vibration and Shock in Damped Mechanical Systems*, New York: Wiley, 1968, Chapters 2 and 3.

2 J. C. Snowdon and G. G. Parfitt, Isolation From Mechanical Shock With One- and Two-Stage Mounting Systems, *Journal Acous. Soc. Am.*, **31**(7), 967–976 (1959).

3 D. S. Steinberg, *Vibration Analysis For Electronic Equipment*, New York: Wiley, 1973, pp. 354–356.

4 P. B. Lindley, *Engineering Design With Natural Rubber*, Technical Bulletin No. 8, 1964, London: The Natural Rubber Producers Research Association.

5 D. Schiff, N. Jones, and S. Fox, Design and Test of a Spacecraft Instrument Shock Isolator, *Shock Vibration Bull.*, Bulletin 57, Part 3, pp. 29–45, (January, 1987).

7

FATIGUE

Fatigue is a failure mode in which a structural element loses its mechanical integrity owing to a large number of stress cycles in which the individual stress magnitudes are too small to affect structural integrity if only a few cycles are imposed on the element. In some applications, such as aerospace applications, fatigue is most commonly the result of exposure to a vibration environment, either a sinusoidal vibration environment at a single or a few frequencies or a random-vibration environment such as discussed in Chapter 4. The smaller the stress amplitudes of the vibration environment, the larger the number of cycles that can be tolerated without failure of the structural member. For some materials, as the stress amplitude decreases, a fatigue threshold stress value is reached below which fatigue failures will not occur even for an infinite number of cycles. The fatigue life of a structural element may be estimated in units of numbers of cycles or in units of time, for example, minutes, provided the vibration environment is adequately specified and the stress response of the element can be determined. In this chapter, high-cycle fatigue is addressed, in which the strain cycles are confined to the elastic range of the material.

7.1 FATIGUE CURVES

The stress cycles to which a structural element is exposed may include a spectrum of alternating stress amplitudes superimposed on a steady stress. Basic fatigue data are often displayed on a plot of cyclic stress amplitude S versus the number of cycles to failure N for a specified value of the stress ratio R. The stress ratio is the algebraic ratio of the minimum stress to the

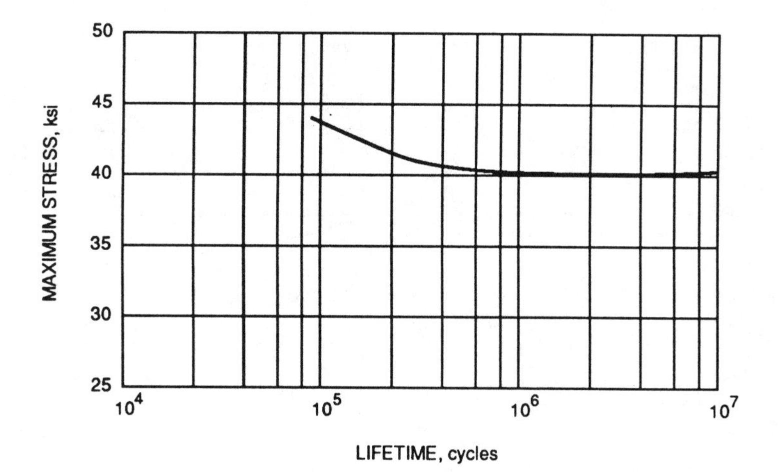

FIGURE 7.1 Fatigue curve for ZK60A-F magnesium alloy extrusion showing fatigue threshold at 40 ksi.

maximum stress in one cycle, with tensile stress being considered positive and compressive stress negative:

$$R = S_{\min}/S_{\max} \tag{7-1}$$

where $S_{\min}$ is the lowest algebraic value of stress in the stress cycle and $S_{\max}$ is the highest algebraic value of stress in the stress cycle. When R is not specified, it is usually assumed to be zero.

Figure 7.1 shows an S–N curve for a magnesium alloy with $R = 0.25$ and with a fatigue threshold at 40 ksi (40,000 psi).[1] Figure 7.2 shows S–N curves

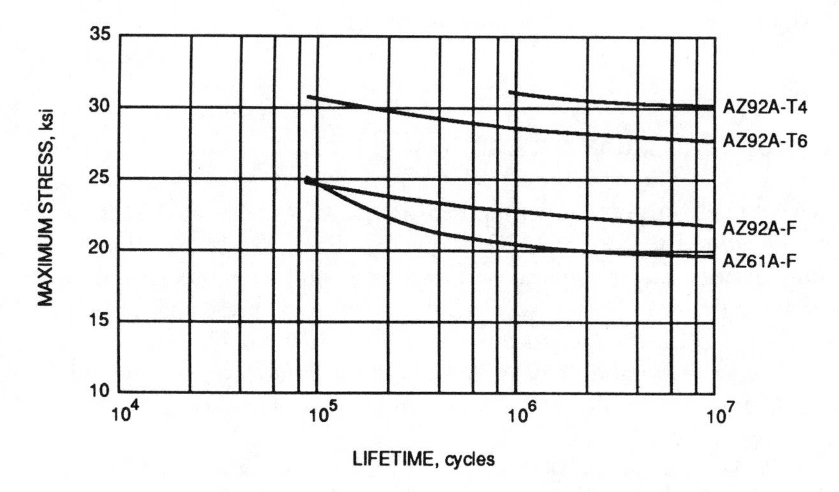

FIGURE 7.2 Fatigue curves for magnesium alloy with different tempers.

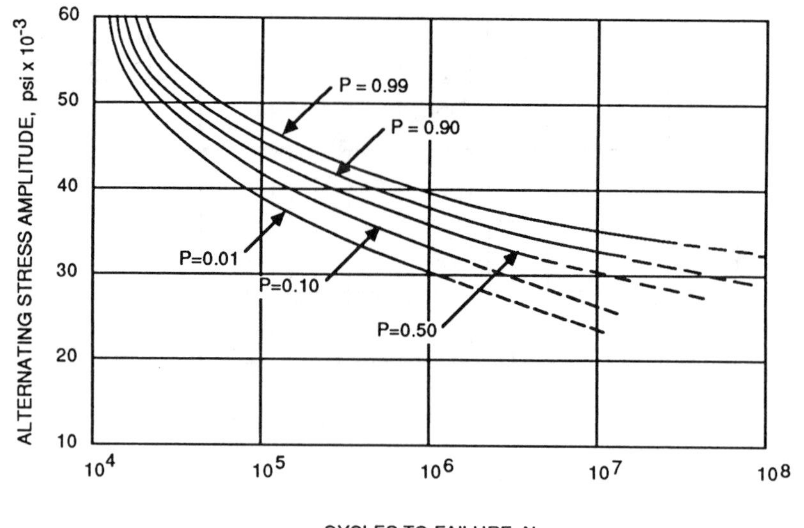

FIGURE 7.3 *S–N–P* curves for 7075-T6 aluminum alloy; *P* is the probability of failure.

for four other magnesium alloys with $R = 0.25$, with no clearly defined fatigue threshold up to 10^7 cycles. These *S–N* curves are interpreted as representing a 0.5 probability of failure at the indicated number of cycles for a specified maximum stress. In some cases several values of probability of failure P may be shown with a separate *S–N* curve for each P value, as shown in the *S–N–P* curves for the aluminum alloy of Fig. 7.3. In this figure, as the probability of failure decreases, the number of cycles to failure decreases for a given stress level.[2]

7.2 NONZERO MEAN STRESS

When a steady stress is superimposed on a cyclic stress, the resultant mean stress is nonzero. [See Eq. (7-2).] The modified Goodman relationship is commonly used under conditions of high-cycle fatigue to predict failure when the fatigue stressing is uniaxial. This methodology is presented in Table 7-1.[2] Stress amplitudes that produce negative quantities on the left-hand side of Table 7-1 result in more than N cycles without failure; stress amplitudes that yield zero values result in failure at N cycles; and stress amplitudes that produce positive values on the left-hand side of the table will cause failure in less than N cycles. If the values of σ_y, σ_u, and σ_N are known for the material and σ_{max} and σ_m can be estimated for the structure in the specified vibration environment, Table 7-1 allows a failure prediction to be made.

TABLE 7-1 Failure Prediction for Nonzero Mean Stress

Failure Is Predicted to Occur in N Cycles or Less if	When the Following Conditions Are Met
1. $\sigma_{max} - 2\sigma_m - \sigma_y \geq 0$	1. $-\sigma_y \leq \sigma_m \leq \sigma_N - \sigma_y$
2. $\sigma_{max} - \sigma_m - \sigma_N \geq 0$	2. $\sigma_N - \sigma_y \leq \sigma_m \leq 0$
3. $\sigma_{max} - (1 - r)\sigma_m - \sigma_N \geq 0$	3. $0 \leq \sigma_m \leq (\sigma_y - \sigma_N)/(1 - r)$
4. $\sigma_{max} - \sigma_y \geq 0$	4. $(\sigma_y - \sigma_N)/(1 - r) \leq \sigma_m \leq \sigma_y$

where σ_{max} = maximum stress in the cycle (psi)
σ_m = mean stress = $(\sigma_{max} + \sigma_{min})/2$ (psi)
σ_{min} = minimum stress in the cycle (psi)
σ_y = yield tensile strength of the material (psi)
σ_N = stress value for failure to occur at N simple, completely reversed cycles (psi)
$r = \sigma_N/\sigma_u$
σ_u = ultimate tensile strength of the material (psi)

7.3 MULTIAXIAL STRESS

If a multiaxial state of stress is present in addition to a nonzero mean stress, a failure prediction may be made by examining the stress conditions on the three principal axes. In this case, the four equations of Table 7-1 are expanded to 12 equations by letting $\sigma_{max} = \sigma_{i\,max}$ and $\sigma_m = \sigma_{i\,m}$, where the subscript $i = 1, 2, 3$ denotes the three principal normal stresses. Whichever one of these 12 equations proves to be most constraining will govern the design. This approach is known as the maximum-normal-stress multiaxial-fatigue-failure theory and represents a combination of the maximum-normal-stress theory for static stresses with the Goodman linear relationships. If the maximum shearing stress theory for static stresses is used instead of the maximum normal stress theory, the same 12 equations described above may be used by substituting $2\tau_{i\,max}$ for $\sigma_{i\,max}$, and $2\tau_{i\,m}$ for $\sigma_{i\,m}$, where the multiaxial principal shearing stresses are $\tau_i = \pm(\sigma_2 - \sigma_3)/2$, $\tau_2 = \pm(\sigma_3 - \sigma_1)/2$, and $\tau_3 = \pm(\sigma_1 - \sigma_2)/2$. If the distortion-energy theory for static stress is used instead of the maximum-normal-stress theory, the failure equations in Table 7-1 may be used by substituting $\sigma = [(\sigma_1 - \sigma_2)^2 + (\sigma_2 - \sigma_3)^2 + (\sigma_3 - \sigma_1)^2]^{1/2}/2^{1/2}$ for σ_{max} and σ_m, where the maximum values of the principal normal stresses are used to calculate σ_{max} and the mean values of the principal normal stresses are used to calculate σ_m. In general, the maximum-normal-stress multiaxial-fatigue-failure theory works best for brittle materials that exhibit a ductility of less than 5% elongation in 2 in.; and the distortion-energy theory and the maximum-shearing-stress theory work best for ductile materials that exhibit a ductility of 5% or more elongation in 2 in.

7.4 STRESS EQUATIONS

When sufficient fatigue test data are available for a material, a regression analysis may be applied to yield an equation that allows the calculation of the number of cycles to failure N_f. This equation usually expresses $\log N_f$ as a function of R or the mean stress S_m; and S_{max} or the equivalent stress S_{eq}, where

$$S_m = \tfrac{1}{2}(S_{max} + S_{min}) \tag{7-2}$$

and

$$S_{eq} = S_{max}(1 - R)^n \tag{7-3}$$

and where R, S_{max}, and S_{min} are defined in Eq. (7-1), and n is a constant provided by the data.

Table 7-2 provides stress equations for some alloys of aluminum, steel, magnesium, and titanium at room temperature.[1] These data are referenced to a specified value of the theoretical elastic stress concentration factor K_t defined as

$$K_t = (\text{actual maximum stress})/(\text{nominal stress}) \tag{7-4}$$

and caused by geometrical or microstructural discontinuities. This definition of K_t is valid only for stress levels within the elastic range. A few stress concentration factors are presented in Figs. 7.4–7.8. Many more are provided in Ref. 3. The "fatigue stress concentration factor," or the "fatigue strength reduction factor," K_f, is defined as the ratio of the effective fatigue stress at the root of the geometric discontinuity to the nominal fatigue stress calculated as if there were no stress concentration effect:

$$K_f = (\text{effective fatigue stress})/(\text{nominal fatigue stress}) \tag{7-5}$$

The notch sensitivity index is defined as

$$q = (K_f - 1)/(K_t - 1) \tag{7-6}$$

and

$$K_f = q(K_t - 1) + 1 \tag{7-7}$$

where $0 \le q \le 1$.

Both K_f and q are functions of the material as well as geometry and type of loading. K_f is smaller than K_t, with the difference increasing as K_t increases. Figure 7.9 shows values of q as a function of notch radius for an aluminum alloy and a range of steel alloys subjected to axial, bending, and torsional loading. K_f may be used as a stress concentration factor, multiplying

TABLE 7-2 Fatigue Stress Equations

$$\log N_f = a_1 - a_2 \log[S_{max}(1 - R)^n - a_3]$$

See Eqs. (7-1)–(7-3) for definitions of symbols in stress equations.

where K_t = theoretical elastic stress concentration factor, Eq. (7-4)
σ_u = ultimate tensile strength (ksi)
σ_y = yield tensile strength (ksi)
ksi = 1,000 psi

All data are at room temperature.
All data are for the longitudinal direction unless otherwise noted.
Some steels are presented for more than one temper, resulting in different values of tensile strength and stress equation constants.

Material and Form	K_t	σ_u (ksi)	σ_y (ksi)	a_1	a_2	a_3	n
		Aluminum					
2014-T6	1.0	67–78	60–72	21.49	9.44	0	0.67
Wrought products	1.6	72	64	10.65	4.02	20.2	0.55
	2.4	72	64	10.59	4.36	11.7	0.52
	3.4	75	67	8.35	3.10	10.6	0.52
2024-T3	1.0	73	54	11.1	3.97	15.8	0.56
0.090 in. Sheet	1.5	76	—	7.5	2.13	23.7	0.66
	2.0	73	—	9.2	3.33	12.3	0.68
	4.0	67	—	8.3	3.30	8.5	0.66
	5.0	62	—	8.9	3.73	3.9	0.56
2024-T4	1.0	69–85	45–65	20.83	9.09	0	0.52
Wrought products	1.6	73	49	12.25	5.16	18.7	0.57
	2.4	73	49	14.33	6.35	3.2	0.48
	3.4	74–84	—	8.14	2.76	11.6	0.52
2219-T851	1.0	68	52	—	—	—	—
2.00 in. Plate	2.0	94	—	7.92	2.69	16.0	0.64
	3.2	92	—	8.46	2.83	3.93	0.76
	5.0	91	—	8.76	3.05	0	0.722
6061-T6	1.0	45	40	20.68	9.84	0	0.63
Wrought products							
7049-T73	1.0	78	70	9.95	3.62	24.2	0.57
Forgings	2.4	95	—	10.6	4.18	0	0.80
7050-T7351X	1.0	72–79	62–69	10.5	3.79	16	0.55
Extrusions	3.0	72–79	62–69	7.73	2.58	0	0.56
7050-T7451	1.0	79	72	9.73	3.24	15.5	0.63
1.0 in. Plate	3.0	75–81	65–72	10.0	3.96	0	0.64

TABLE 7-2 *(Continued)*

Material and Form	K_t	σ_u (ksi)	σ_y (ksi)	a_1	a_2	a_3	n
7050-T7452	1.0	76–81	66–72	7.06	1.89	30	0.60
Forgings	3.0	73–81	59–72	8.21	2.96	5	0.68
7050-T74	1.0	74–81	68–71	16.8	6.97	0	0
Forgings	3.0	77–81	68–71	10.5	4.14	0	0.629
7050-T7651X	1.0	84–90	75–81	11.8	4.38	12	0.61
Extrusions	3.0	78–90	68–81	8.22	2.90	5	0.57
7075-T6	1.0	82	72	18.21	7.73	10	0.62
Rolled bar	1.6	99	—	8.28	2.62	15	0.53
	3.4	97	—	9.19	3.60	5	0.39
7075-T6	1.0	82	76	14.86	5.80	0	0.49
0.090 in. Sheet	1.5	87	—	9.57	3.52	18.7	0.49
	2.0	88	—	7.50	2.46	18.6	0.54
	4.0	82	—	10.2	4.63	5.3	0.51
	5.0	77	—	7.51	2.92	6.7	0.58
7475-T61, -T761	1.0	75–81	66–76	16.9	7.03	0	0
Sheet	3.0	75–82	67–76	13.4	6.29	0	0
7475-T7351	1.0	70	60	17.42	7.56	0	0.40
Plate	3.0	70–72	60–63	8.46	3.21	7.5	0.72
		Magnesium					
AZ31B-F	1.0	38	26	7.13	2.20	12.9	0.56
Forging				(R between -1.0 and -0.5)			
(transverse)							
	1.0	38	26	8.87	3.26	15.0	0.33
				(R between 0.0 and $+0.5$)			
	3.3	38	26	8.24	4.34	0	0
HK31A-H24	1.0	39.1	28.3	6.85	2.16	14.7	0.47
0.050 in. Sheet							
(transverse)							
ZK60A-T5	1.0	47.5	40.9	7.56	2.73	23.7	0.40
Extruded bar	2.4	63.7	40.9	5.51	1.36	13.2	0.42
0.50 in. dia.	3.4	58.2	40.9	9.27	4.13	5.3	0.46
(longitudinal)							
		Steel					
4130	1.0	117	99	9.65	2.85	61.3	0.41
0.075 in. Sheet				(for $-0.60 \leq R \leq +0.02$)			
	1.0	117	99	9.27	3.57	43.3	0
				(for $R = -1.00$)			
	1.5	123	—	7.94	2.01	61.3	0.88
	2.0	120	—	17.1	6.49	0	0.86
	4.0	120	—	12.6	4.69	0	0.63

TABLE 7-2 *(Continued)*

Material and Form	K_t	σ_u (ksi)	σ_y (ksi)	a_1	a_2	a_3	n
	5.0	120	—	12.0	4.57	0	0.56
	1.0	180	174	20.3	7.31	0	0.49
	2.0	180	174	8.87	2.81	41.5	0.46
	4.0	180	174	12.4	4.45	0	0.60
4340	1.0	125	—	14.96	6.46	60.0	0.70
$1\frac{1}{8}$ in.-dia. rolled bar	3.3	150	—	9.75	3.08	20.0	0.84
	1.0	158	147	13.51	5.01	80.0	0.75
	3.3	190	—	7.90	2.00	40.0	0.60
	1.0	208–221	189–117	9.31	2.73	93.4	0.59
	3.3	251	—	7.52	1.96	31.2	0.65
	1.0	266–291	232	11.62	3.75	80	0.44
	2.0	390	—	9.46	2.65	50	0.64
	3.0	352	—	7.14	1.74	56.4	0.51
300M	1.0	290	242	10.58	3.02	75.0	0.39
Forging	2.0	456	—	12.87	5.08	55.0	0.36
	3.0	435	—	9.52	3.00	25.0	0.50
	5.0	379	—	9.61	3.04	10.0	0.52
Custom 450 (H900)	1.0	192	188	—	—	—	—
$1\frac{1}{16}$-in.-dia. bar	3.0	304	—	9.64	3.21	39.28	0.65
Custom 450 (H1050)	1.0	156	151	—	—	—	—
$1\frac{1}{16}$-in.-dia. bar	3.0	244	—	9.59	3.15	33.23	0.607
Custom 455 (H950)	1.0	245	242	38.1	15.7 (for $R = -1.0$)	0	0
$1\frac{1}{16}$-in.-dia. bar	1.0	245	242	82.9	34.8 (for $R = +0.026$)	0	0
	1.0	245	242	85.9	34.7 (for $R = +0.50$)	0	0
	3.0	361	—	7.42	1.90	47.34	0.515
Custom 455 (H1000)	1.0	214	209	—	—	—	—
$1\frac{1}{16}$-in.-dia. bar	3.0	335	—	12.37	4.44	21.43	0.561
PH13-8Mo (H1000)	1.0	205	197	16.32	5.75	92.6	0.64
4 in. × 5 in. Forged bar	3.0	205	197	9.38	3.07	30.0	0.69
7 in. × 7 in. Hand forged	1.0	210	204	18.12	6.54	0	0.11
15-5PH (H1025)	1.0	163	159	—	—	—	—
2 in. × 6 in. bar	3.0	278	—	19.69	9.14	18.16	0.595

TABLE 7-2 *(Continued)*

Material and Form	K_t	σ_u (ksi)	σ_y (ksi)	a_1	a_2	a_3	n
			Steel				
PH15-7Mo (TH1050)	1.0	201	196	23.24	8.32	0	0.47
0.025 in. Sheet	4.0	201	196	10.42	3.91	32	0.58
17-4PH (H900)	1.0	202	195	30.6	11.2	0	0.52
1 in. and $1\frac{1}{8}$-in. -dia. bar	3.0	202	195	9.10	2.79	48.4	0.67
0.787-in.-dia. bar	4.0	207	—	9.03	2.91	26.1	0.51
17-4PH	1.0	165	161	—	—	—	—
(H1025) 2 in. × 6 in. bar	3.0	280	—	21.60	9.24	0	0.581
17-4PH (H1100) 0.787-in.-dia. bar	4.0	151	—	14.6	5.56	0	0.69
			Titanium				
Ti-6Al-4V	1.0	137	129	19.18	7.55	0	0
Annealed					(for $R = -1.0$)		
1.25-in.-dia. bar	1.0	137	129	5.70	0.94	82.3	0
					(for $-0.02 \le R \le +0.12$)		
	1.0	137	129	7.08	2.18	99.6	0
					(for $+0.15 \le R \le +0.37$)		
1.00-in.-dia. bar	2.43	150	143	24.1	10.7	0	0.49
Extrusion	1.0	143	127	24.8	9.6	0	0
	2.8	143	127	14.8	5.8	0	0.50
Ti-8Al-1Mo-1V	1.0	147.2	135.6	10.57	3.46	66.7	0.61
0.050 in. Sheet (long-transverse)	2.6	147.2	135.6	14.49	5.90	12.7	0.55
Ti-6Al-4V	1.0	166–177	153–167	14.29	4.91	30.6	0.42
0.063 in. Sheet	2.8	166–177	153–167	10.87	3.80	24.0	0.50
Ti-6Al-4V	1.0	155	145	24.6	9.35	0	0.48
1.00 in. Plate	3.0	187	—	14.4	5.51	0	0.58
Ti-13V-11Cr-3Al	1.0	138.5	132.8	10.15	3.41	52.2	0.97
0.043 in. Sheet Annealed	3.0	138.5	132.8	21.93	11.03	0	0.53
			Inconel 718				
0.066 in. Sheet	1.0	197	164	15.51	5.52	20	0.53
(long-transverse)	3.0	197	164	7.24	1.64	31.8	0.62

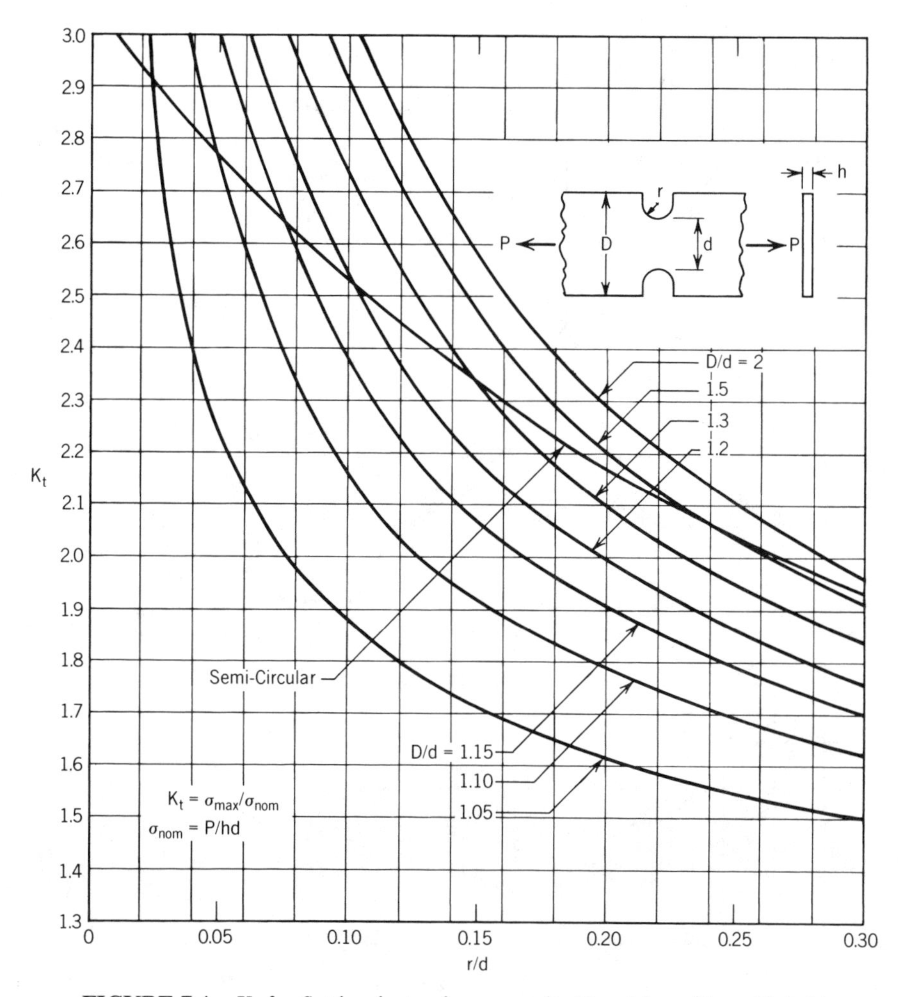

FIGURE 7.4 K_t for flat bar in tension, opposite U notches. (From Ref. 3.)

the applied nominal cyclic stress to obtain the effective cyclic stress for the notched structural element in fatigue analysis.

7.5 CONSTANT-LIFE FATIGUE DIAGRAM

Another method of presenting multiparameter fatigue life when there are adequate data is the constant-life fatigue diagram. Figure 7.10 shows this diagram for AISI4340 steel, unnotched and notched ($K_t = 3.3$), with data taken from $1\frac{1}{8}$-inch-diameter rolled bar. (This steel has an ultimate tensile strength ranging from 125 to 260 ksi, depending on the tempering process.)

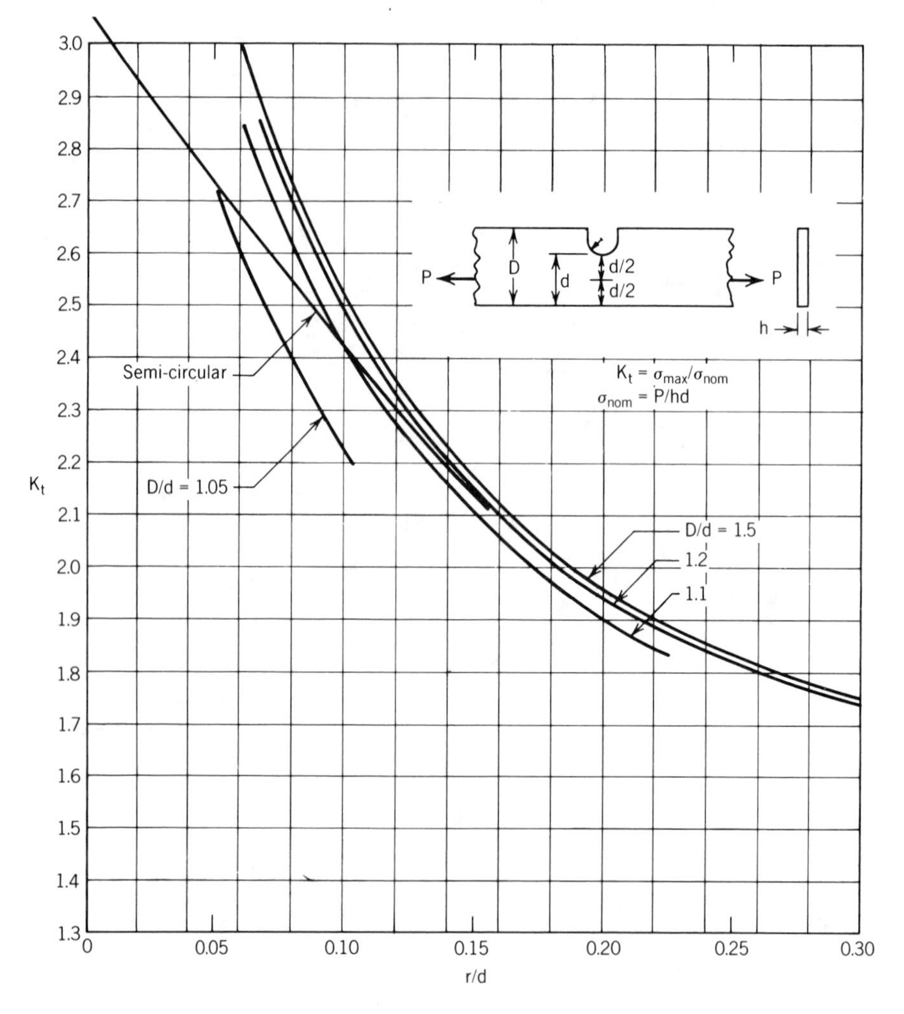

FIGURE 7.5 K_t for flat bar in tension, U notch on one side. (From Ref. 3.)

The parameter A is defined as

$$A = (S_{max} - S_{min})/(S_{max} + S_{min}) = (1 - R)/(1 + R) \qquad (7\text{-}8)$$

Given any two of the parameters R (or A), S_{max}, S_{min}, S_m, a point on the diagram will be defined and the number of cycles to failure may be estimated by interpolating between the constant-cycle number curves. The ultimate tensile strength is the point on the mean stress axis where the constant-cycle number curves converge to a single stress value. Many constant-life fatigue diagrams have been generated for aluminum and steel alloys and other metals.[1]

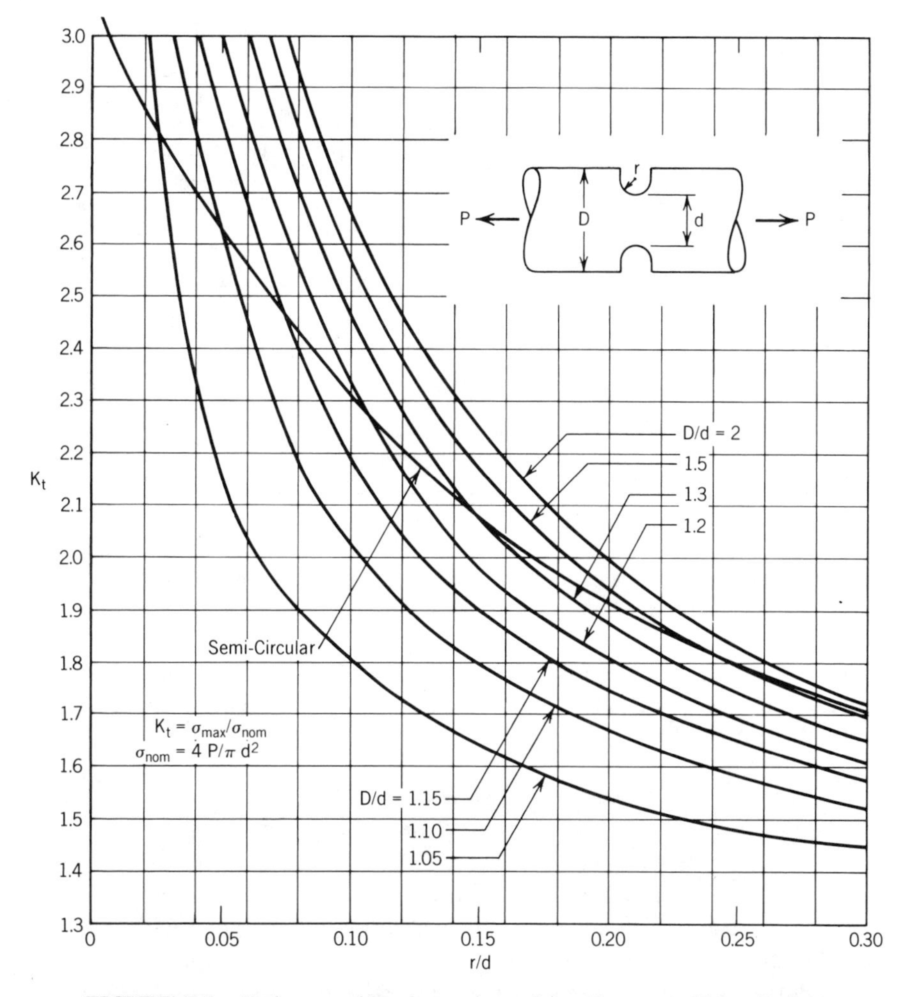

FIGURE 7.6 K_t for round bar in tension, with a U groove. (From Ref. 3.)

7.6 CUMULATIVE FATIGUE DAMAGE

In most cases where fatigue is a significant failure mode, the alternating stress amplitude will vary throughout the duration of the vibration environment. Miner's theory[4] is used to estimate fatigue life in this type of situation. It is based on the concept of cumulative damage, where a fraction of the fatigue life is used up by each stress level, with the fraction being proportional to the number of cycles at each stress level. When these fractions at each stress level add up to unity, fatigue failure will occur. This is represented mathematically as follows. If N_i is the number of cycles to fatigue failure at stress level S_i, and n_i is the total number of cycles actually experienced at stress level S_i, then the

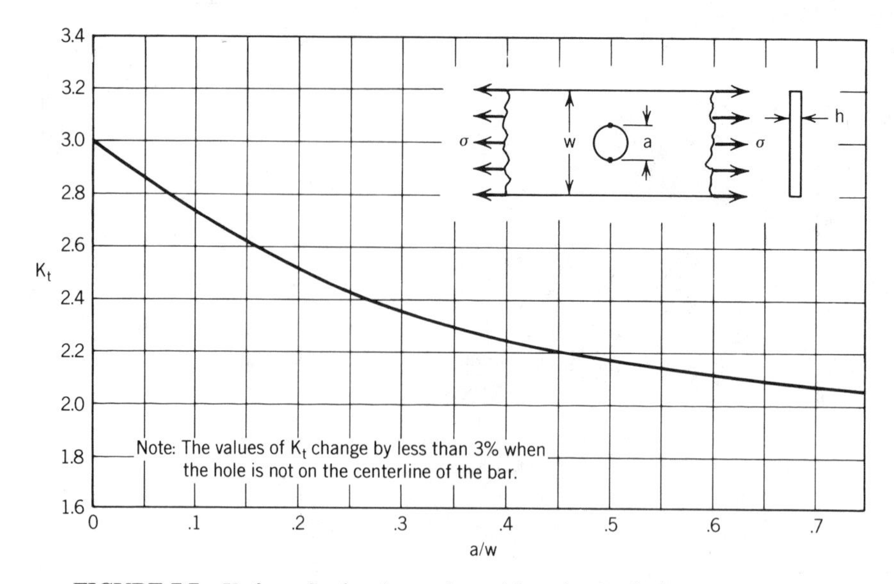

FIGURE 7.7 K_t for a flat bar in tension with a circular hole. (From Ref. 3.)

fraction of the fatigue life used up at stress level S_i is n_i/N_i. If these fractions are added up for all of the stress levels imposed by the environment, the total fraction of fatigue life used up is

$$R = \sum_i \frac{n_i}{N_i} \tag{7-9}$$

When $R \geq 1$, fatigue failure will occur. Equation (7-9) is only approximate since the order in which stresses occur is also important. For SAE1030 steel, when low stresses are followed by high stresses, failure occurs for $R = 1.1\text{–}1.3$; when high stresses are followed by low stresses, failure occurs for $R \simeq 0.8$.[5] In general, when stress amplitudes are mixed in a random way, the failure range for R is $0.6 < R < 1.6$. As an example, for aerospace electronic systems a value of $R \leq 0.7$ is often used; in a manned space mission, a conservative value of $R \leq 0.3$ may be used.

7.7 FATIGUE LIFE

If the fatigue environment consists primarily of a single stress amplitude S and frequency F, the fatigue life is

$$t = N/F \text{ sec} \tag{7-10}$$

where N is the number of cycles to fatigue failure at stress level S.

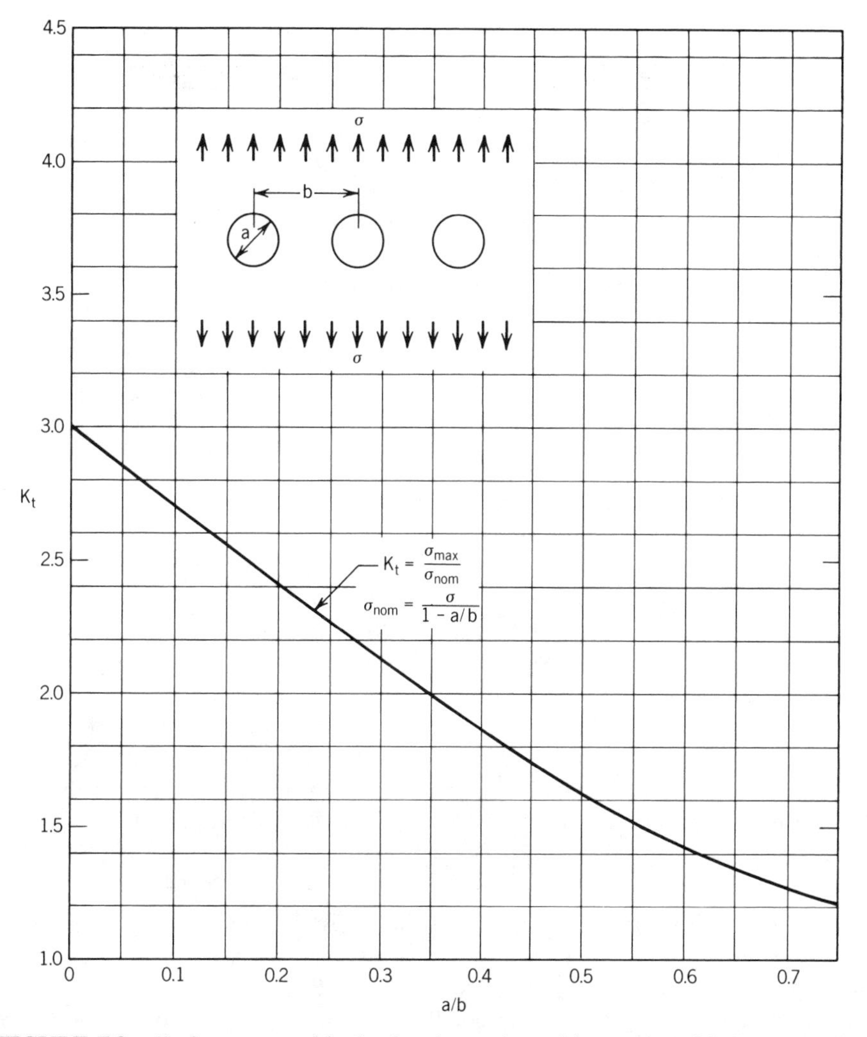

FIGURE 7.8 K_t for a very wide flat bar in tension with a series of holes across its width. (From Ref. 3.)

In many cases, a log S vs log N plot of fatigue data will result in a straight line,[6] as in Fig. 7.11. In this case, two different points in the line are related by the equation

$$N_1 S_1^p = N_2 S_2^p \tag{7-11}$$

where N_1 = number of cycles to fatigue failure at stress level S_1
N_2 = number of cycles to fatigue failure at stress level S_2
$1/p$ = slope of the S–N line on a log S vs log N plot

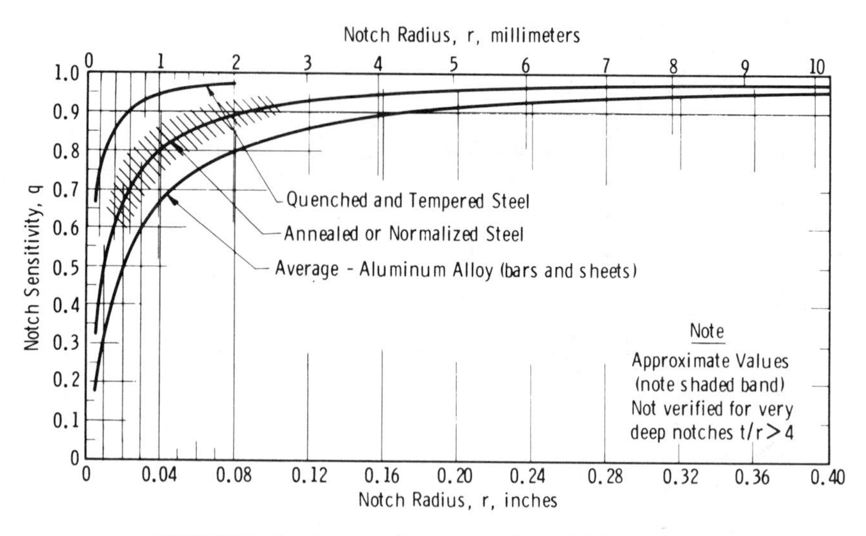

FIGURE 7.9 Average fatique notch sensitivity index.

The value of p may be obtained from the S–N data for a specific material in a structural application. For a printed circuit board (PCB) with connectors, lead wires, and solder joints a value of $p = 6.4$ may be used. Equation (7-11) may be useful in estimating fatigue life, as shown in the following example.

Extensive testing of PCBs indicates that at least 10^7 cycles may be tolerated before fatigue failure occurs if the peak displacement at the center of the PCB is limited to

$$d_{max} = 0.003b/L^2 \text{ in.} \tag{7-12}$$

where b is the short side of a rectangular PCB (in.) and L is the maximum length of a component mounted on the PCB (in.). See Section 4.3. For values of actual displacement d greater than d_{max}, fatigue failure will occur before 10^7 cycles. Assuming that the stress S is proportional to the displacement d, Eq. (7-11) may be written as

$$N_1 = N_2(S_2/S_1)^p = N_2(d_{max}/d)^p \tag{7-13}$$

where N_1 = number of cycles to fatigue failure for actual displacement d
N_2 = 10^7 cycles corresponding to d_{max}
d_{max} = value calculated from Eq. (7-12)
d = actual displacement as calculated in Section 4.3

The fatigue life is then found by substituting Eq. (7-13) into Eq. (7-10):

$$t = 10^7(0.003b/dL^2)^p/F \text{ sec} \tag{7-14}$$

where F, in this example, is the fundamental natural frequency of the PCB.

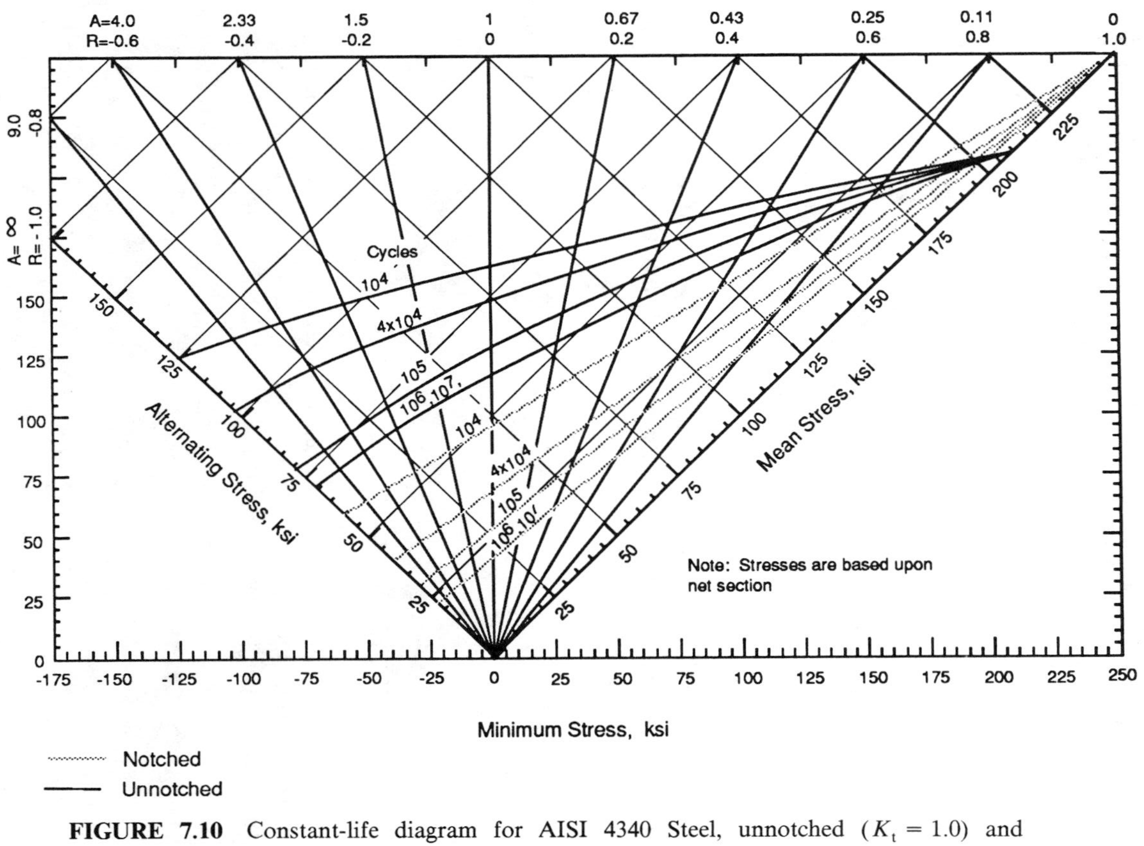

FIGURE 7.10 Constant-life diagram for AISI 4340 Steel, unnotched ($K_t = 1.0$) and notched ($K_t = 3.3$).

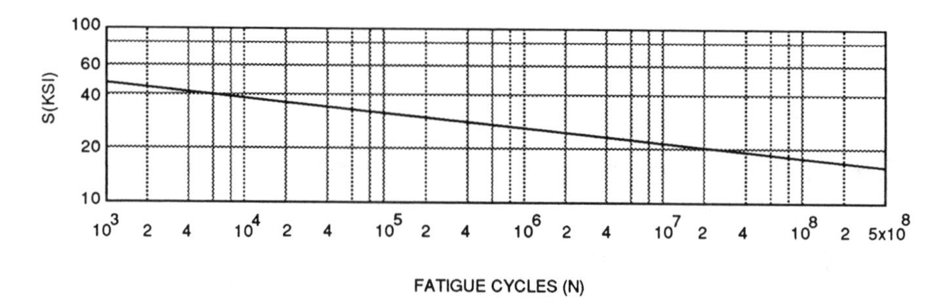

FIGURE 7.11 log–log S–N curve for 6061-T6 aluminum alloy.

For a frequency spectrum of fatigue cycles, the fatigue life is given by

$$t = \sum_i \frac{n_i}{F_i} \text{ sec} \tag{7-15}$$

where n_i is the total number of cycles actually experienced at stress level S_i and F_i is the frequency of vibration at stress level S_i (see Chapter 4). Equation (7-12) is subject to the condition [see Eq. (7-9)] that

$$R = \sum \frac{n_i}{N_i} = 1 \tag{7-16}$$

7.8 SAFETY FACTORS (FS)

In beamlike or barlike structural elements where the stresses are known or predictable, safety factors may be estimated for steady and alternating stress environments.[3] The following sections deal with ductile materials.

7.8.1 Steady Stress

For axial tension

$$\text{FS} = \sigma_{\text{y}}/\sigma_0 \tag{7-17}$$

where σ_{y} is the yield strength of the material (psi) and σ_0 is the steady axial stress (psi).

For bending, subject to the constraint $L_{\text{b}}\sigma_{\text{y}} < \sigma_{\text{UT}}$,

$$\text{FS} = L(X)_{\text{b}} \cdot \sigma_y/\sigma_{\text{b}} \tag{7-18}$$

TABLE 7-3 $L(X)$ Values For Bending and Torsion of Bars and Tubes[a]

Cross Section	Stress Mode	$X = 1/4$	$X = 1/2$	$X = 1$
Solid rectangle	Bending	1.22	1.375	1.50
Solid circle	Bending	1.25	1.50	1.70
Solid circle	Torsion	—	—	1.33
Circular tube $D_i/D_o = 0.75$[b]	Bending	1.23	1.34	1.44

[a]These L values apply only for materials where the stress–strain diagram becomes horizontal after the yield point.
[b]D_i is the inside diameter of tube; D_o is the outside diameter of tube.

and for torsion, subject to the constraint $L_s\sigma_y < \sigma_{UT}$,

$$FS = L(X)_s \cdot \sigma_y/3^{1/2}\tau_0 \tag{7-19}$$

where σ_b = steady bending stress (psi)
τ_0 = steady shear stress (psi)
$L(X)$ = limit design factor = $A(X)/B$ (subscript b for bending; subscript s for torsion)
$A(X)$ = load or moment required to cause yielding from the outer fiber of the element to a depth of X toward the center of the element, where $X = \frac{1}{4}, \frac{1}{2}$, or 1 (complete yielding) of the distance from the extreme fiber to the center
B = load or moment required to cause yielding only at the extreme fiber

Some values of L are provided in Table 7-3 and Eqs. (7-20) and (7-21); these apply only to materials in which the stress–strain diagram becomes horizontal after the yield point. These L values may be used for low- and medium-carbon steel. For other materials where stress continues to increase with strain beyond the yield point, a value of L closer to unity should be used. For complete yielding throughout the cross section ($X = 1$) in circular tubes,

$$L_b = (16/\pi)\left[1 - (D_i/D_o)^3\right]\left[1 - (D_i/D_o)^4\right]^{-1}, \quad \text{in bending} \tag{7-20}$$

$$L_s = \tfrac{4}{3}\left[1 - (D_i/D_o)^3\right]\left[1 - (D_i/D_o)^4\right]^{-1}, \quad \text{in torsion} \tag{7-21}$$

For combined stress,

$$FS = \sigma_y \Big/ \left[(\sigma_0 + \sigma_b/L_b)^2 + 3(\tau_0/L_s)^2\right]^{1/2} \tag{7-22}$$

7.8.2 Alternating Stress

For an alternating stress

$$FS = [\sigma_f]/[K_f \sigma_a], \quad \text{for axial or bending loads} \tag{7-23}$$

where σ_f = fatigue limit in axial or bending test (psi)
σ_a = alternating stress amplitude (psi)
K_f = fatigue notch factor, Eq. (7-7)

$$FS = \sigma_f/\left(3^{1/2}[q(K_s - 1) + 1]\tau_a\right), \quad \text{for torsion of a round bar} \tag{7-24}$$

where K_s = stress concentration factor for shear stress
τ_a = alternating shear stress amplitude
q = fatigue notch sensitivity index, Eq. (7-6)

For combined stress,

$$FS = \sigma_f \Big/ \left\{ (K_f \sigma_a)^2 + 3[q(K_s - 1) + 1]^2 \tau_a^2 \right\}^{1/2} \tag{7-25}$$

7.8.3 Combined Alternating and Steady Stresses

For combined alternating and steady stresses, with both axial loading and bending, subject to the constraint $L_b \sigma_y < \sigma_{UT}$,

$$FS = \left[\sigma_0/\sigma_y + (\sigma_b/L_b \sigma_y) + (K_f \sigma_a/\sigma_f)\right]^{-1} \tag{7-26}$$

For combined alternating and steady stresses in torsion

$$FS = (1/3^{1/2})\left\{\tau_0/L_s \sigma_y + [q(K_s - 1) + 1]\tau_a/\sigma_f\right\}^{-1} \tag{7-27}$$

where $L_s \sigma_y < \sigma_{UT}$.

For combined alternating and steady stresses of both axial and torsion loads,

$$FS = \left(\{FS[\text{Eq. (7-26)}]\}^{-2} + \{FS[\text{Eq. (7-27)}]\}^{-2}\right)^{-1/2} \tag{7-28}$$

REFERENCES

1 MIL-HDBK-5E, June, 1987, "Military Standardization Handbook: Metallic Materials and Elements for Aerospace Vehicle Structures," Philadelphia, PA: Naval Publications and Forms Center.

2 M. Kutz (Ed.), *Mechanical Engineer's Handbook*, New York: Wiley, 1986, Section 19.5.

3 R. E. Peterson, *Stress Concentration Factors*, New York: Wiley, 1974.

4 M. A. Miner, "Cumulative Damage in Fatigue," *J. Appl. Mech.* **12** (Sept. 1945).

5 S. H. Crandall, *Random Vibration*, New York: Technology Press, Wiley, 1958.

6 D. S. Steinberg, *Vibration Analysis For Electronic Equipment*, New York: Wiley, 1973, p. 449.

8

FRACTURE

8.1 FRACTURE FORMULAS AND PARAMETERS

Fracture can occur in brittle materials in which failure is due to the rapid propagation of a crack across a loaded component. However, ductile materials can also fail by brittle fracture in the presence of cracks or flaws in the material or of welds and residual stresses due to machining and other causes. Fracture control is concerned with avoiding fracture by controlling crack size and stresses to lie below the critical level for fracture for the material in the loaded component.

When a crack is present in a material, there are three different ways in which a stress may be applied to cause the crack to grow.[1] As shown in Fig. 8.1, in the crack-opening mode (Mode I), the stress causes the crack surfaces to move apart. In the edge-sliding or forward-sliding mode (Mode II), the crack surfaces slide over each other in a direction normal to the leading edge of the crack. In the side-sliding, parallel-shear, or tearing mode (Mode III), the crack surfaces slide over each other in a direction parallel to the leading edge of the crack.

Fracture is predicted to occur when

$$\sigma, \tau \geq K_c / C(\pi a)^{1/2} \text{ ksi} \tag{8-1}$$

where σ = gross section tensile stress (ksi), Mode I
τ = gross section shear stress (ksi), Modes II and III
K_c = critical stress intensity [ksi (in.)$^{1/2}$]
C = dimensionless constant, dependent on type of loading and geometry, provided in Table 8-2
a = crack length (in.)

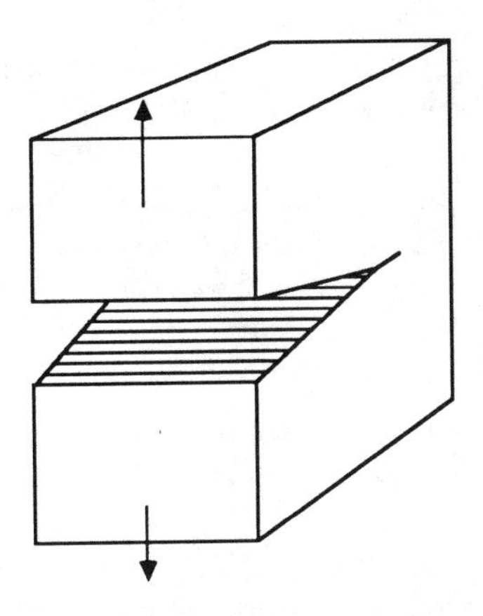

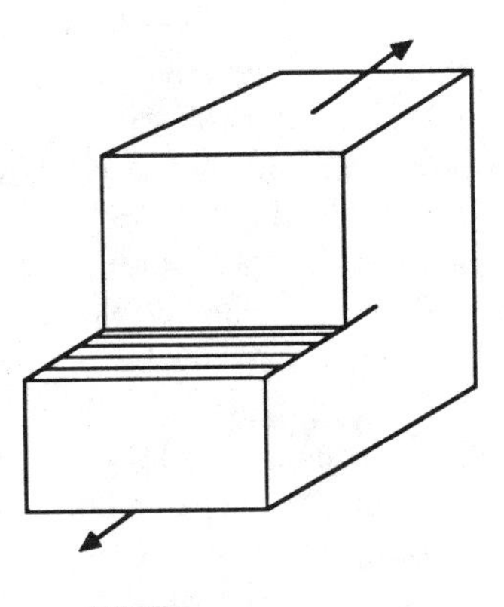

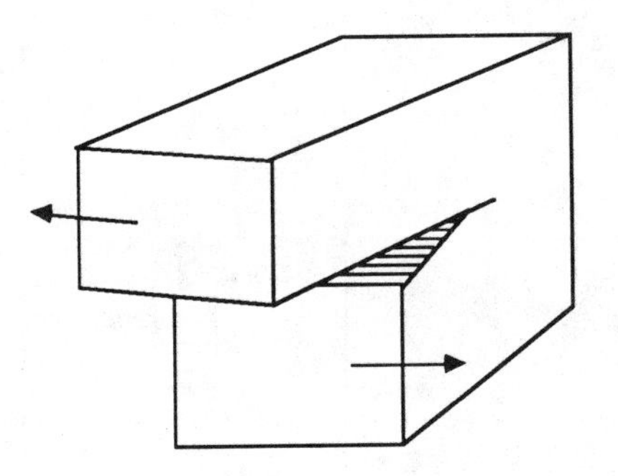

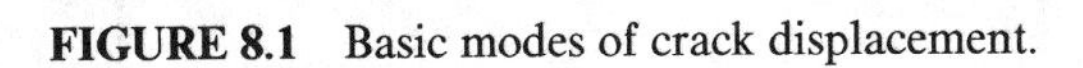
FIGURE 8.1 Basic modes of crack displacement.

For a given material, the critical stress intensity K_c in the crack opening mode (Mode I) decreases to a lower limiting value K_{IC} as the state of strain near the crack approaches the condition of plane strain. This basic material property, K_{IC}, is called the "plane strain fracture toughness" for the material, and is provided in tables. For plane strain conditions to exist in the material, the material thickness B must equal or exceed a lower limit:

$$B \geq 2.5(K_{IC}/\sigma_y)^2 \text{ in.} \tag{8-2}$$

where σ_y is the yield strength of the material (ksi).

When Eq. (8-2) does not apply, a condition of plane stress may exist near the crack, and the value of K_c under plane stress conditions may be considerably larger than the value of K_{IC}. Although it would be conservative to use the value of K_{IC} as a failure criterion in this case, it may lead to overdesign. The methods of elastic–plastic fracture mechanics required to obtain a more efficient design are beyond the scope of this book.

In applications, such as aerospace applications, a plane strain threshold stress intensity factor ΔK_0, which is a property of the material, may be used when the stresses in a component are low. The structure is predicted not to fail by fracture if

$$\sigma_T \leq \Delta K_0/2 \text{ ksi(in.)}^{1/2} \tag{8-3}$$

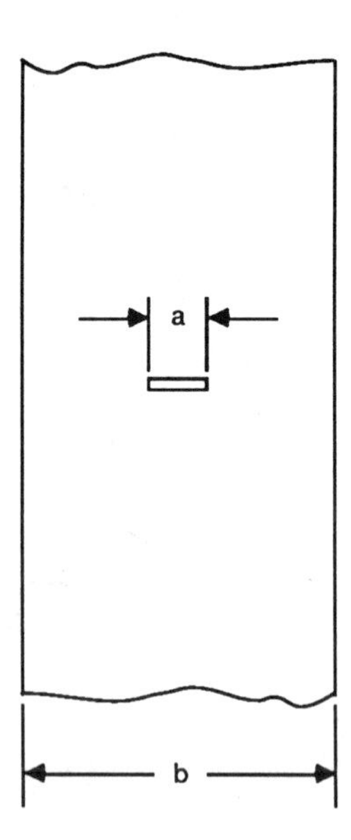

FIGURE 8.2 Central through-the-thickness crack.

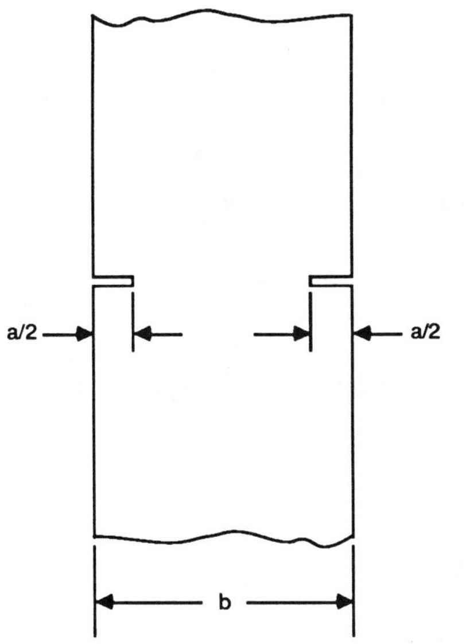

FIGURE 8.3 Double-edge notch.

where σ_T is the gross section tensile stress in the vicinity of the crack (ksi) and ΔK_0 is the plane strain threshold stress intensity factor [ksi(in.)$^{1/2}$].

Table 8-1 presents values of σ_y, K_{IC}, ΔK_0, and K_c for some commonly used aerospace structural materials.[1-3] Table 8-2 provides values of the constant C for use in Eq. (8-1) for a few commonly occurring crack configurations.[1]

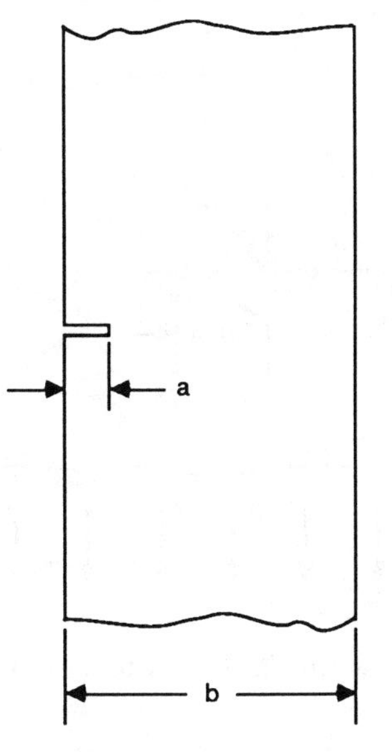

FIGURE 8.4 Single-edge notch.

8.2 CRACK GROWTH RATE AND PART LIFE

In the plane strain case, the critical crack size at which fracture occurs is, from Eq. (8-1),

$$a_{cr} = (K_{IC}/C\sigma_{max})^2/\pi \text{ in.} \tag{8-4}$$

where σ_{max} is the maximum value of the nominal stress to which the component will be subjected (ksi). The initial crack length a_0, due to preexisting flaws in the material, may be estimated by observation or from prior experience. The rate at which the crack grows due to sustained cyclic loading is[3]

$$\frac{da}{dN} = C_R(\Delta K)^n \text{ in. per cycle} \tag{8-5}$$

where a = instantaneous crack length (in.), $a_0 \leq a \leq a_{cr}$
N = number of loading cycles
C_R = empirical parameter, dependent on material properties, frequency, and mean load
n = slope of the $\ln(da/dN)$ vs $\ln(\Delta K)$ plot

and

$$\Delta K = C(\Delta\sigma)(\pi a)^{1/2} \text{ ksi(in.)}^{1/2} \tag{8-6}$$

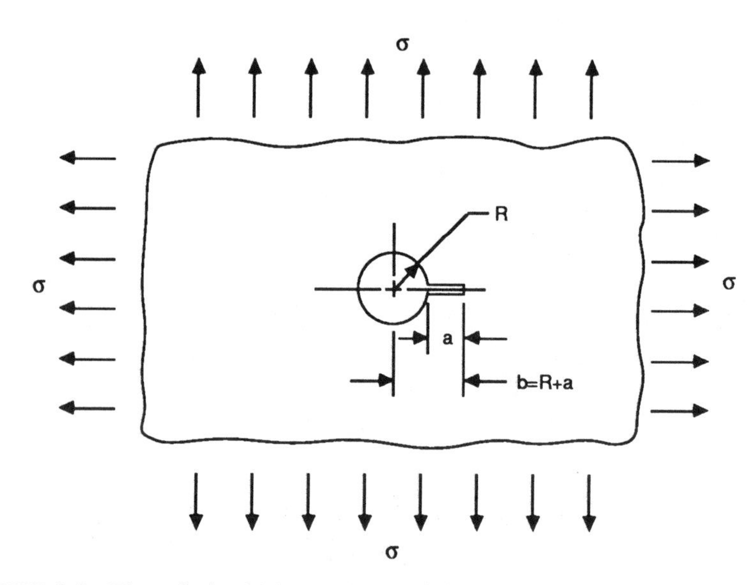

FIGURE 8.5 Through-the-thickness circular hole with a crack, equal biaxial tensile stresses.

TABLE 8-1 Fracture Parameters[a, b]

Material	σ_y (ksi)	K_{IC} [ksi(in.)$^{1/2}$]	ΔK_0 [ksi(in.)$^{1/2}$]	K_c^c [ksi(in.)$^{1/2}$]
Aluminum				
2014-T6 forged	64	28	—	—
2024-T351 plate	43–56	27–43	—	—
2219-T87 plate	51–53	25–34	3.5	40
2219-T62 plate	36–37	—	3.5	35
6061-T6 plate	34–35	—	3.5	40
7075-T6	85	30	—	—
7075-T651 plate	75–81	25–28	—	—
7075-T351 plate	58–66	28–32	—	—
7075-T73 extrusion	56	19–22	3.5	40
Titanium				
3Al-2.5V tubing, annealed, relieved	—	—	3.5	50
6Al-4V plate, annealed	121–143	38–77	6.0	80
Steel and high temperature alloys				
300M	262–270	—	4.0	61
4340, 500°F temper, plate	217–238	45–57	6.0	90
4340, 800°F temper, forged	197–211	72–83	—	—
AISI 301, 302, 303, 304, 321, 347 annealed	30	—	15.0	100
15-5 plate	155	—	10.0	95
21-6-9 tubing	57	—	10.0	100
AM355	155–165	—	10.0	95
D6AC, 1000°F temper, plate	217	93	—	—
A538	250	100	—	—
A286 STA	85–95	—	15.0	100–150
Inconel 718 STA	145–156	—	15.0	115
PH13-8Mo forging	205–212	49–104	—	—
Beryllium	30	—	5.0	18
Weld joints				
3Al-2.5V titanium tube	—	—	3.0	40
6Al-4V titanium annealed, GTA weld	—	—	7.0	50
304 stainless steel, EB weld	—	—	8.0	80
EF nickel/304L interface, transverse to interface	—	—	8.0	60
EF nickel/304L interface, parallel to interface	—	—	8.0	80
Inconel 718 EB weld, STA	—	—	10.0	75–120

Source: Refs. 1–3.

[a] Material properties are at room temperature with tension in the longitudinal direction and crack opening in Mode I.

[b] σ_y is the tensile yield strength (ksi); K_{IC} is the plane strain fracture toughness [ksi(in.)$^{1/2}$]; ΔK_0 is the plane strain threshold intensity factor [ksi(in.)$^{1/2}$]; K_c is the critical stress intensity [ksi(in.)$^{1/2}$]; and ksi = 1000 psi.

[c] K_c values are dependent on specimen thickness and may differ from values in the table.

TABLE 8-2 Values of C for Use in Eq. (8-1)

Description of Crack/Stress Configuration	a/b				
	0.0	0.2	0.4	0.6	0.8
1. Through-the-thickness crack (see Fig. 8.2)					
Modes I and II	1.00	1.03	1.11	1.31	1.83
Mode III	1.00	1.02	1.08	1.22	1.58
2. Double-edge notch (see Fig. 8.3)					
Modes I and II	1.14	1.19	1.14	1.24	1.59
Mode III	1.00	1.02	1.08	1.21	1.59
3. Single-edge notch (see Fig. 8.4)					
Mode I	1.12	1.37	2.12	4.02	12.0
4. Through-the-thickness circular hole with a crack under equal biaxial tensile stresses (see Fig. 8.5)					
Mode I	2.24	1.86	1.46	1.13	0.90

Source: Ref. 1.

as obtained from Eq. (8-1), where $\Delta\sigma$ is the cyclic minimal stress range, $\sigma_{max} - \sigma_{min}$ (ksi). In Eq. (8-6), ΔK is the stress-intensity-factor range.

The number of cycles to failure N_f is found by substituting Eq. (8-6) into Eq. (8-5), integrating Eq. (8-5), and solving for N_f:

$$N_f = \frac{1}{C_R(\Delta\sigma)^n(\pi)^{n/2}} \int_{a_0}^{a_{cr}} \frac{da}{C^n a^{n/2}} \text{ cycles} \tag{8-7}$$

If C does not change much with a, an average value of C may be used and Eq. (8-7) becomes

$$N_f = \left[\frac{1}{C_R(\Delta\sigma)^n(\pi)^{n/2}C^n(n/2-1)}\right]\left[\left(\frac{1}{a_0}\right)^{(n/2-1)} - \left(\frac{1}{a_{cr}}\right)^{(n/2-1)}\right] \tag{8-8}$$

Values of n and C_R for some materials are presented in Table 8-3.

The values of N_f found from Eqs. (8-7) or (8-8) determine part life based on an estimate of number of cycles per unit time (e.g., per day or per year).

8.3 FRACTURE CONTROL

Fracture control is concerned with the criteria and procedures required to prevent metallic-component structural failures due to the presence of flaws or defects in the material. This is often accomplished by development of a

TABLE 8-3 Crack-Growth-Rate Parameters[a]

Material	C_R	n
Aluminum		
2219-T62 plate	8.0×10^{-9}	2.79
2219-T87 plate	2.19×10^{-9}	3.30
6061-T6 plate	19.4×10^{-9}	2.64
7075-T73 extrusion	10.65×10^{-9}	2.67
Steel		
15-5	12.48×10^{-9}	1.98
21-6-9 stainless tubing	0.50×10^{-9}	2.9
300H	4.195×10^{-9}	2.184
AISI 302, 304, 316, 321, 347, 348	0.4127×10^{-9}	2.891
AH355	12.5×10^{-9}	1.98
4340	0.745×10^{-9}	2.737
Heat-resistant alloy		
A286,STA, LT orientation	0.052×10^{-9}	3.09
A286,STA, TL orientation	0.00267×10^{-9}	3.97
Inconel 718 STA	0.40×10^{-9}	2.7
Beryllium	0.422×10^{-9}	3.086
Weld joint		
3Al-2.5V titanium tube	1.0×10^{-9}	3.3
6Al-4V titanium annealed, GTA weld	1.13×10^{-9}	2.95
304L stainless steel, EB weld	0.82×10^{-9}	2.89
EF nickel/304L, normal or parallel to interface	0.40×10^{-9}	2.9
Inconel 718 EB weld, STA	0.359×10^{-9}	3.46
Inconel 718 FB weld, STA	0.163×10^{-9}	3.46

[a] For use in Eq. (8-5): $da/dN = C_R(\Delta K)^n$ in. per cycle with C_R in in. per cycle and ΔK in ksi(in.)$^{1/2}$.

fracture control plan that provides for analysis of components to determine which are fail-safe and which are fracture critical and require a safe-life evaluation involving applicable elements of analysis, inspection, test, and materials handling. Pressurized structures (e.g., tanks) that contain nonhazardous fluids do not require a safe-life analysis if they are demonstrated to meet the designer/verification criteria of developing a pressure-limiting leak at a safe pressure margin below the design burst pressure. Structural elements that are fail-safe or whose fractured parts will be contained by a surrounding structure (see Section 12.4) may be exempt from fracture control. In some cases, small parts, such as screw heads or nuts, may also be exempt from fracture control.

Structural elements whose potential failure mode is ductile failure and not brittle fracture may be exempt from fracture control. The ductile-failure-mode criteria relate the fracture toughness of the material to the load, thickness and stress:

$$K_{IC}/\sigma_0 > 2pt^{1/2}\ (\text{in.})^{1/2} \tag{8-9}$$

where σ_0 is the operating stress level at maximum design load including stress concentration factors (ksi) and t is the material thickness at the location of σ_0. Note: This criterion is not applicable to round bars, bolts, or trunnions with undefined thickness. The proof test factor p is given by

$$p \geq (\text{FS})(\sigma_y/\sigma_u) \text{ dimensionless} \tag{8-10}$$

where FS is the ultimate safety factor, σ_y is the tensile yield stress of the material (ksi), and σ_u is the tensile ultimate stress of the material (ksi), or (see Section 8.2)

$$p \geq (a_{cr}/a_0)^{1/2} \text{ dimensionless} \tag{8-11}$$

Equation (8-10) applies to single-use structures with a low number of cyclic loadings; Eq. (8-11) applies to multiple-use structures subjected to a high number of cyclic loadings. In Eqs. (8-9)–(8-11), the following conditions apply:

$$p \geq 1.1 \tag{8-12}$$

$$p\sigma_0 \leq \sigma_y \text{ except in areas of stress concentration where local yielding is not detrimental}$$

Also, if Eq. (8-11) is used, it is necessary to specify nondestructive examination (NDE) requirements that are adequate to exclude flaws larger than a_0.

Parts that come under fracture control must have a safe-life which is demonstrated by analysis or test or both. The safe-life is defined as a multiple of the service life to which the part will be exposed. For example, a safe-life may be specified as at least four times the service life:

$$N_f \geq 4N_s \text{ cycles} \tag{8-13}$$

where N_s is the number of cycles in the part service life and N_f, estimated from Eqs. (8-7) or (8-8), must satisfy Eq. (8-13). The validity of this procedure depends on the reliable determination of the original flaw size a_0 in the structural material, as can be seen from Eqs. (8-7) and (8-8).

REFERENCES

1 M. Kutz (Ed.), *Mechanical Engineers Handbook*, New York: Wiley, 1986, p. 261ff.

2 MIL-HDBK-5E, June 1987, *Metallic Materials and Elements for Aerospace Vehicle Structures*, Philadelphia, PA: Naval Publications and Forms Center.

3 A. G. Atkins and Y. W. Mai, *Elastic and Plastic Fracture*, 1985 New York: Halsted Press, Wiley, 1985.

9

ELASTIC INSTABILITY

Under some loading conditions the load that will cause failure of a structural element is determined by the stiffness of the element and not by the strength of the material. Structural failures discussed in preceding chapters were due to stresses exceeding the ultimate tensile, shear, or compressive strength of the material or to fatigue or fracture. This chapter deals with failures that can occur with stresses much less than the ultimate material strength, and with the initial application of the stress, before fatigue or fracture can occur. This type of failure is called elastic instability, or buckling, and occurs when the element/load configuration results in a bending moment or a twisting moment that is proportional to the deformation of the element, while the stresses are in the elastic range. The load that just causes buckling is called the critical load (for buckling) P_b and it is proportional to the stiffness of the element, which is given by (EI/L) for beams and $Eh^3/12(1 - \nu^2)$ for plates. See Section 2, Chapter 2. For small values of stiffness, buckling will occur before stresses reach the material ultimate strength values.

Table 9-1, Cases 1–13, presents buckling critical loads P_b for slender columns for various end conditions,[1] subject to the following assumptions:

1. The column is nominally straight and the load is axial and concentric with the longitudinal axis of the unloaded column.
2. The column adjusts itself to shortening only by bending and not by twisting. This assumption is valid for closed cross sections.
3. No local buckling (buckling of any part of the cross section) occurs before the whole column has developed its full strength.

These assumptions do not apply to Cases 14 and 15 of Table 9-1.

TABLE 9-1 Buckling Critical Loads for Axially, End-Loaded Beams

Cases 1–5. Uniform beams

P_b = Buckling critical load (lbf)
L = Length of beam (in.)
E = Modulus of elasticity of beam material (psi)
I = Area moment of inertia of beams cross section about central axis perpendicular to plane of buckling (in.^{4})
F = Free end
P = Pinned (hinged) end
C = Clamped (fixed) end
G = Guided end

Case Number	End Conditions	P_b
1	G–C	$\pi^2 EI/L^2$
2	F–C	$\pi^2 EI/4L^2$
3	P–P	$\pi^2 EI/L^2$
4	P–C	$2.048\pi^2 EI/L^2$
5	C–C	$4\pi^2 EI/L^2$

Cases 6–13. Symmetrically tapered beams.

$P_b = KEI/L^2$ (lbf)
I = Area moment of inertia of cross section of untapered, middle portion of beam (in.^{4})
I_0 = Area moment of inertia of beam cross section at ends of beam (in.^{4})
I_x = Area moment of inertia of beam cross section in tapered portion of beam a distance x in. from virtual beam vertex (in.^{4})
a = Length of untapered portion of beam (in.)
b = Length of taper to virtual vertex (in.)
$(L - a)/2$ = Length of each tapered portion of beam (in.)
L = Total length of beam (in.)

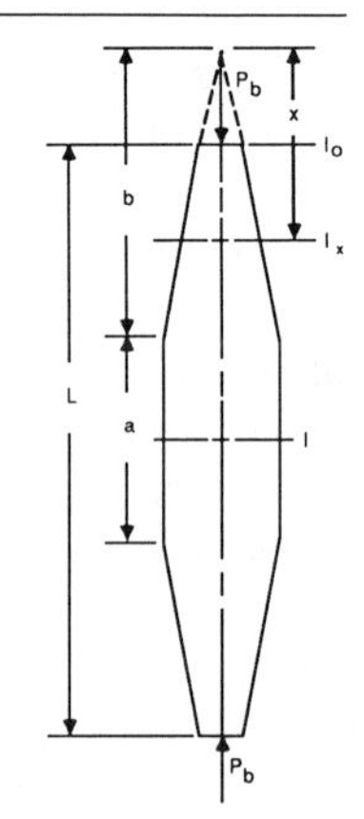

Case Number	Taper Conditions	End Conditions	a/L	I_0/I = 0.10	0.2	0.4	0.6	0.8
				K Values for $P_b = KEI/L^2$				
6	$I_x = I(x/b)$	P–P	0	6.48	7.01	7.86	8.61	9.27
			0.2	7.58	7.99	8.59	9.12	9.53
			0.4	8.63	8.90	9.19	9.55	9.68
			0.6	9.46	9.73	9.70	9.76	9.82
			0.8	9.82	9.82	9.83	9.85	9.86
7	$I_x = I(x/b)$	C–C	0	—	20.36	26.16	31.04	35.40
			0.2	—	22.36	27.80	32.20	36.00
			0.4	—	23.42	28.96	32.92	36.36
			0.6	—	25.44	30.20	33.80	36.84
			0.8	—	29.00	33.08	35.80	37.84

TABLE 9-1 *(Continued)*

Case Number	Taper Conditions	End Conditions	a/L	K Values for $P_b = KEI/L^2$, I_0/I = 0.10	0.2	0.4	0.6	0.8
8	$I_x = I(x/b)^2$	P–P	0	5.40	6.37	7.61	8.51	9.24
			0.2	6.67	7.49	8.42	9.04	9.50
			0.4	8.08	8.61	9.15	9.48	9.70
			0.6	9.25	9.44	9.63	9.74	9.82
			0.8	9.79	9.81	9.84	9.85	9.86
9	$I_x = I(x/b)^2$	C–C	0	—	18.94	25.54	30.79	35.35
			0.2	—	21.25	27.35	32.02	35.97
			0.4	—	22.91	28.52	32.77	36.34
			0.6	—	24.29	29.69	33.63	36.80
			0.8	—	27.67	32.59	35.64	37.81
10	$I_x = I(x/b)^3$	P–P	0	5.01	6.14	7.52	8.50	9.23
			0.2	6.32	7.31	8.38	9.02	9.50
			0.4	7.84	8.49	9.10	9.46	9.69
			0.6	9.14	9.39	9.62	9.74	9.81
			0.8	9.77	9.81	9.84	9.85	9.86
11	$I_x = I(x/b)^3$	C–C	0	—	18.48	25.32	30.72	35.32
			0.2	—	20.88	27.20	31.96	35.96
			0.4	—	22.64	28.40	32.72	36.32
			0.6	—	23.96	29.52	33.56	36.80
			0.8	—	27.24	32.44	35.60	37.80
12	$I_x = I(x/b)^4$	P–P	0	4.81	6.02	7.48	8.47	9.23
			0.2	6.11	7.20	8.33	9.01	9.49
			0.4	7.68	8.42	9.10	9.45	9.69
			0.6	9.08	9.38	9.62	9.74	9.81
			0.8	9.77	9.80	9.84	9.85	9.86
13	$I_x = I(x/b)^4$	C–C	0	—	18.23	25.23	30.68	35.33
			0.2	—	20.71	27.13	31.94	35.96
			0.4	—	22.49	28.33	32.69	36.32
			0.6	—	23.80	29.46	33.54	36.78
			0.8	—	27.03	32.35	35.56	37.80

14. Uniform cantilever beam of narrow rectangular section under an end load P_b applied at a distance a above ($a > 0$) or below ($a < 0$) the centroid of the section

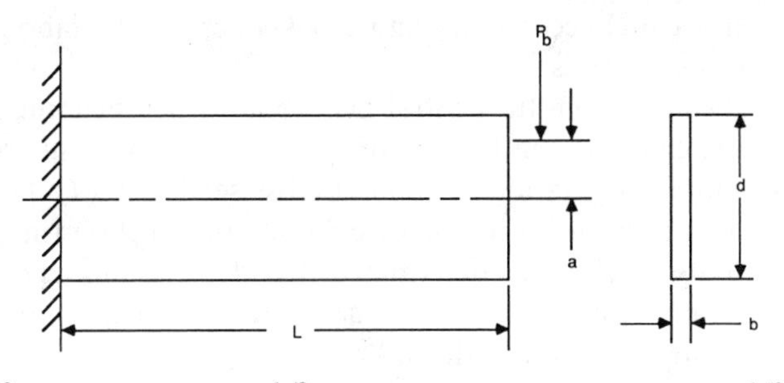

$$P_b = \alpha b^3 (d/L^2)[(1 - 0.63b/d)EG]^{1/2}\{1 - (\beta a/L)[E/G(1 - 0.63b/d)]^{1/2}\} \text{ lbf}$$
$$\alpha = 0.669, \quad \beta = 0.5$$

TABLE 9-1 *(Continued)*

For a load W uniformly distributed along the beam, the buckling critical load is $W_b = 3P_b$ lbf

15. Uniform beam of narrow rectangular section under a center load P_b applied at a distance a above $(a > 0)$ or below $(a > 0)$ the centroid of the section. Ends of beam simply supported and constrained against twisting

P_b same as Case 14, with $\alpha = 2.82$ and $\beta = 1.74$
For a load W uniformly distributed along the beam, the buckling critical load is

$$W_b = 1.67P_b \text{ lbf}$$

If the ends of the beam are fixed and the load is applied at the centroid of the middle cross section, P_b is the same as Case 14 with $\alpha = 4.43$ and $a = 0$

Other formulas have been developed for intermediate length columns, two of which follow.[2,3] The same three assumptions cited previously apply, plus the assumption that the column cross section is constant along its length.

The secant formula for intermediate length columns is

$$P_b = A\sigma_y / \left\{1 + (ec/r^2)\sec\left[(KL/2r)(P_b/AE)^{1/2}\right]\right\} \text{ lbf} \tag{9-1}$$

where P_b = critical buckling load (lbf)
A = cross-sectional area of column (in.^{2})
σ_y = yield strength of column material (psi)
e = equivalent eccentricity, being that eccentricity in a perfectly straight column that would cause the same amount of bending as the actual eccentricity and crookedness in the fabricated column, dimensionless
c = distance from the central axis about which bending occurs to the extreme fiber on the compression side of the bent column (in.)
r = radius of gyration of column cross section $= (I/A)^{1/2}$ (in.)
I = area moment of inertia of column cross section (in.^{4})
K = fixity coefficient determined by the column end condition = (effective column length)/(actual column length), dimensionless
L = actual column length (in.)
E = modulus of elasticity (psi)

TABLE 9-2 Values of K = (Effective Column Length) / (Actual Column Length)

End Conditions	K
One end fixed, one end free	2.0
Both ends fixed	0.5
Both ends pinned	0.875
Both ends riveted	0.75

The factor (ec/r^2), called the eccentric ratio, accounts for the column crookedness due to fabrication and the unintentional eccentricity of loading, and is generally taken as

$$(ec/r^2) = 0.25 \tag{9-2}$$

Values of K are given in Table 9-2. The factor (KL/r) is often called the slenderness ratio.

The AISI formula for compression members[4] is

$$\begin{aligned} \sigma_a &= 0.522\sigma_y - 0.0132(\sigma_y KL/r)^2/E \text{ psi for } KL/r < \pi(2E/\sigma_y)^{1/2} \\ \sigma_a &= 5.15E/(KL/r)^2 \text{ psi for } KL/r \geq \pi(2E/\sigma_y)^{1/2} \end{aligned} \tag{9-3}$$

where σ_a is the allowable average compression stress over the column cross section (psi) and the other quantities are as previously defined.

Figure 9.1 shows[5] the allowable column stress σ_a for 6061-T6 and 2024-T3 aluminum alloy round tubes as a function of slenderness ratio KL/r. Figure 9.2 presents[5] the crushing (or crippling) stress σ_c as a function of D/t, the ratio of tube diameter to wall thickness, for the same tubes. The crushing stress is the upper limit of the column stress at which local buckling failure will occur. For given values of KL/r and D/t, failure will occur at the lower of the two stress values σ_a and σ_c. For small values of KL/r and large values of D/t, failure will be by local buckling at a stress value σ_c. For large values of KL/r and small values of D/t, failure will be by column buckling at a stress value σ_a.

Table 9-3 contains buckling critical stresses and loads for some plates and shells for various configurations of loads and boundary conditions. Note that actual buckling loads are sometimes significantly less than the theoretical buckling loads indicated in the tables. This can be caused by changes in the amount of actual end constraint in the case of columns and by geometrical irregularities in the case of flat plates and shells, as indicated in Cases 8, 12, and 13 of Table 9-3. The critical loads and stresses shown in Table 9-3 should be considered as a theoretical upper limit that is more closely approached as the actual shape more perfectly approximates the ideal geometrical form.

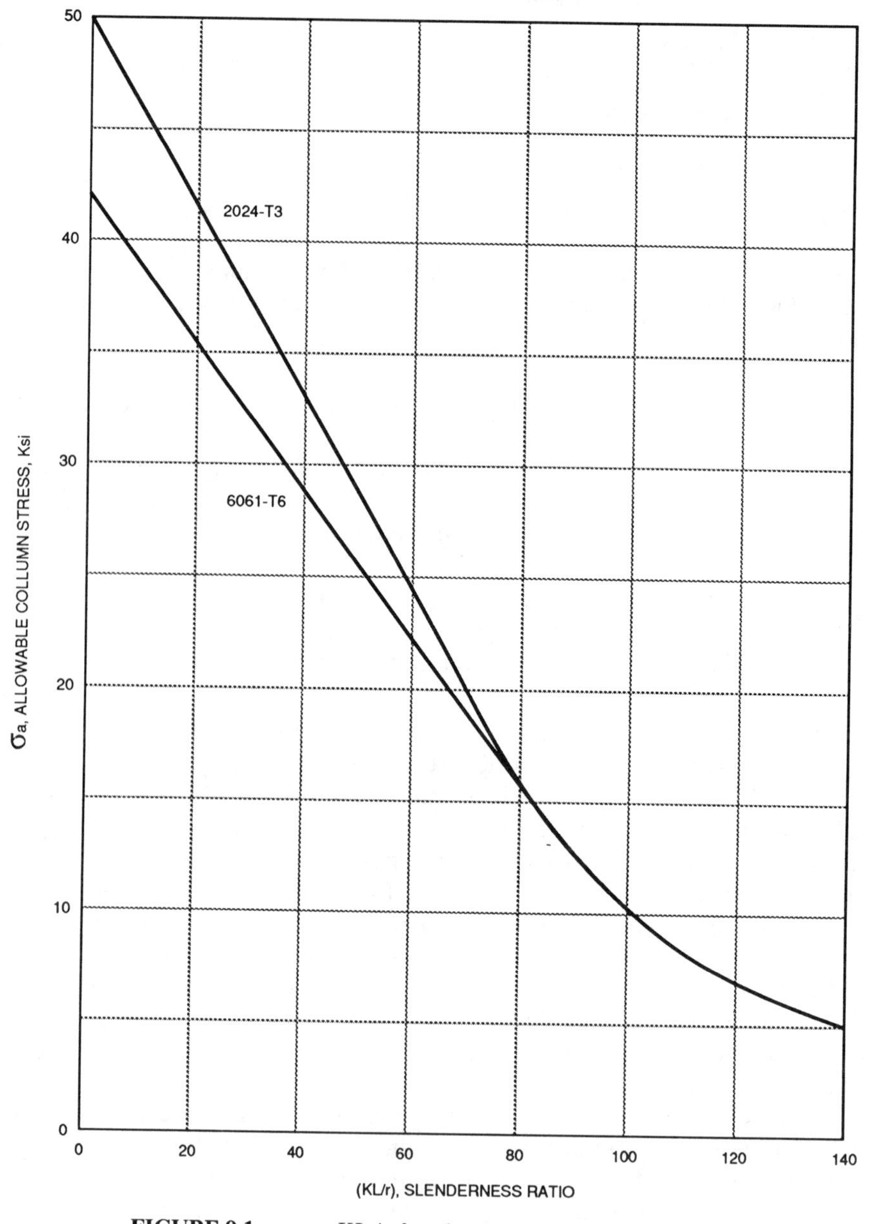

FIGURE 9.1 σ_a vs KL/r for aluminum alloy round tubing.

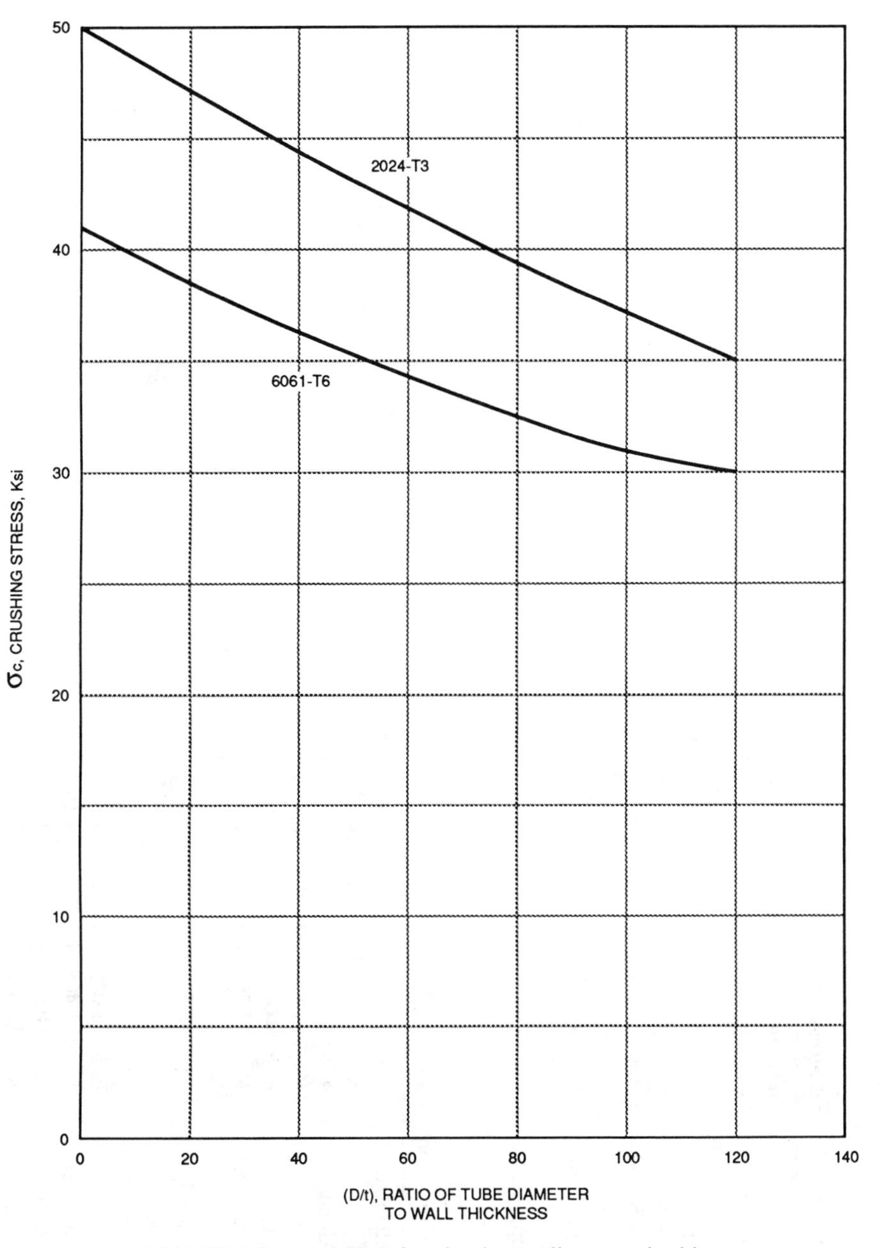

FIGURE 9.2 σ_c vs D/t for aluminum alloy round tubing.

TABLE 9-3 Buckling Critical Stresses and Loads for Plates and Shells

Cases 1–6. Flat rectangular plate under uniform compression on two opposite edges b

σ_b = Critical unit compressive stress = $KE(t/b)^2/(1-\nu^2)$ psi
E = Modulus of elasticity (psi)
ν = Poisson's ratio
t = Thickness (in.)
b = Length of stressed edges (in.)
a = Length of unstressed edges (in.)
K = Constant given in table, function of a/b
F = Free edge
S = Simply supported edge
C = Clamped edge

b/t > 10

Case Number	Edge Conditions																		
1	All edges	a/b	0.2	0.3		0.4	0.6	0.8	1.0	1.2	1.4	1.6	1.8	2.0	2.2	2.4	2.7	3.0	∞
	S	K	22.2	10.9		6.92	4.23	3.45	3.29	3.40	3.68	3.45	3.32	3.29	3.32	3.40	3.32	3.29	3.29
2	All edges	a/b	1	2		3	∞												
	C	K	7.7	6.7		6.4	5.73												
3	Edges b, S	a/b	0.4	0.5		0.6	0.7	0.8	1.0	1.2	1.4	1.6	1.8	2.1	∞				
	Edges a, C	K	7.76	6.32		5.80	5.76	6.00	6.32	5.80	5.76	6.00	5.80	5.76	5.73				
4	Edges b, S																		
	One edge a, S	a/b	0.5	1.0		1.2	1.4	1.6	1.8	2.0	2.5	3.0	4.0	5.0					
	One edge a, F	K	3.62	1.18		0.934	0.784	0.687	0.622	0.574	0.502	0.464	0.425	0.416					
5	Edges b, S																		
	One edge a, C	a/b	1.0	1.1	1.2	1.3	1.4	1.5	1.6	1.7	1.8	1.9	2.0	2.2	2.4				
	One edge a, F	K	1.40	1.28	1.21	1.16	1.12	1.10	1.09	1.09	1.10	1.12	1.14	1.19	1.21				
6	Edges b, C	a/b	0.6	0.8		1.0	1.2	1.4	1.6	1.7	1.8	2.0	2.5	3.0					
	Edges a, S	K	11.0	7.18		5.54	4.80	4.48	4.39	4.39	4.26	3.99	3.72	3.63					

Cases 7 and 8. Flat rectangular plate under uniform shear on all edges

τ_b = critical unit shear stress = $KE(t/b)^2/(1 - \nu^2)$

Case Number	Edge Conditions											
7	All edges	a/b	1.0	1.2	1.4	1.5	1.6	1.8	2.0	2.5	3.0	∞
	S	K	7.75	6.58	6.00	5.84	5.76	5.59	5.43	5.18	5.02	4.40
8[a]	All edges	a/b	1	2	∞							
	C	K	12.7	9.5	7.4							

Cases 9 and 10. Flat rectangular plate under concentrated center loads on two opposite edges a

P_b = critical load = $KEt^3/(1 - \nu^2)b$ lbf

Case Number	Edge Conditions	
9	All edges S	$K = \pi/3$ for $a/b > 2$
10	Edges b, S; Edges a, C	$K = 2\pi/3$ for $a/b > 2$

Case 11. Curved panel under uniform compression on curved edges b; all edges S

σ_b = initial unit compressive stress (psi)

b = width of panel measured on arc (in.)

r = radius of curvature of panel (in.)

t = thickness of panel (in.)

For $a > b$, $\sigma_b = E\{[12(1 - \nu^2)(t/r)^2 + (\pi t/b)^4]^{1/2} + (\pi t/b)^2\}/6(1 - \nu^2)$ psi

For $b/r < 1/2$ and $a \simeq b$, $\sigma_b = 0.6Et/r$ psi

TABLE 9-3 *(Continued)*

Case Number	Edge Conditions
Case 12. Circular tube under uniform axial compression on edges; ends not constrained. σ_b = critical unit compressive stress = $Et/r[3(1 - \nu^2)]^{1/2}$ psi[b] r = radius of tube (in.) t = thickness of tube wall (in.) Most accurate for tubes with length > $10(rt)^{1/2}$ in.	
Case 13. Truncated conical shell under axial load, ends held circular P_b = critical buckling load (lbf) $P_b = 2\pi Et^2 \cos^2\alpha/[3(1 - \nu^2)]^{1/2}$ theoretical $P_b \simeq 0.3(2\pi Et^2 \cos^2\alpha)$ actual	

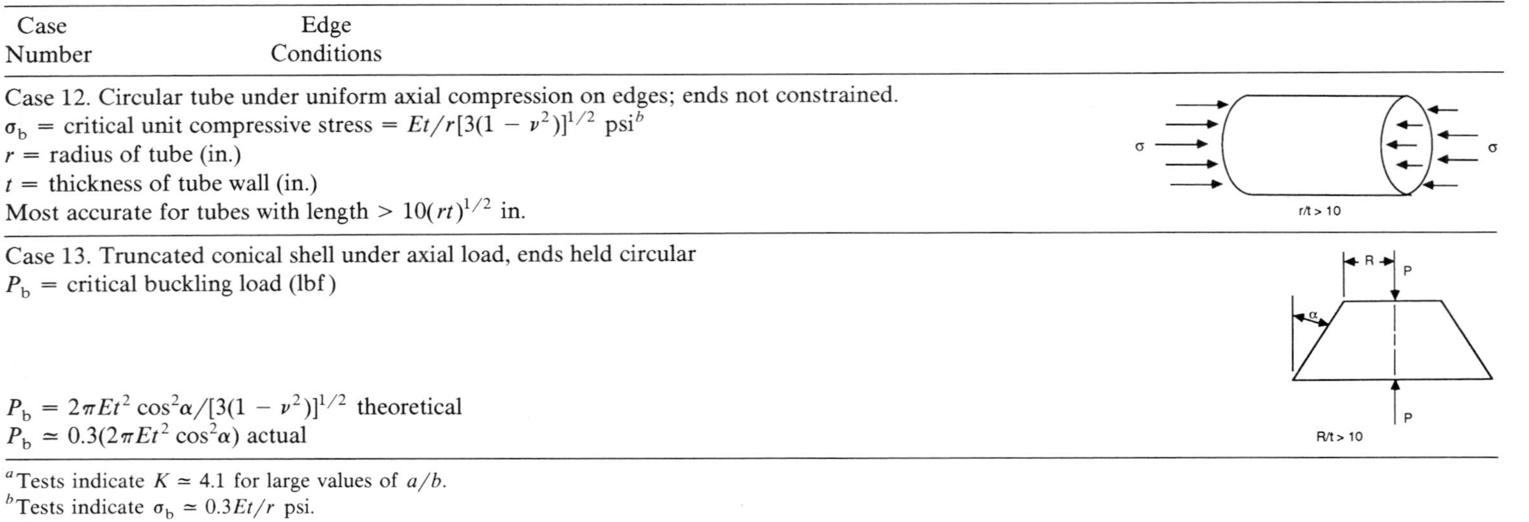

[a]Tests indicate $K \simeq 4.1$ for large values of a/b.
[b]Tests indicate $\sigma_b \simeq 0.3Et/r$ psi.

REFERENCES

1 R. J. Roark and W. C. Young, *Formulas For Stress and Strain*, 5th Ed., New York: McGraw-Hill, 1985, Tables 34 and 35.

2 R. J. Roark and W. C. Young, in Ref. 1, p. 422.

3 R. H. Perry, *Engineering Manual*, New York: McGraw-Hill, 1967, pp. 6-44 and 6-45.

4 R. N. White and C. G. Salman, *Building Structural Design Handbook*, New York: Wiley, 1987, pp. 649 and 650.

5 MIL-HDBK-5E, "Metallic Materials and Elements for Aerospace Vehicle Structures," 1 June 1987.

10

STRUCTURAL ANALYSIS OF MOUNTED HOUSINGS

This chapter reviews the analysis of housings secured to mounting surfaces and subjected to acceleration environments due to vibration, shock, and quasistatic load factors, such as can be induced during lift-off and landing. The analysis includes the attachment to the mounting surface, preloading, maximum material stresses, buckling of the housing walls, and containment of fractured internal components.

10.1 ATTACHMENT TO MOUNTING SURFACES

The maximum acceleration due to the specified acceleration environments may occur in any direction relative to the housing/mounting surface orientation, and the combined tension/shear stresses on the bolts are analyzed by taking the acceleration vector in both directions along each of three orthogonal axes, one axis and one direction at a time, in order to identify the worst-case acceleration orientation. These axes are chosen to be parallel to an axis of symmetry or parallel to the edges of the housing if it is rectangular.

Load-path redundancy is incorporated into the analysis by assuming that the worst-case bolt is missing. The worst-case bolt is the one that would most weaken the attachment (most compromise mechanical integrity) for the acceleration direction under investigation. If more than 10 mounting bolts are used, the two worst-case bolts may be assumed missing. If the remaining bolts will sustain the load with the required margin of safety, the mounting has adequate load-path redundancy.

Consider a rectangular housing that is bolted to a mounting surface, Fig. 10.1, with the bolt pattern shown in Fig. 10.2. The housing weight is W lbf and the maximum acceleration is G in units of $g = 386$ in./sec^2. With acceleration in the $+Z$ direction and the two worst-case bolts missing, the tensile force per remaining bolt due to acceleration is

$$F_T(+Z) = GW/15 \text{ lbf} \tag{10-1}$$

Acceleration in the $\pm X$ and $\pm Y$ directions subjects the bolts to both a shear load and a tensile load. The shear load is

$$F_S(\pm X, \pm Y) = GW/15 \text{ lbf} \tag{10-2}$$

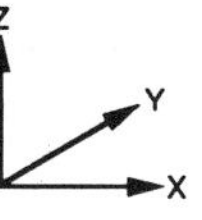

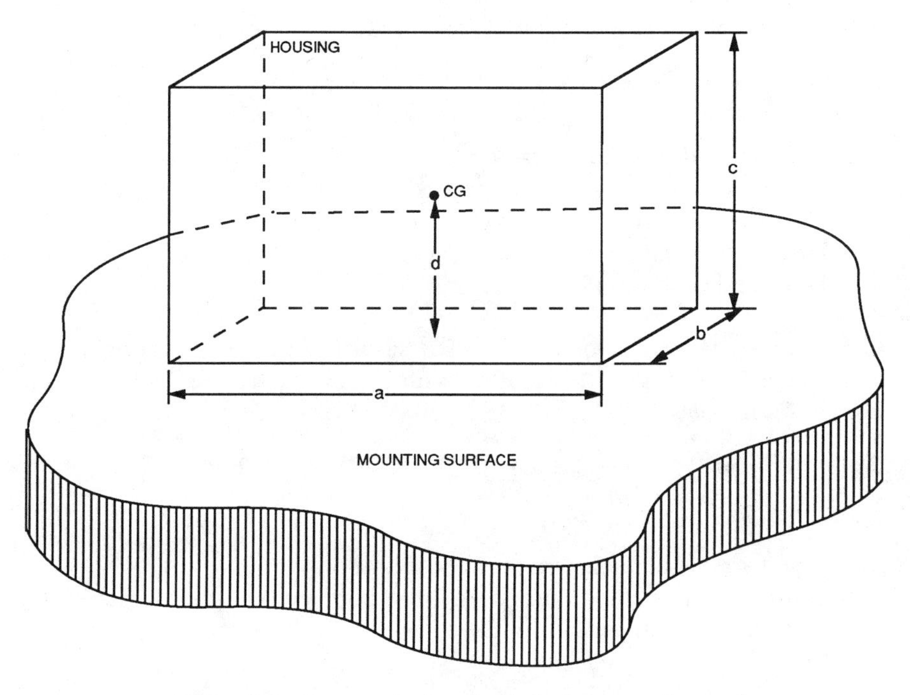

FIGURE 10.1 Rectangular housing bolted to mounting surface.

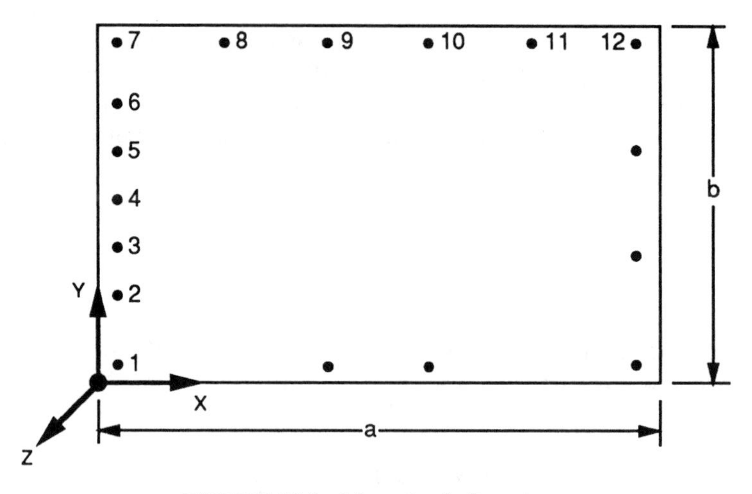

FIGURE 10.2 Mounting bolt pattern.

The tensile load is found by equating moments. The overturning moment is

$$M = GWd \text{ in.} \cdot \text{lbf} \tag{10-3}$$

where d is the height of the housing CG (center of gravity) above the mounting surface. This overturning moment must be counteracted by the bolts. For acceleration in the $+X$ direction, the tensile force per bolt is

$$F_T(+X) = M / \sum_i n_i (a - x_i) \text{ lbf} \tag{10-4}$$

where M is found from Eq. (10-3) and
n_i = number of bolts having coordinate X_i
x_i = distance from origin in X direction (in.)
a = length of housing in X direction (in.)
i = bolt number shown in Fig. 10.2

In Eq. (10-4), the axis of rotation for the overturning moment is the right edge of the housing at $x = a$ in Fig. 10.2. With the two worst-case bolts missing, Eq. (10-4) becomes

$$F_T(+X) = M / [5(a - x_1) + (a - x_8) + 2(a - x_9) + 2(a - x_{10}) + (a - x_{11}) + 4(a - x_{12})] \text{ lbf} \tag{10-5}$$

where the two missing bolts are at x coordinate x_1.

Similarly, for acceleration in the $-X$ direction, and the axis of rotation at $x = 0$, the tensile load per bolt is

$$F_T(-X) = M / [7x_1 + x_8 + 2x_9 + 2x_{10} + x_{11} + 2x_{12}] \text{ lbf} \tag{10-6}$$

where the two missing bolts are at x coordinate x_{12}.

For acceleration in the $+Y$ direction, and the axis of rotation at $y = b$, the tensile load per bolt is

$$F_T(+Y) = M/[2(b - y_1) + (b - y_2) + 2(b - y_3) + (b - y_4) + 2(b - y_5) + (b - y_6) + 6(b - y_7)] \text{ lbf} \quad (10\text{-}7)$$

where the two missing bolts are at y coordinate y_1.

For acceleration in the $-Y$ direction, and the axis of rotation at $y = 0$, the tensile load per bolt is

$$F_T(-Y) = M/[4y_1 + y_2 + 2y_3 + y_4 + 2y_5 + y_6 + 4y_7] \text{ lbf} \quad (10\text{-}8)$$

where the two missing bolts are at y coordinate y_7.

The maximum preload P_{min} to be applied to the bolts should equal or exceed the following three values:

$$P_{min} = 0.25F_{UT} \text{ lbf} \quad (10\text{-}9)$$

where F_{UT} is the ultimate tensile strength of the bolt (psi);

$$P_{min} = 1.15 \max F_T \text{ lbf} \quad (10\text{-}10)$$

where max F_T is maximum tensile working load on bolt (psi);

$$P_{min} = \max F_S/\mu \text{ lbf} \quad (10\text{-}11)$$

where max F_S is the maximum shear load on bolt (psi) and μ is the worst-case (lowest) coefficient of friction between housing and mounting surface, and equals 0.5 for an aluminum–aluminum interface.

Equation (10-9) is a rule-of-thumb minimum preload to prevent loosening of the bolt in service. Equation (10-10) provides a safety factor of 1.15 to prevent joint separation under the maximum tensile working load. Equation (10-11) provides enough friction to prevent relative transverse motion between the housing and the mounting surface during vibration.

Under maximum working load, the total tensile load in the bolt (Section 10.2) is

$$F_T(\max) = P + F_T R/(1 - R) \text{ lbf} \quad (10\text{-}12)$$

where R = spring constant of joint, Section 10.2
P = preload on bolt (lbf)
F_T = maximum tensile working load on bolt (lbf)

The maximum tensile stress in the bolt is

$$\sigma = F_T(\max)/A_T \text{ psi} \quad (10\text{-}13)$$

where $F_T(\max)$ is obtained from Eq. (10-12) and A_T is the basic minor area of bolt (in.^{2}). The maximum shear stress in the bolt is

$$\tau = F_S/KA_S \text{ psi} \tag{10-14}$$

where F_S is obtained from Eq. (10-2).
A_S = basic shank area of bolt (in.^{2})
K = shear coefficient = $6(1 + \nu)/(7 + 6\nu)$ for a circular bolt section = 0.884
ν = Poisson's ratio = 0.27 for a stainless steel bolt

The principal stresses in the bolt are

$$\sigma_1 = \sigma/2 + \left(\tau^2 + \sigma^2/4\right)^{1/2} \text{ psi} \tag{10-15}$$

and

$$\sigma_2 = \sigma_3 = \sigma/2 - \left(\tau^2 + \sigma^2/4\right)^{1/2} \text{ psi} \tag{10-16}$$

and the maximum shear stress in the screw is

$$\tau_{\max} = \left(\tau^2 + \sigma^2/4\right)^{1/2} \text{ psi} \tag{10-17}$$

The margins of safety for ultimate strength are

$$\text{MS}_T = \sigma_{UT}/\sigma_1 \text{FS}_U - 1 \tag{10-18}$$

in tension and

$$\text{MS}_S = \tau_U/\tau_{\max} \text{FS}_U - 1 \tag{10-19}$$

in shear, where FS_U is the safety factor for ultimate strength. The margin of safety need only be equal to or greater than zero when expressed in the above form.

Typical safety factors for metallic flight structures are shown in Table 10-1.[1]

Note that even if the bolt preload P is greater than the value of $P_{\min}$ given by Eq. (10-11), it is still advisable to require $\text{MS}_S \geq 0$ in Eq. (10-19).

TABLE 10-1 Quasistatic and Random Vibration Safety Factors FS for Metallic Flight Structures

Category	Yield	Ultimate
Nontested structures	1.25	2.0
Tested structures	1.1	1.4

10.2 PRELOADING OF THREADED FASTENERS

When threaded fasteners, for example, bolts or screws, are used to connect a housing to a mounting surface, they are tightened to subject the fasteners to tensile forces along their longitudinal axes. The extent to which a fastener is tightened is expressed in terms of the preload P, but is controlled and is usually measured in terms of the tightening torque T. The preload is required to keep the joint from separating when subjected to working loads, which would lead to more rapid fatigue and failure. Under working loads, the total load on the bolt is[2]

$$F = P + WR/(1 + R) \text{ lbf} \tag{10-20}$$

where F = total load on bolt (lbf)
P = preload (lbf)
W = working load (lbf)
R = spring constant, or joint stiffness

The spring constant R may be expressed as

$$R = AE(\text{bolt})/AE(\text{plate}) \tag{10-21}$$

where AE = (effective load bearing area) × (modulus of elasticity of material)
For the bolt,

$$AE(\text{bolt}) = (\pi/4)(D)^2 E_b \text{ lbf} \tag{10-22}$$

where D is the shank diameter of bolt (in.) and E_b is the modulus of elasticity of bolt material (psi). For the plate, the effective load bearing area A_p is

$$A_p = (\pi/4)\left[(D_W + 0.1L)^2 - D_H^2\right] \text{ in.}^2 \tag{10-23}$$

where D_W = diameter of washer or effective diameter of bolt head (in.)
D_H = diameter of bolt hole (in.)
L = length of joint or thickness of plate(s) being compressed

and

$$\text{AE(plate)} = A_p E_p \text{ lbf} \tag{10-24}$$

where E_p is the modulus of elasticity of plate material (psi). Table 10-2 presents values of R for some steel bolts with steel, aluminum, and magnesium plates.

The maximum allowable total load on the bolt F_m will be determined by the bolt ultimate tensile strength and the required factor of safety:

$$F_m = F_U/\text{SF lbf} \tag{10-25}$$

TABLE 10-2 *R* Values for Steel Bolts in Various Plates

Bolt Type[a]	Bolt Shank Diameter (in.)	Bolt Head Diameter (in.)	*R* Values[e]		
			Steel Plate[b]	Aluminum Plate[c]	Magnesium Plate[d]
NAS-03	0.189	0.375	0.24	0.94	1.51
04	0.249	0.438	0.45	1.25	2.03
05	0.312	0.500	0.62	1.70	2.76
06	0.374	0.563	0.78	2.18	3.50
07	0.437	0.625	0.94	2.57	4.18
08	0.499	0.750	0.77	2.13	3.45
09	0.562	0.875	0.67	1.84	2.98
10	0.624	0.938	0.77	2.13	3.45
12	0.749	1.063	0.95	2.62	4.23
14	0.874	1.250	0.92	2.54	4.10
16	0.999	1.438	0.90	2.49	4.00
18	1.124	1.625	0.89	2.46	3.95
20	1.249	1.812	0.89	2.46	3.95
AN-03	0.189	0.335	0.45	1.25	2.02
04	0.249	0.398	0.62	1.70	2.75
05	0.312	0.460	0.84	2.32	3.74
06	0.374	0.523	1.06	2.92	4.73
07	0.437	0.585	1.21	3.33	5.40
08	0.499	0.710	0.95	2.60	4.20
09	0.562	0.835	0.77	2.14	3.44
10	0.624	0.898	0.88	2.42	3.90
12	0.749	1.023	1.12	3.10	5.00
14	0.874	1.210	1.04	2.84	4.60
16	0.999	1.398	1.03	2.81	4.54
18	1.124	1.585	0.97	2.67	4.30
20	1.249	1.772	0.90	2.48	4.00

Source: Ref. 2

[a] $E = 28 \times 10^6$ psi for bolt material.
[b] $E = 29.5 \times 10^6$ psi for steel plate.
[c] $E = 10.5 \times 10^6$ psi for aluminum plate.
[d] $E = 6.5 \times 10^6$ psi for magnesium plate.
[e] R values without washers and for a zero-degree cone angle in the flange.

where F_U is the bolt ultimate tensile strength (lbf) and SF is the factor of safety. See Table 10-1 for values of safety factors. Ultimate tensile strength can be found from

$$F_U = \sigma_T A \text{ lbf} \tag{10-26}$$

where σ_T is the ultimate tensile strength of bolt material (psi) and A is the basic minor area of bolt (in.^{2}). Values of A are provided in Table 10-3.

From Eqs. (10-20)–(10-26), the maximum preload that can be allowed and still meet the factor of safety requirements is

$$P_m = \sigma_T A/\text{SF} - WR/(1 + R) \text{ lbf} \tag{10-27}$$

TABLE 10-3 Tensile Areas A for Bolts

Diameter (in.)	Size	A (in.^{2})
0.112	4–40	0.0050896
0.138	6–32	0.0076821
0.164	8–32	0.012233
0.190	10–32	0.018074
0.250	1/4–28	0.033394
0.312	5/16–24	0.053666
0.375	3/8–24	0.082397
0.438	7/16–20	0.11115
0.500	1/2–20	0.15116
0.562	9/16–18	0.19190
0.625	5/8–18	0.24349
0.750	3/4–16	0.35605
0.875	7/8–14	0.48695
1.000	1–12	0.63307
1.125	1-1/8–12	0.82162
1.250	1-1/4–12	1.0347
1.375	1-3/8–12	1.2724
1.500	1-1/2–12	1.5345

where

$$R = (E_b/E_p)D^2/\left[(D_W + 0.1L)^2 - D_H^2\right] \tag{10-28}$$

The ratio of bolt working load to preload W/P is an important design factor. As this ratio increases toward unity, the joint becomes more efficient but the margin of error in the estimate of P decreases and accurate determination of P becomes increasingly important to avoid separation of the joint.

The tightening torque T is[3]

$$T = \tfrac{1}{2}P\left[1.25D\mu_N + D_p \tan(\beta + \phi)/\cos\alpha\right] \text{ in.} \cdot \text{lbf} \tag{10-29}$$

where μ_N = coefficient of friction between nut and washer
D_p = pitch diameter of the bolt threads (in.)
β = helix angle of thread = $\sin^{-1}(1/\pi n D_p)$
n = number of threads per inch
$\phi = \tan^{-1}\mu_T$
μ_T = coefficient of friction between threads
$\alpha = \frac{1}{2}$ of thread profile angle = 30° for basic thread form

Table 10-4 presents values of D and D_p for various bolt sizes.

TABLE 10-4 Shank Diameter and Pitch Diameter for Bolts

Size	Shank Diameter D (in.)	Pitch Diameter D_p (in.)
2–56	0.0860	0.0744
4–40	0.1120	0.0958
6–32	0.1380	0.1177
8–32	0.1640	0.1437
10–24	0.1900	0.1629
12–24	0.2160	0.1889
1/4–20	0.2500	0.2175
5/16–18	0.3125	0.2764
3/8–16	0.3750	0.3344
7/16–14	0.4375	0.3911
1/2–13	0.5000	0.4500
9/16–12	0.5625	0.5084
5/8–11	0.6250	0.5660
3/4–10	0.7500	0.6850
7/8–9	0.8750	0.8028
1–8	1.0000	0.9188
1-1/8–7	1.1250	1.0322
1-1/4–7	1.2500	1.1572
1-3/8–6	1.3750	1.2667
1-1/2–6	1.5000	1.3917
1-3/4–5	1.7500	1.6201

For steel bolts, typical values of the friction coefficients are $\mu_N \simeq 0.14$ for nut friction and $\mu_T \simeq 0.12$ for thread friction.

Torqueing of the bolt is one of the largest sources of error in attempting to control the preload. Typical accuracy ranges for various types of torqueing tools are presented in Table 10-5.

For steel bolts, the following equations provide an approximate estimate of the torque required to obtain a specified preload:

For #4, 6, 8, and 10 bolt sizes,

$$T = P/(50 - 2.5N) \text{ in.} \cdot \text{lbf} \tag{10-30}$$

where T = tightening torque (in.· lbf)
P = preload (lbf)
N = bolt number = 4, 6, 8, or 10

For $\frac{1}{4}$ to $\frac{7}{16}$ in. diameter bolts,

$$T = P/(33 - 48D) \text{ in.} \cdot \text{lbf} \tag{10-31}$$

TABLE 10-5 Accuracy Ranges for Torqueing Tools

Type of Tool	Accuracy of Range (Percentage of Full Scale)
Impact wrench	10–30
Gearhead air-powered wrench	10–20
Mechanical multiplier	5–20
Bar torque wrench	3–15
Hydraulic wrench	3–10
Worm-gear torque wrench	0.25–5
Digital torque wrench	0.25–1

Source: Ref. 4

where D is the bolt diameter (in.).

$$\begin{aligned} &\text{For } \tfrac{1}{2} \text{ in. bolts, } T = 0.086P \text{ in.} \cdot \text{lbf} \\ &\text{For } \tfrac{5}{8} \text{ in. bolts, } T = 0.107P \text{ in.} \cdot \text{lbf} \\ &\text{For } \tfrac{3}{4} \text{ in. bolts, } T = 0.127P \text{ in.} \cdot \text{lbf} \\ &\text{For } \tfrac{7}{8} \text{ in. bolts, } T = 0.147P \text{ in.} \cdot \text{lbf} \\ &\text{For 1 in. bolts, } T = 0.167P \text{ in.} \cdot \text{lbf} \end{aligned} \tag{10-32}$$

Table 10-6 provides safe maximum torque values for nonlubricated bolts of various materials, with the exception of nylon where the breaking torque is presented.

10.3 BUCKLING OF HOUSING WALLS

With reference to Fig. 10.1, assume acceleration in the $\pm X$ direction, with the entire overturning moment M of Eq. (10-3) supported by the front and back walls, which are placed in shear as shown in Fig. 10.3. Then the shear stress on each wall is[5]

$$\tau = M/2act \text{ psi} \tag{10-33}$$

where t is the wall thickness (in.). The critical buckling stress due to shear is, from Table 9-3,

$$\tau_b = KE(t/c)^2/(1 - \nu^2)\text{psi} \tag{10-34}$$

where E is Young's modulus of wall material, ν is Poisson's ratio for wall material, and K is a function of a/c given in Table 9-3. The edge conditions on the walls allow the appropriate selection from Cases 7 and 8 of Table 9-3. The buckling stress ratio for shear is

$$R_S = \tau/\tau_b \tag{10-35}$$

TABLE 10-6 Torque Values (in. · lbf) for Various Bolt Materials[a]

Bolt		Stainless Steel		Silicon		Aluminum	
Size	Monel	316	18–8	Bronze	Brass	2024-T4	Nylon[b]
2–56	2.5	2.6	2.5	2.3	2.0	1.4	0.44
2–64	3.1	3.2	3.0	2.8	2.5	1.7	—
3–48	4.0	4.0	3.9	3.6	3.2	2.1	—
3–56	4.5	4.6	4.4	4.1	3.6	2.4	—
4–40	5.3	5.5	5.2	4.8	4.3	2.9	1.19
4–48	6.7	6.9	6.6	6.1	5.4	3.6	—
5–40	7.8	8.1	7.7	7.1	6.3	4.2	—
5–44	9.6	9.8	9.4	8.7	7.7	5.1	—
6–32	9.8	10.1	9.6	8.9	7.9	5.3	2.14
6–40	12.3	12.7	12.1	11.2	9.9	6.6	—
8–32	20.2	20.7	19.8	18.4	16.2	10.8	4.3
8–36	22.4	23.0	22.0	20.4	18.0	12.0	—
10–24	25.9	23.8	22.8	21.2	18.6	13.8	6.61
10–32	34.9	33.1	31.7	29.3	25.9	19.2	8.2
1/4–20	85.3	78.8	75.2	68.8	61.5	45.6	16.0
1/4–28	106	99.0	94.0	87.0	77.0	57.0	20.8
5/16–18	149	138	132	123	107	80.0	34.9
5/16–24	160	147	142	131	116	86.0	—
3/8–16	266	247	236	219	192	143	—
3/8–24	294	271	259	240	212	157	—
7/16–14	427	393	376	249	317	228	—
7/16–20	451	418	400	371	327	242	—
1/2–13	584	542	517	480	422	313	—
1/2–20	613	565	541	502	443	328	—
9/16–12	774	713	682	632	558	413	—
9/16–18	855	787	752	697	615	456	—
5/8–11	1330	1160	1110	1030	907	715	—
5/8–18	1480	1300	1240	1150	1020	798	—
3/4–10	1830	1580	1530	1420	1250	980	—
3/4–16	1790	1560	1490	1380	1220	958	—
7/8–9	2770	2430	2330	2140	1900	1490	—
7/8–14	2750	2420	2320	2130	1890	1490	—
1–8	4130	3595	3440	3180	2810	2200	—
1–14	3730	3250	3110	2880	2540	1990	—
1-1/8–7	5990	5180	4960	4600	4040	3180	—
1-1/8–12	5640	4900	4680	4330	3820	3010	—
1-1/4–7	7520	6550	6280	5820	5140	4030	—
1-1/4–12	6900	6050	5760	5360	4730	3700	—
1-1/2–6	12800	11200	10700	9870	8720	6840	—
1-1/2–12	10100	8780	8440	7810	6900	5400	—

[a] Values are for nonlubricated bolts.

[b] Safe maximum torque values are listed for all materials except nylon, where the breaking torque is given.

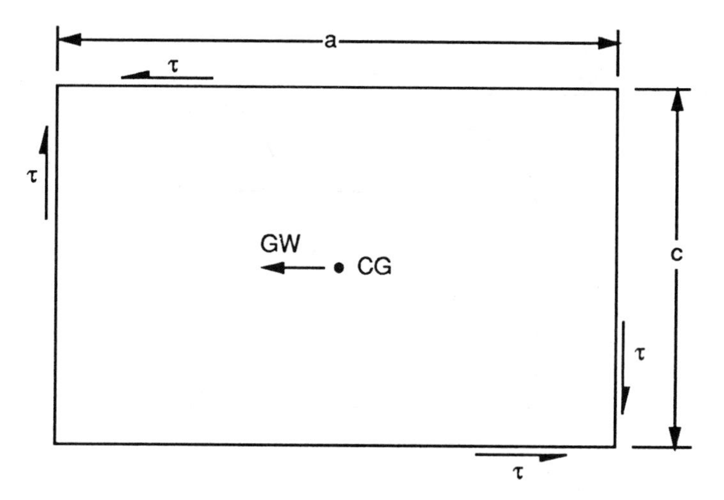

FIGURE 10.3 Front and back walls in shear.

and the margin of safety is

$$MS_S = 1/R_S - 1 = (\tau_b/\tau) - 1 \tag{10-36}$$

Next, consider a rectangular-shaped housing that is bolted to a mounting surface at the two ends of its longest dimension, Fig. 10.4. When acceleration loads act along the X direction, the housing will bend in the direction of the load and will also rotate around its mounting axis as shown in Fig. 10.5. From Table 4-1, Case 8, the maximum material stress due to bending of the side plate is

$$\sigma = GwL^2c/8I \text{ psi} \tag{10-37}$$

which occurs at midspan, $Z = L/2$,

where G = acceleration along the X axis, in units of $g = 386$ in./sec^2

w = weight per unit length of housing = W/L (lbf/in.)

W = total weight of housing (lbf)

L = length of housing (in.)

c = distance from neutral axis of housing (parallel to Z axis) to outside surface of side plate (in.)

I = area moment of inertia of housing about neutral axis (in.^{4})

The critical buckling stress for a compressive bending load on the side plate is, from Table 9-3,

$$\sigma_b = KE(t_2/h')^2/(1 - \nu^2) \text{ psi} \tag{10-38}$$

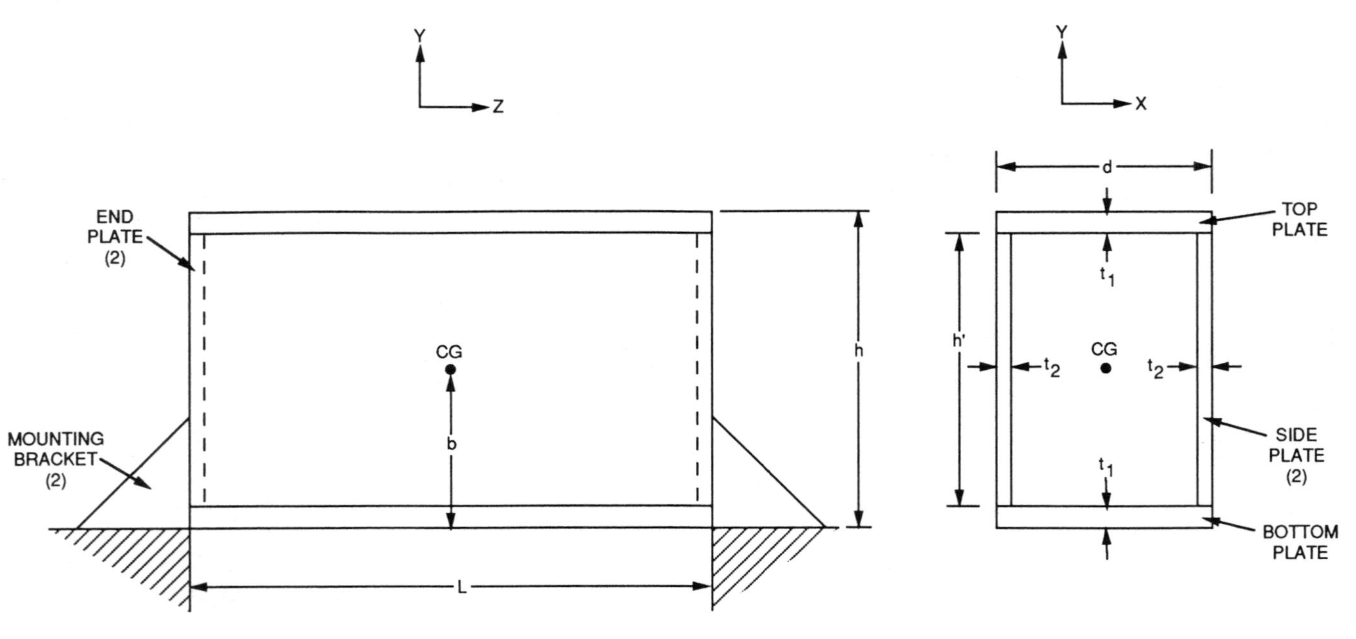

FIGURE 10.4 Rectangular housing supported at ends.

where h' = height of side plates = $(h - 2t_1)$ (in.)
t_2 = thickness of side plates (in.)
t_1 = thickness of top and bottom plates (in.)
K = function of (L/h'), given in Table 9-3
E = Young's modulus of side plate material (psi)
ν = Poisson's ratio for side plate material

The manner in which the side plates are connected to the bottom plate, the top plate, and the end plates determines the edge conditions (clamped, simply supported, or free) and allows the appropriate selection from Cases 1–6 of Table 9-3, and the determination of K based on the ratio L/h'. The buckling stress ratio in bending is then

$$R_b = \sigma/\sigma_b \tag{10-39}$$

The maximum material stress due to shear in the side plate, caused by torsion of the housing as shown in Fig. 10.5, can be found from the shear flow equation.[6] Assuming the top and bottom plates have equal effective stiffness

$$k_1 = dt_1\eta_1 \text{ in.}^2 \tag{10-40}$$

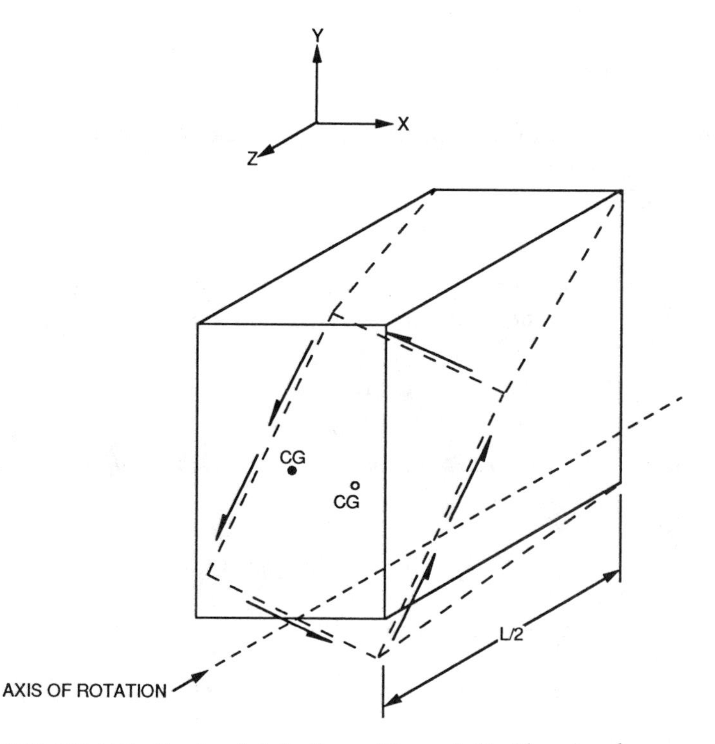

FIGURE 10.5 Torsional deflection, with one-half of housing shown.

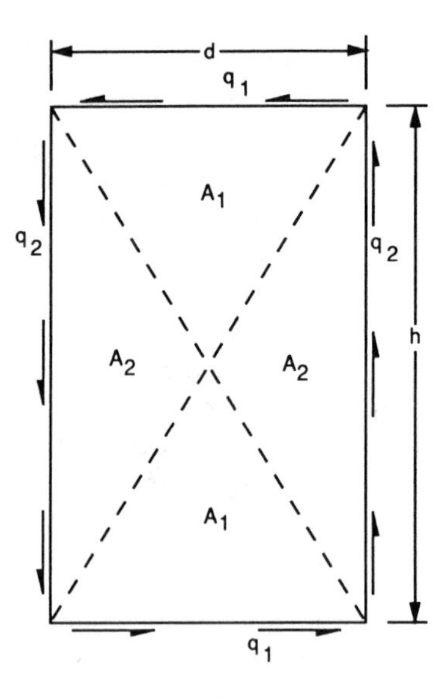

FIGURE 10.6 Shear flow and shear center in housing.

and the two side plates have equal effective stiffness

$$k_2 = h't_2\eta_2 \text{ in.}^2 \tag{10-41}$$

the shear center will be at the center of the housing cross section, due to symmetry, as shown by the intersection of the two dashed lines in Fig. 10.6. η_1 and η_2 are the attachment efficiency factors (see Section 3.5). The torque causing torsion and shear is

$$T = Gwb \text{ in.} \cdot \text{lbf} \tag{10-42}$$

where b is the height of the CG above axis of rotation (in.) and G and W have been previously defined. The shear flow equation is

$$T = 2\sum_i A_i q_i / \text{in.} \cdot \text{lbf} \tag{10-43}$$

Since $A_1 = \frac{1}{2}(h/2)d = hd/4$ in.^{2} and $A_2 = \frac{1}{2}(d/2)h = hd/4$ in.^{2}, Eqs. (10-42) and (10-43) may be combined to yield

$$(q_1 + q_2) = GWb/hd \text{ lbf/in.} \tag{10-44}$$

The shear flow in each plate is proportional to shear stiffness:

$$q_1/q_2 = k_1/k_2 = dt_1\eta_1/h't_2\eta_2 \tag{10-45}$$

For welded joints between plates, $\eta = 1.0$; for bolted joints, $\eta = 0.25$–0.50.

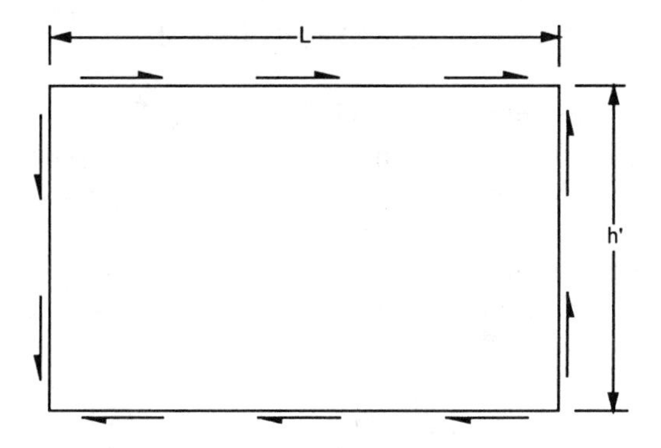

FIGURE 10.7 Shear flow in side plate.

Equations (10-40), (10-41), (10-44), and (10-45) yield

$$q_1 = (GWb/hd)/(1 + h't_2\eta_2/dt_1\eta_1) \text{ lbf/in.} \quad (10\text{-}46)$$

and

$$q_2 = (GWb/hd)/(1 + dt_1\eta_1/h't_2\eta_2) \text{ lbf/in.} \quad (10\text{-}47)$$

The shear stress in the side plate is

$$\tau = q_2/t_2 \text{ psi} \quad (10\text{-}48)$$

The critical buckling stress due to shear in the side plate (Fig. 10.7) is

$$\tau_c = KE(t_2/h')^2/(1 - \nu^2) \text{ psi} \quad (10\text{-}49)$$

where K is a function of L/h' given in Table 9-3. The edge conditions on the side plate allow the appropriate selection from Cases 7 and 8 of Table 9-3, and the determination of the value of K based on the ratio L/h'. The buckling stress ratio in shear for the side plate is

$$R_s = \tau/\tau_c \quad (10\text{-}50)$$

The margin of safety for buckling, taking into account both the bending and shear stresses in the side plate, is

$$\text{MS} = 1/(R_b^2 + R_s^2)^{1/2} - 1 \quad (10\text{-}51)$$

10.4 CONTAINMENT OF FRACTURED INTERNAL COMPONENTS

If there are any internal parts that may be able to break loose in the acceleration environment, it is necessary that the housing have adequate strength to contain these parts and prevent rupture or penetration of a housing wall. Candidates for consideration as loose parts include components, such as transformers, that may be mounted with only one screw and do not have load path redundancy; and components that are mounted to circuit boards by means of lead wires which may suffer fatigue failure in a vibration environment. If there is more than one candidate for consideration, the worst case is the heaviest, with the smallest contact area on impact, and the farthest distance to travel (under acceleration) to the target wall. The worst-case target wall is the largest, thinnest wall. However, these guidelines may not be readily applicable, and several different cases may require evaluation. In the following it is assumed that a quasistatic acceleration load causes the loose part to move relative to a housing wall for a distance X under acceleration G, expressed in units of $g = 386$ in./sec^2.

Assume that a part having a mass of W lbf becomes loose and is accelerated under a quasistatic acceleration load factor G a distance of X in. For impact at the center (worst-case location) of a rectangular housing wall the following cases may be applied.[7]

1. Concentric, circular, uniformly loaded impact area of radius $R_0 \ll b$. All four edges of housing wall simply supported. Max σ at center. Max y at center. At center,

$$\sigma_0 = (3W/2\pi t^2)[(1 + \nu)\ln(2b/\pi R) + \beta] \text{ psi} \qquad (10\text{-}52)$$

and

$$y_0 = \alpha W b^2/Et^3 \text{ in.} \qquad (10\text{-}53)$$

where σ_0 and y_0 are stress and deflection under 1-g acceleration normal to plate and

W = weight of impacting object under 1-g acceleration (lbf)
t = housing wall thickness (in.)
ν = Poisson's ratio for housing wall material
b = short dimension of housing wall (in.)
a = long dimension of housing wall (in.)
$R = (1.6R_0^2 + t^2)^{1/2} - 0.675t$ for $R_0 < 0.5t$; $R = R_0 > 0.5t$ (in.)
R_0 = actual radius of impact area (in.)
E = Young's modulus for housing wall material (psi)
α = function of the ratio a/b, Table 10-7
β = function of the ratio a/b, Table 10-7

TABLE 10-7 Values of α and β for Simply Supported, Rectangular Wall, Case 1. ($\nu = 0.3$)

	a/b						
	1.0	1.2	1.4	1.6	1.8	2.0	∞
β	0.435	0.650	0.789	0.875	0.927	0.958	1.000
α	0.1267	0.1478	0.1621	0.1715	0.1770	0.1805	0.1851

TABLE 10-8 Values of α, β, and γ for Fixed, Rectangular Wall, Case 2. ($\nu = 0.3$)

	a/b						
	1.0	1.2	1.4	1.6	1.8	2.0	∞
β	−0.238	−0.078	0.011	0.053	0.068	0.067	0.067
γ	0.7542	0.8940	0.9624	0.9906	1.0000	1.004	1.008
α	0.0611	0.0706	0.0754	0.0777	0.0786	0.0788	0.0791

2. Concentric, circular, uniformly loaded impact area of radius $R_0 \ll b$. All four edges of housing wall fixed. Max y at center. At center, Eqs. (10-52) and (10-53) apply with Table 10-8 for determination of α and β. At center of long edge,

$$\sigma_0 = \gamma W/t^2 \text{ psi} \tag{10-54}$$

with Table 10-8 for determination of γ.

3. Concentric, rectangular, uniformly loaded impact area. All four edges of housing wall simply supported. Max σ at center. At center,

$$\sigma_0 = \beta W/t^2 \text{ psi} \tag{10-55}$$

where β is a function of b_1/b, a_1/b, and a/b, Table 10-9, and

b_1 = short dimension of impact area (in.)
a_1 = long dimension of impact area (in.)
b = short dimension of housing wall (in.)
a = long dimension of housing wall (in.)

For impact at the center (worst-case location) of a circular housing wall, the following cases may be applied.[8]

4. Concentric, uniformly loaded impact area of radius $R_0 \ll a$. Edge of housing wall simply supported. Max σ at center, max Y at center. At center,

$$\sigma_0 = (3W/2\pi t^2)[(1+\nu)\ln(a/R) + 1] \text{ psi} \tag{10-56}$$

TABLE 10-9 Values of β for Simply Supported, Rectangular Wall, Case 3 ($\nu = 0.3$)

	$a = b$						$a = 1.4b$						$a = 2b$					
	a_1/b						a_1/b						a_1/b					
b_1/b	0	0.2	0.4	0.6	0.8	1.0	0	0.2	0.4	0.8	1.2	1.4	0	0.4	0.8	1.2	1.6	2.0
0	—	1.82	1.38	1.12	0.93	0.76	—	2.0	1.55	1.12	0.84	0.75	—	1.64	1.20	0.97	0.78	0.64
0.2	1.82	1.28	1.08	0.90	0.76	0.63	1.78	1.43	1.23	0.95	0.74	0.64	1.73	1.31	1.03	0.84	0.68	0.57
0.4	1.39	1.07	0.84	0.72	0.62	0.52	1.39	1.13	1.00	0.80	0.62	0.55	1.32	1.08	0.88	0.74	0.60	0.50
0.6	1.12	0.90	0.72	0.60	0.52	0.43	1.10	0.91	0.82	0.68	0.53	0.47	1.04	0.90	0.76	0.64	0.54	0.44
0.8	0.92	0.76	0.62	0.51	0.42	0.36	0.90	0.76	0.68	0.57	0.45	0.40	0.87	0.76	0.63	0.54	0.44	0.38
1.0	0.76	0.63	0.52	0.42	0.35	0.30	0.75	0.62	0.57	0.47	0.38	0.33	0.71	0.61	0.53	0.45	0.38	0.30

and

$$y_0 = \left[Wa^2/16\pi D\right]\left[(3 + \nu)/(1 + \nu)\right] \text{ in.} \tag{10-57}$$

where a is the radius of housing wall (in.) $D = Et^3/12(1 - \nu^2)$, and other quantities are defined following Eq. (10-53).

5. Concentric, uniformly loaded impact area of radius $R_0 \ll a$. Edge of housing wall fixed. Max σ at center, max y at center. At center,

$$\sigma_0 = (3W/2\pi t^2)(1 + \nu)\ln(a/R) \text{ psi} \tag{10-58}$$

and

$$y_0 = Wa^2/16\pi D \text{ in.} \tag{10-59}$$

The stress and deflection of the housing wall due to the mass W being accelerated a distance of X in. under an acceleration of G is[9]

$$\sigma = \sigma_0(2GX/y_0)^{1/2} \text{ psi} \tag{10-60}$$

and

$$y = (2GXy_0)^{1/2} \text{ in.} \tag{10-61}$$

where σ_0 and y_0 are the values obtained in the preceding work, G is acceleration in units of $g = 386$ in./sec^2, and X is the distance traveled under quasistatic acceleration G prior to impact.

If the mass W is initially in contact with the wall and the quasistatic acceleration G is suddenly applied, the stress and deflection of the housing wall is

$$\sigma = 2G\sigma_0 \text{ psi} \tag{10-62}$$

and

$$y = 2Gy_0 \text{ in.} \tag{10-63}$$

The margin of safety is

$$\text{MS}_\text{v} = \sigma_{\text{UT}}/\sigma \text{FS}_\text{U} - 1 \tag{10-64}$$

where σ_{UT} is the ultimate tensile strength of wall material (psi).

REFERENCES

1 *NASA, Marshal Space Flight Center*, "Flight Equipment Requirements for Safety Critical Structures," *JA*-418, 29 October 1986.

2 C. C. Osgood, "How Elasticity Influences Bolted-Joint Design," *Machine Design*, February 24, 1972, pp. 92–95 and March 9, 1972, pp. 104–107.

3 F. Yeaple, "Bolt Torque Equations Predict Stresses," *Product Engineering*, October 1978, pp. 28–33.

4 *Machine Design*, November 13, 1986, pp. 24–26.

5 R. J. Roark and W. C. Young, *Formulas For Stress and Strain*, 5th Ed., New York: McGraw Hill, 1975, p. 82.

6 D. H. Allen and W. E. Haisler, *Introduction To Aerospace Structural Analysis*, New York: Wiley, 1985.

7 R. J. Roark and W. C. Young, in Ref. 5, Table 26.

8 R. J. Roark and W. C. Young, in Ref. 5, Table 24.

9 R. J. Roark and W. C. Young, in Ref. 5, p. 575.

11

VENTING

When space vehicles are launched from earth, their rapid ascent results in a reduction of ambient air pressure from 1 atm to essentially zero pressure during time periods on the order of 1 min. It is important to design a subsystem so that the air inside the housing may be vented (may escape) rapidly enough to avoid excessive pressure differences between the internal air pressure in the housing and the lower external air pressure. If this pressure difference becomes too large, it could structurally damage or rupture the housing, which is at the same time being subjected to shock and vibration loads due to launch and separation. The maximum pressure differential supported by the housing may be constrained to remain within safe structural design limits by proper sizing of vents that allow air to escape from the housing.

11.1 VENT AREA

During ascent, the air pressure, P_2, external to the housing (downstream of the vents) is continuously decreasing. Figure 11.1 shows a typical ambient air pressure profile for a launch. The rate of change of the air pressure P_1 inside the housing (upstream of the vent) will be bounded by two theoretical extremes: the pressure values based on the assumption of isentropic (reversible adiabatic) flow out of the housing where the air inside the housing exchanges no heat with its surroundings[1-3] and the pressure values based on the assumption of reversible isothermal flow out of the housing, where the air inside the housing maintains a constant temperature. Both of these cases are

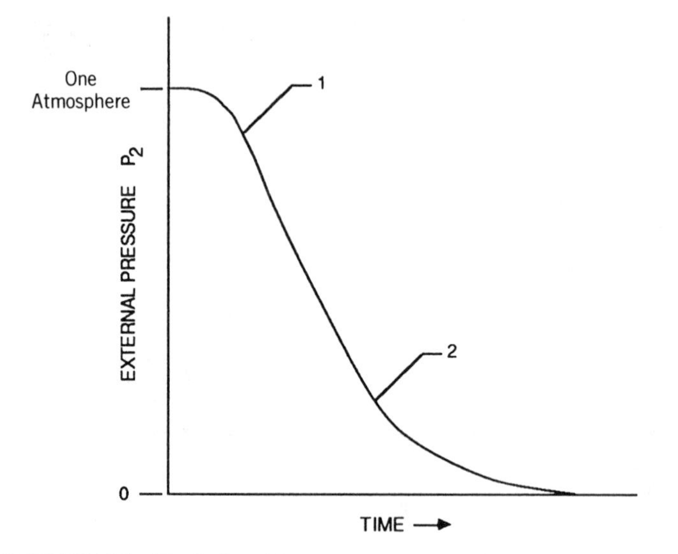

FIGURE 11.1 Typical ambient pressure profile for a launch vehicle.

presented here. However, the isothermal case is recommended for vent design for two reasons: it is more conservative in that for a specified maximum allowable pressure differential it requires larger vent capacity than the isentropic case and for most subsystems of interest the housing and its internal components have a large surface area compared to the contained air volume, allowing ready heat transfer with the air and tending to keep the air at a constant temperature.

For isothermal venting, the rate of change of air pressure inside the housing, for subcritical flow, is given by

$$\dot{P}_1 = -(CAP_0/\rho_0 V_0)(P_2/P_1)[2\rho_1 P_1 \ln(P_1/P_2)]^{1/2} \text{ psi/sec} \quad (11\text{-}1a)$$

or

$$\dot{P}_1 = -P_2(CA/V_0)[2(P_0/\rho_0)\ln(P_1/P_2)]^{1/2} \text{ psi/sec} \quad (11\text{-}1b)$$

where $\dot{P}_1 = dP_1/dt$ (psi/sec)
$P_1 = P_1(t)$ = air pressure inside housing at time t (psi)
$P_2 = P_2(t)$ = air pressure external to housing at time t (psi)
P_0 = initial air pressure inside housing at time $t = 0$ (psi)
ρ_0 = initial air density inside housing at time $t = 0$ (lbm/in.^{3})
$\rho_1 = \rho_1(t)$ = air density inside housing at time t (lbm/in.^{3})
V_0 = volume available for air inside housing (in.^{3})
C = vent discharge coefficient, dimensionless
A = vent area (in.^{2})

Equations (11-1) apply when the flow is subcritical:

$$(P_2/P_1)_{crit} = [2/(\gamma + 1)]^{\gamma/(\gamma-1)} < P_2/P_1 \leq 1 \qquad (11\text{-}2)$$

where $\gamma = C_P/C_V$ = ratio of specific heat at constant pressure to specific heat at constant volume, dimensionless. For air, $\gamma = 1.4$ and $(P_2/P_1)_{crit} = 0.52828$.

When the flow is critical,

$$(P_2/P_1) \leq (P_2/P_1)_{crit} \qquad (11\text{-}3)$$

and Eq. (11-1a) becomes, for critical flow,

$$\dot{P}_1 = -(CAP_0/\rho_0 V_0)(P_2/P_1)_{crit}[2\rho_1 P_1 \ln(P_1/P_2)_{crit}]^{1/2} \text{ psi/sec} \qquad (11\text{-}4)$$

For air in critical flow,

$$\dot{P}_1 = -0.42201(CAP_0/\rho_0 V_0)(2\rho_1 P_1)^{1/2} \qquad (11\text{-}5)$$

For isentropic venting, the rate of change of air pressure inside the housing is given, for subcritical flow, by

$$\dot{P}_1 = -\left(CA\gamma P_0^{1/\gamma}/\rho_0 V_0\right) P_1^{(\gamma-1)/\gamma} \times \left\{[2\gamma\rho_1 P_1/(\gamma - 1)]\left[(P_2/P_1)^{2/\gamma} - (P_2/P_1)^{(\gamma+1)/\gamma}\right]\right\}^{1/2} \text{ psi/sec} \qquad (11\text{-}6)$$

where all the quantities have been previously defined and $\gamma = 1.4$ for air. Equations (11-2) and (11-3) apply to Eq. (11-6) for the subcritical and critical flows, with $(P_2/P_1)_{crit}$ replacing P_2/P_1 for critical flow. For air, with $(P_2/P_1)_{crit} = 0.52828$,

$$\dot{P}_1 = -0.958625\left(CAP_0^{1/\gamma}/\rho_0 V_0\right) P_1^{(\gamma-1)/\gamma}(\rho_1 P_1)^{1/2} \text{ psi/sec} \qquad (11\text{-}7)$$

for critical flow.

Equations (11-1), (11-5), (11-6), and (11-7) are differential equations in the dependent variable P_1 for subcritical and critical flow of air out of the housing for the isothermal and isentropic cases, respectively. The use of subcritical or critical flow equations are governed by the logic statements of Eqs. (11-2) and (11-3) based on the instantaneous pressure ratio P_2/P_1. Since $P_2(t)$ is specified and all of the other quantities in these equations are specified, the values of $P_1(t)$ may be found by numerical integration, yielding the pressure differential as a function of time:

$$\Delta P(t) = P_1(t) - P_2(t) \text{ psi} \qquad (11\text{-}8)$$

In many launch situations, the external-pressure-versus-time curve of Fig. 11.1 is nearly linear and has its maximum rate of decrease $P_{2(\max)}$ from point 1 to point 2. In this case, an approximate value may be calculated for the required vent area A by setting

$$\dot{P}_1(t_2) = \dot{P}_{2(\max)} \text{ psi/sec} \tag{11-9}$$

and using

$$P_1(t_2) - P_2(t_2) = \Delta P_{\max} \text{ psi} \tag{11-10}$$

in Eq. (11-1b), where $\Delta P_{\max}$ is the maximum safe pressure differential that can be supported by the housing:

$$CA = -\left[aV_0/(2P_0/\rho_0)^{1/2}\right]\Big/P_2(t_2)\{\ln[1 + \Delta P_{\max}/P_2(t_2)]\}^{1/2} \text{ in.}^2 \tag{11-11}$$

where

$$a = -(P_1 - P_2)/(t_2 - t_1) \text{ psi/sec} \tag{11-12}$$

is the slope of the linear part of the curve in Fig. 11.1. In using Eq. (11-11), the constraint of Eqs. (11-3)–(11-5) must be observed. The required value of A may also be found by successive approximations.

For square-edged and sharp-edged orifices, the vent discharge coefficient C can vary from 0.60 to 0.84, while for nozzles $C = 0.95$. Porous metal plugs are often used to eliminate electromagnetic interference (EMI), which can occur with orifices. For these porous plugs there are manufacturers' data available on pressure drop versus air flow rate.

11.2 MAXIMUM SAFE PRESSURE DIFFERENTIAL

The maximum safe pressure differential $\Delta P_{\max}$ may be found by determining the margin between the maximum allowable loads and the actual loads on the housing due to shock and vibration, using the methods provided in preceding chapters. The pressure differential will result in an externally directed, uniformly distributed static load normal to every housing surface across which a pressure differential exists. The material stresses due to these loads may be calculated using Tables 4-6–4-8 of Chapter 4, where the maximum material stress σ_m due to a 1-g acceleration normal to a (flat-plate) housing wall is provided. In these tables, σ_m is directly proportional to the weight density of the wall material times the wall thickness:

$$\sigma_m = \text{constant} \cdot q = \text{constant} \cdot \rho t \text{ psi} \tag{11-13}$$

where q = 1-g load per unit area of wall (psi)
ρ = weight density of wall material (lbf/in.^{3})
t = wall thickness (in.)

Then the added material stress due to a pressure differential ΔP is given by

$$\sigma_p = (\Delta P) \cdot \sigma_m / q \text{ psi} \tag{11-14}$$

The maximum safe pressure differential is then

$$\Delta P_{max} = (\sigma_{allow} - \sigma_{pk}) q / \sigma_m \text{ psi} \tag{11-15}$$

where σ_{allow} is the maximum allowable wall material stress including safety factors (psi) and σ_{pk} is the peak wall material stress due to shock or vibration (psi).

11.3 PRESSURE-RESPONSE EQUATIONS

This section presents the pressure-response equations (time to reach a specified pressure in a single volume such as a tank) for perfect gases vented from a tank to a fixed lower-pressure external environment, and for perfect gases charged into a tank from a fixed higher-pressure external environment. The cases covered are isothermal and isentropic venting and charging for monatomic and diatomic gases, both subcritical and critical flows through the orifices. For large gas volumes and small surface-to-volume ratios such as found in tanks, the isentropic case will provide a more accurate analysis than the isothermal case.

The time required for the gas in the tank to go from an initial pressure P_i to a final pressure P_f is[1,4]

$$t = C(\tau_f - \tau_i) \text{ sec} \tag{11-16}$$

where t is the time (sec) for pressure change to occur and C and τ are specified in the following sections. The quantity C has the units of seconds, and the quantity τ is dimensionless. The subscript f indicates the final state and the subscript i the initial state.

11.3.1 Venting to a Fixed External Pressure

In this section, the external pressure P_2 is constant and the pressure inside the tank is $P_1 \geq P_2$. For a monatomic gas, $\gamma = \frac{5}{3}$, and the flow regimes are defined as follows [see Eq. (11-2)]:

$$\text{Subcritical flow:} \quad 0.487139 \leq P_2/P_1 \leq 1 \tag{11-17}$$

$$\text{Critical flow:} \quad 0 \leq P_2/P_1 \leq 0.487139 \tag{11-18}$$

For a diatomic gas, $\gamma = \frac{7}{5}$, and the flow regimes are

$$\text{Subcritical flow:} \quad 0.528282 \leq P_2/P_1 \leq 1 \tag{11-19}$$

$$\text{Critical flow:} \quad 0 \leq P_2/P_1 \leq 0.528282 \tag{11-20}$$

where P_2 is the constant pressure downstream of the vent and P_1 is the gas pressure in the tank, upstream of the vent. When both critical and subcritical flow occur, the critical flow time must be added to the subcritical flow time to obtain the total flow time.

Isothermal Venting

For isothermal venting,

$$C = V/C_1 C_D A_T R (T_1)^{1/2} \text{ sec} \tag{11-21}$$

where V = housing gas volume (in.^{3})
C_D = vent discharge coefficient
A_T = vent throat area (in.^{2})
R = gas constant = 7.16×10^6 in.^{2}/sec$^2 \cdot {}^\circ$R $\cdot$ (MW)
MW = molecular weight of the gas
T_1 = tank and gas temperature, constant (°R)

and

$$C_1 = (\gamma/R)^{1/2}[2/(\gamma+1)]^{\beta} \tag{11-22}$$

where $\beta = (\gamma+1)/2(\gamma-1)$
$\gamma = C_p/C_v$ = ratio of specific heat at constant pressure to specific heat at constant volume.

For a monatomic gas in subcritical flow,

$$\tau = 0.32246 - 0.54127\left[(P_2/P_1)^{-0.4} + 2\right]\left[(P_2/P_1)^{-0.4} - 1\right]^{0.5} \tag{11-23}$$

where the ratio P_2/P_1 meets the constraint of Eq. (11-17).
For a monatomic gas in critical flow,

$$\tau = \ln(P_2/P_1) \tag{11-24}$$

where the ratio P_2/P_1 meets the constraint of Eq. (11-18).

For a diatomic gas in subcritical flow, where Eq. (11-19) applies,

$$\tau = 0.31249 - 0.60388\left[3(P_2/P_1)^{-4/7} - 4(P_2/P_1)^{-2/7} + 4\right] \times\left[(P_2/P_1)^{-2/7} - 1\right] \quad (11\text{-}25)$$

For a diatomic gas in critical flow, where Eq. (11-20) applies,

$$\tau = \ln(P_2/P_1) \quad (11\text{-}26)$$

Isentropic Venting

For isentropic venting,

$$C = V/\gamma C_1 C_D A_T R(T_i)^{1/2}(P_2/P_i)^{\alpha} \text{ sec} \quad (11\text{-}27)$$

where T_i = initial tank gas temperature (°R)
P_i = initial tank gas pressure (psi)
$\alpha = (\gamma - 1)/2\gamma$

and the other quantities have been defined in Section 11.4.1.

For a monatomic gas in subcritical flow, where Eq. (11-17) applies,

$$\tau = 5.3175 - 0.81190\Big((P_2/P_1)^{-0.2}\left[(P_2/P_1)^{-0.4} - 1\right]^{0.5} + \ln\Big\{(P_2/P_1)^{-0.2} + \left[(P_2/P_1)^{-0.4} - 1\right]^{0.5}\Big\}\Big) \quad (11\text{-}28)$$

For a monatomic gas in critical flow, where Eq. (11-18) applies,

$$\tau = 5(P_2/P_1)^{0.2} \quad (11\text{-}29)$$

For a diatomic gas in subcritical flow, where Eq. (11-19) applies,

$$\tau = 7.0482 - 0.4529\Big(\left[(P_2/P_1)^{-2/7} + 1.5\right](P_2/P_1)^{-1/7}\left[(P_2/P_1)^{-2/7} - 1\right]^{0.5} + 0.75\ln\Big\{(P_2/P_1)^{-2/7} - 0.5 + (P_2/P_1)^{-1/7}\left[(P_2/P_1)^{-2/7} - 1\right]^{0.5}\Big\}\Big) \quad (11\text{-}30)$$

For a diatomic gas in critical flow, where Eq. (11-20) applies,

$$\tau = 7(P_2/P_1)^{1/7} \quad (11\text{-}31)$$

11.3.2 Charging from a Fixed External Pressure

In this section, the external pressure P_2 is constant and the pressure inside the tank is $P_1 \leq P_2$.

For isothermal charging,

$$C = V(T_2)^{1/2}/C_1CART_i \text{ sec} \tag{11-32}$$

where T_2 is the temperature (°R) of charging gas external to the tank and all of the other quantities have been previously defined.

For isentropic charging,

$$C = V/C_1C_DA\gamma R(T_2)^{1/2} \text{ sec} \tag{11-33}$$

For either isothermal or isentropic charging a perfect gas, the following equations apply:

Subcritical flow, monatomic gas:

$$\tau = 1.2990 - 1.6238\left[1 - (P_1/P_2)^{0.4}\right]^{1/2} \tag{11-34}$$

Subcritical flow, diatomic gas:

$$\tau = 1.2679 - 1.8116\left[1 - (P_1/P_2)^{2/7}\right]^{1/2} \tag{11-35}$$

Critical flow (any perfect gas):

$$\tau = P_1/P_2 \tag{11-36}$$

In using Eqs. (11-34)–(11-36), the logic Eqs. (11-17)–(11-20) must be applied, and when both critical and subcritical flows occur, the critical flow time must be added to the subcritical flow time to obtain a total flow time.

REFERENCES

1 *Aerospace Fluid Component Designer's Handbook*, Volume I, Revision D, February 1970, Sections 3.10.5 and 3.10.6, NTISAD-874542.

2 S. Eskinazi, *Principles of Fluid Mechanics*, Boston, MA: Allyn and Bacon, 1962.

3 C. E. Kirby and G. W. Ivey, "Determination of Flow Areas for Series-Parallel Compartment Venting to Satisfy Pressure Differential Requirements," AIAA Paper No. 72-707, AIAA 5th Fluid and Plasma Dynamics Conference, Boston, MA, June 26–28, 1972.

4 H. T. Yang, Formulas for venting or charging gas from a single volume, *AIAA J.* **24** (11) 1709–1711 (October, 1986).

12

THERMAL ANALYSIS

Various systems, including, for example, aerospace ones, have many components that will exhibit reduced reliability, reduced lifetime, malfunction, or failure if they are exposed to excessively high or low temperatures. These components usually have specified upper and lower temperature limits within which they can operate without performance or lifetime degradation, and a specified broader temperature range to which they can be exposed in the nonoperating condition without suffering any deleterious effects. These temperature limits require that the subsystem design configuration take into account the thermal requirements and that thermal analysis demonstrates conformance to the specified limits. In addition to the thermal requirements of components within the subsystem, in aerospace applications there are often temperature limits imposed on the external surface of the subsystem housing and on the heat flow (power) dumped into or drained out of the satellite heat sink when the subsystem may be directly exposed to solar insolation or directly radiating to space.

This chapter deals with the subsystem thermal analysis resulting from the preceding requirements. Only steady-state thermal analysis is addressed, in which the heat sources, temperatures, and heat flows are constant and independent of time. Worst-case steady-state thermal conditions usually impose the most severe constraints on subsystem thermal design, both in the maximum-heat-sink-temperature/maximum-subsystem-power operating condition and in the minimum-heat-sink-temperature/zero-subsystem-power nonoperating condition.

12.1 HEAT SOURCES AND SINKS

Heat sources within a subsystem are due to the heat generated within semiconductors, resistors, transformers, and other electrical and electromechanical elements when the subsystem is operating. Usually a "worst-case" analysis is done that results in an upper-limit estimate of power dissipation by each of the elements in the subsystem. Heat sources external to the subsystem in aerospace applications for example, include heat transfer by conduction and radiation from other systems, satellite structures, and satellite heat sinks that are at a higher temperature than the subsystem under analysis. If any of these external systems are at a lower temperature, they act as heat sinks. If any part of the subsystem has a direct view into space, it will radiate to space, which acts as a heat sink, or may be exposed to the sun, which will act as a heat source.

12.2 RADIATION

The power radiated into space by a surface is

$$Q = \varepsilon \sigma A T^4 \quad \mathrm{W} \tag{12-1}$$

where ε = emissivity = ratio of the radiation emitted by the surface to the radiation that would be emitted by a black body of the same area at the same temperature, dimensionless ($0 \leq \varepsilon \leq 1$)
σ = Stefan–Boltzmann constant [$= 3.66 \times 10^{-11}$ W/in.^{2} (K)^4]
A = surface area in.^{2}
T = absolute temperature (K)
W = watt

The power received by a surface exposed to the sun and orbiting the earth outside the earth's atmosphere is

$$Q = \alpha I A \cos\theta \quad \mathrm{W} \tag{12-2}$$

where α = absorptivity = ratio of the radiation absorbed by a surface to the radiation incident on the surface, dimensionless ($0 \leq \alpha \leq 1$)
I = solar constant = 0.873 W/in.^{2}, average value
A = surface area (in.^{2})
θ = angle of incidence, which is the angle between a direct solar ray and the normal to the surface

It should be noted that most solid materials of interest in aerospace applications are essentially opaque to thermal radiation of the wavelength range encountered in these applications, and for such materials the absorptivity is equal to the emissivity:

$$\alpha = \varepsilon \tag{12-3}$$

The net power received by surface 2 from surface 1 at higher temperature, when the surfaces are parallel and of equal area and the smallest surface dimension is much greater than the distance between the surfaces, is

$$Q = A\sigma\left(T_1^4 - T_2^4\right)/\left(1/\varepsilon_1 + 1/\varepsilon_2 - 1\right) \ \ \text{W} \tag{12-4}$$

where the subscripts refer to surfaces 1 and 2 and $T_1 > T_2$.

A radiative heat shield commonly used in aerospace applications is multilayer insulation (MLI), generally composed of a multiplicity of thin plastic metallized sheets of low emissivity, and configured to minimize heat flow by conduction. The power transmitted by radiation between the two outer most surfaces at temperatures T_1 and T_2 when all surfaces have the same emissivity is

$$Q = A\sigma\left(T_1^4 - T_2^4\right)/(2/\varepsilon - 1)(n + 1) \ \ \text{W} \tag{12-5}$$

where n is the number of intermediate sheets between the two outer layers.

Two surfaces may be separated enough that only part of the radiation leaving one surface impinges on the other surface. The angle factor* F_{12} is defined as the fraction of the radiant energy leaving surface 1 that impinges on surface 2. For diffuse radiation, where the radiant energy flux is independent of the direction of emission,

$$A_1 F_{12} = A_2 F_{21} \tag{12-6}$$

where A_1 is the area of surface 1 (in.^{2}) and A_2 is the area of surface 2 (in.^{2}).

If the energy leaving surface 1 impinges on other "surfaces," then

$$\sum_{i=1}^{n} F_{1i} = 1 \tag{12-7}$$

where a "surface" need not be a solid part of an enclosure, but can be an opening to the surroundings, the atmosphere, or space, and may be represented as a black body ($\varepsilon = 1$).

If surface i is concave, then some of the energy leaving one part of the surface may impinge on another part of the surface, and

$$F_{ii} > 0 \tag{12-8}$$

For plane or convex surfaces,

$$F_{ii} = 0 \tag{12-9}$$

*The "angle factor" is also called the "shape factor," "geometrical factor," or "configuration factor."

In solving static (equilibrium) radiation problems for temperatures and heat flows, the following assumptions are commonly made:

1. The radiative surfaces are "gray bodies," with constant emissivity ε and reflectivity $\rho = 1 - \varepsilon$, whose values are independent of surface temperature and independent of the temperature of the sources of the impinging radiation.
2. All of the surfaces are diffuse reflectors where the angular distribution of the reflected radiation follows Lambert's cosine law.
3. Each surface is isothermal with a constant temperature and zero transmissivity (opaque to radiation).

A typical aerospace problem is the calculation of radiative energy exchanges between a subsystem and its surroundings. This may be modeled[1] by defining an enclosure completely surrounding the subsystem (encompassing 4π steradians of solid angle) and consisting of a number of surfaces, some of which are gray bodies and some of which may be openings to space, represented by black bodies. The total number of surfaces in both the surroundings and the enclosed subsystem is n. The algebraic relations between the subsystem and its surroundings are given by

$$J_i = \varepsilon_i \sigma T_i^4 + \rho_i \sum_{j=1}^{n} J_j F_{ij} \quad \text{W/cm}^2, \quad i = 1, 2, \ldots, n \tag{12-10}$$

where J_i = radiosity = total energy per unit area emitted by surface i. The first term on the right-hand side of Eq. (12-10) is the radiant energy density emitted by surface i due to its temperature T_i. The last term in Eq. (12-10) is the reflection by surface i of the energy incident on it from all of the surrounding surfaces as well as from itself, if concave, per Eq. (12-8).

Equation (12-10) may be rewritten as

$$\sum_{j=1}^{n} a_{ij} J_j + K_i = 0, \quad i = 1, 2, \ldots, n \tag{12-11}$$

where $a_{ii} = F_{ii} - 1/\rho_i$
$a_{ij} = F_{ij}, \quad i \neq j$
$K_i = (\varepsilon_i/\rho_i)\sigma T_i^4$

The a_{ij} can be written as a matrix:

$$(a_{ij}) = \begin{bmatrix} (F_{11} - 1/\rho_1) & F_{12} & \cdots & F_{1n} \\ F_{21} & (F_{22} - 1/\rho_2) & \cdots & F_{2n} \\ \vdots & \vdots & & \vdots \\ F_{n1} & F_{n2} & \cdots & (F_{nn} - 1/\rho_n) \end{bmatrix} \tag{12-12}$$

The solution to the set of equations (12-11) can then be represented in matrix form as

$$(J_j) = -(a_{ij})^{-1}(K_i) \tag{12-13}$$

where $(a_{ij})^{-1}$ is the inverse of the (a_{ij}) matrix defined in Eq. (12-12). If the values of F_{ij}, ρ_i, and the surface temperature of all of the surfaces T_i are known, then the solution of Eq. (12-13) is completely determined and may be solved by computer. With the values of J_i calculated, net radiation heat fluxes may be found from

$$Q_i = A_i(\varepsilon_i/\rho_i)(\sigma T_i^4 - J_i) \quad \text{W} \tag{12-14}$$

Alternatively, one surface temperature and at least $(n - 1)$ values of J_i may be specified and the remaining surface temperatures may then be determined. In this case the surface temperatures are found from

$$T_i = \left[(Q_i\rho_i/A_i\varepsilon_i\sigma) + J_i/\sigma\right]^{1/4} \text{ K} \tag{12-15}$$

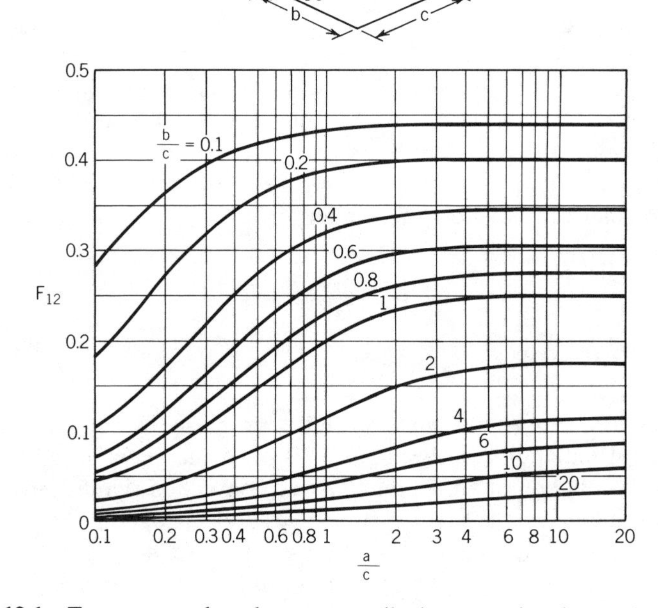

FIGURE 12.1 Two rectangular plates perpendicular to each other with a common edge.

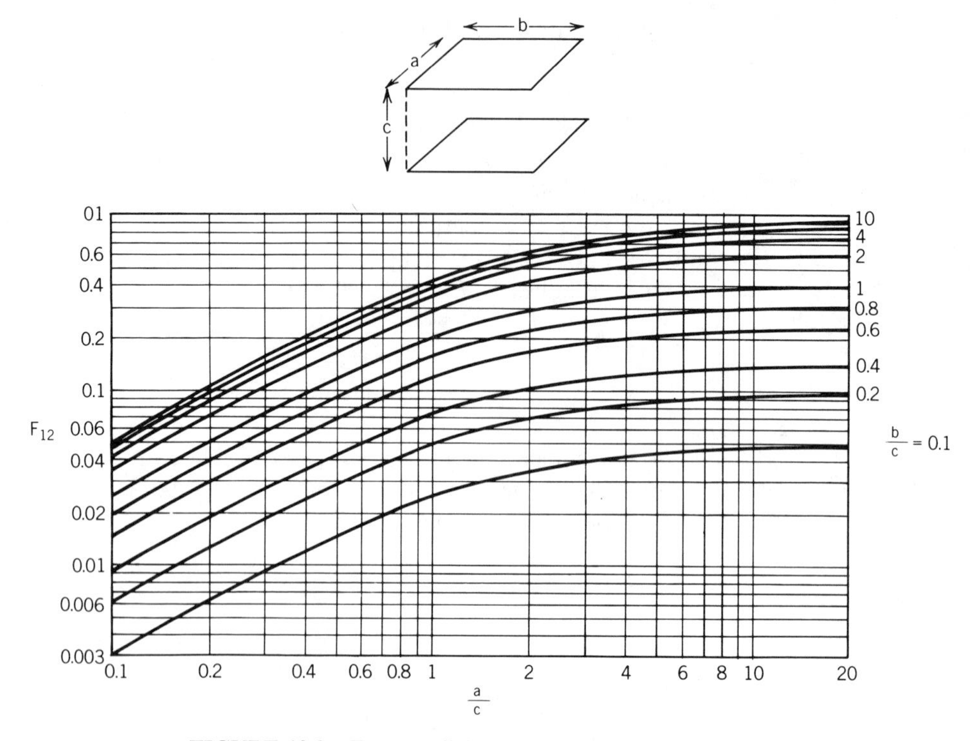

FIGURE 12.2 Two parallel rectangular plates of equal size.

Angle factors F_{12} are presented for some common configurations in Figs. 12.1 through 12.6. Emissivity values for some metallic surfaces are provided in Table 12.1. In many causes, the emissivity of a surface is strongly dependent on the manner of its preparation and on its environmental history, and only rough guidelines can be offered for values of ε.

12.3 CONDUCTION

In most satellite subsystems heat transfer is primarily by conduction through solids and by radiation (see Section 12.2), since the air is usually vented to space. The elimination of air conduction and convection significantly increases temperature increments between heat sources and surrounding structures. The analysis of heat conduction through solids is here limited to the steady-state case, as explained at the beginning of this chapter. The conduction across an interface between two separate solid elements is discussed in Section 12.4.

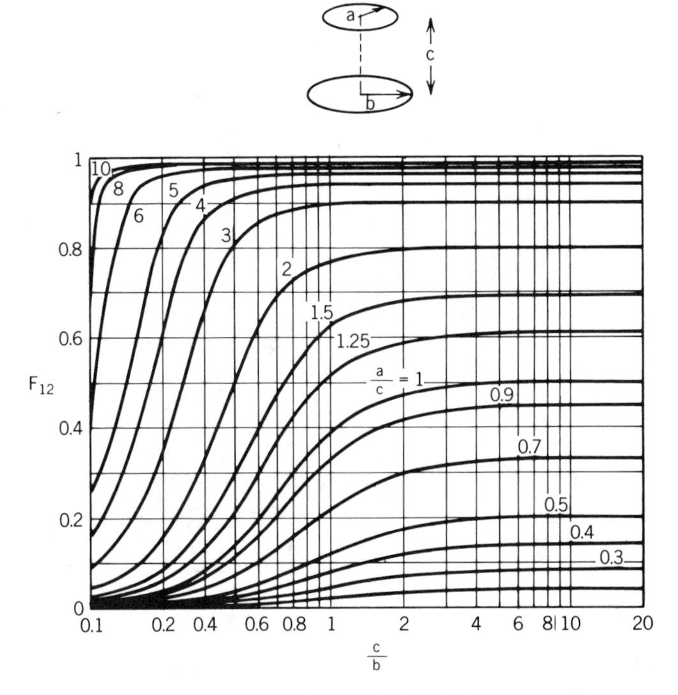

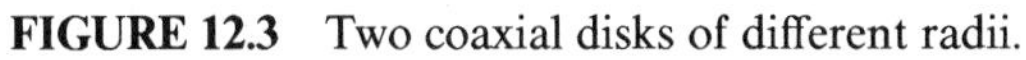

FIGURE 12.3 Two coaxial disks of different radii.

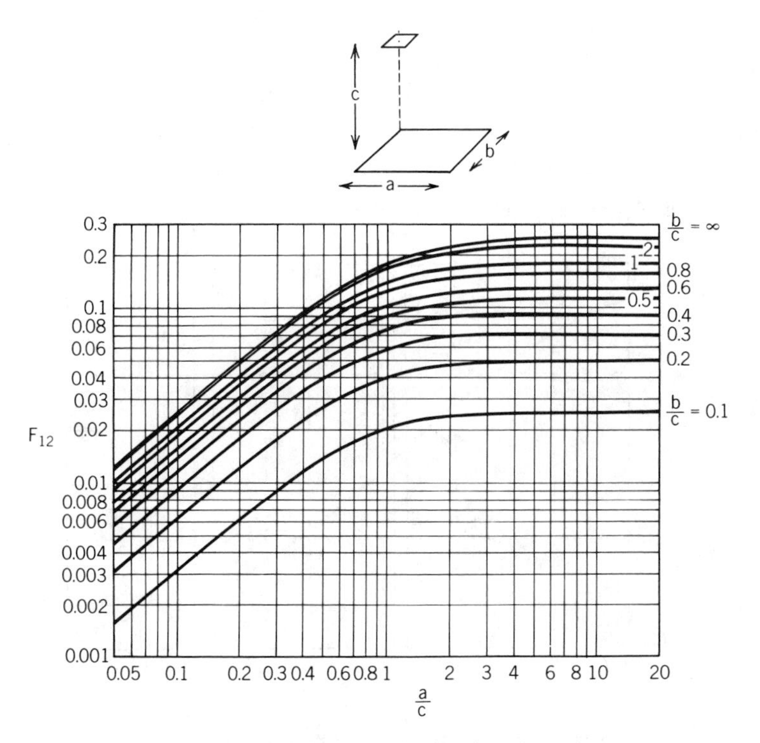

FIGURE 12.4 Rectangle parallel to an infinitesimal surface element over one corner.

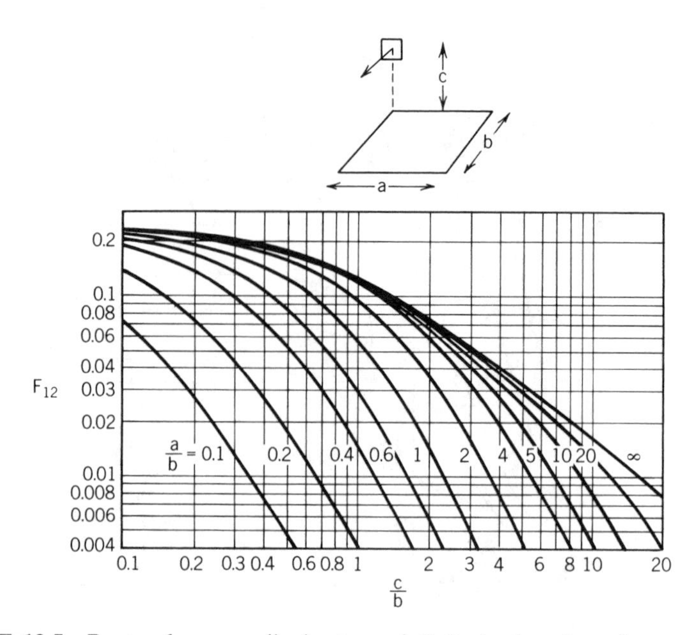

FIGURE 12.5 Rectangle perpendicular to an infinitesimal surface element over one corner.

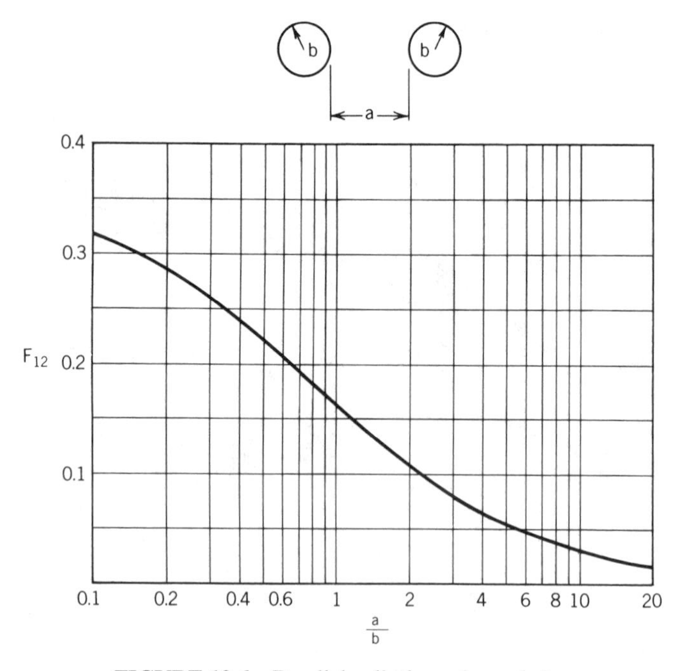

FIGURE 12.6 Parallel cylinders of equal size.

TABLE 12.1 Emissivity Values for Some Metallic Surfaces

Material	Surface Condition	Range of ε
Aluminum	Anodized	0.7–0.9
	Good "as-received" finish, or polished	0.03–0.05
Copper	Colorized	0.2
	Polished	0.02–0.05
	Black oxidized	0.8–0.9
	Rough	0.7
Gold	Polished or buffed	0.02
Iron	Polished	0.1–0.3
	Smooth	0.3–0.6
	Rough	0.9
	Oxidized	0.6–0.95
Steel	Polished	0.1–0.2
	Colorized	0.5
	As-rolled sheet	0.7
	Oxidized	0.3–0.95
Magnesium	Polished, slightly oxidized	0.2
	As received	0.6–0.7
Nickel	Polished	0.04–0.07
	Electroplated, unpolished	0.1
	Oxidized	0.3–0.6
Inconel	Polished	0.1–0.2
	As-received	0.3
	Oxidized	0.2–0.6
Silver	Electrolytic deposit	0.01
	Oxidized	0.02
Tantalum	As-received	0.01–0.03
	Oxidized	0.4
Tin	Coating	0.05–0.08
Titanium	As-received	0.2
	Oxidized	0.3
Tungsten	Unoxidized	0.02–0.03
Zinc	Coating	0.1–0.2

Heat conduction through solids is most simply expressed by the following equations, with reference to Fig. 12.7:

$$A = (KA/L)(T_1 - T_2) \quad \mathrm{W} \tag{12-16}$$

where Q = heat transferred per unit time from surface at temperature T_1 to surface at temperature T_2 (W)

K = thermal conductivity of the solid material (W/in.·°C)

A = cross-sectional area of the solid normal to the direction of heat flow = bw (in.^{2})

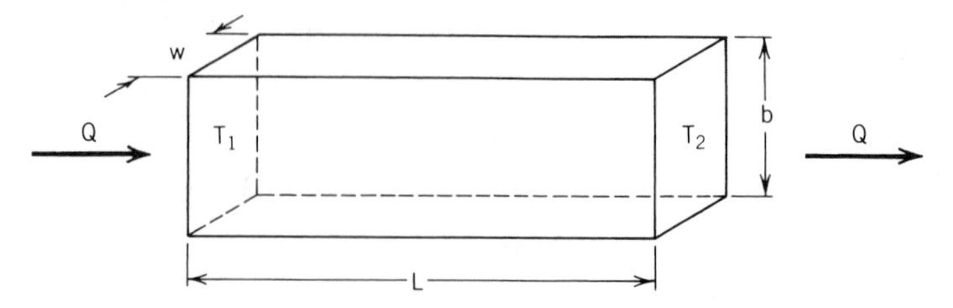

FIGURE 12.7 Heat conduction through a solid.

b = thickness of solid, normal to direction of heat flow (in.)
w = width of solid, normal to direction of heat flow (in.)
L = length of solid parallel to direction of heat flow = heat flow path length (in.)
T_1 = temperature of left-hand surface (°C)
T_2 = temperature of right-hand surface (°C)
$T_1 > T_2$

In Eq. (12-16), it is assumed that the material is homogeneous with a constant K that is independent of temperature, the left-hand surface (T_1) and the right-hand surface (T_2) are both isothermal, there is no heat transfer across any surfaces other than the two ends (T_1 and T_2), and all quantities are independent of time. Under these conditions the thermal gradient across any cross section that is parallel to the two ends is given by

$$\frac{dT}{dL} = \frac{T_1 - T_2}{L} \quad \text{°C/in.} \tag{12-17}$$

and the temperature increment across any path length ΔL is given by

$$\Delta T = \Delta L \left(\frac{dT}{dL} \right) \quad \text{°C} \tag{12-18}$$

The thermal conductance of the solid along the L direction is

$$G = Q/(T_1 - T_2) = KA/L \quad \text{W/°C} \tag{12-19}$$

and the thermal resistance in the same direction is

$$R = 1/G = L/KA \quad \text{°C/W} \tag{12-20}$$

If the solid is linearly tapered in one dimension, as shown in Fig. 12.8,

$$G = K(b_1 - b_2)w/L \ln(b_1/b_2) \quad \text{W/°C} \tag{12-21}$$

where $(b_1 - b_2)/\ln(b_1/b_2)$ is called the log mean width.

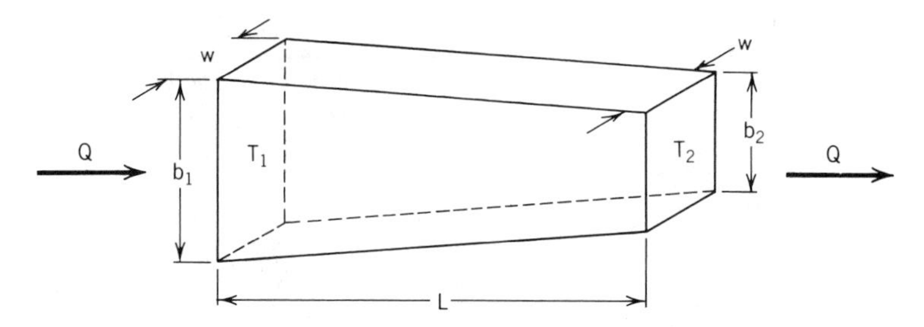

FIGURE 12.8 Solid tapered linearly in one dimension.

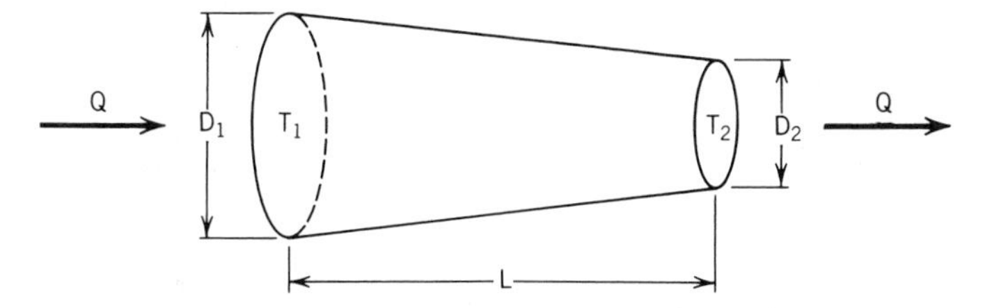

FIGURE 12.9 Truncated right circular cone.

For a circular cylinder with heat flow parallel to the cylindrical axis, Eq. (12-19) applies, with $A = (\pi/4)D^2$, where D is the cylinder diameter and L is its length. For a truncated right circular cone of length L, base diameter D_1, and truncated diameter D_2 (Fig. 12.9), with heat flow parallel to the cone axis,

$$G = (\pi/4)KD_1D_2/L \quad \text{W/°C} \tag{12-22}$$

A formula similar to Eq. (12-22) applies when a solid of any shape cross section is linearly tapered in both dimensions:

$$G = CKd_1d_2/L \quad \text{W/°C} \tag{12-23}$$

where $Cd^2 = A$ = cross-section area (in.^{2})
C = constant = $\pi/4$ for a circular cross section,
= 1 for a square cross section
d = characteristic dimension of the cross section
= diameter for a circular cross section
= side for a square
subscript 1 applies to surface T_1
subscript 2 applies to surface T_2

The values of the thermal conductivity K of a material depend on the material temperatures. However, most of the materials used in satellite subsys-

TABLE 12.2 Thermal Conductivity Values for Some Metals

Material	Thermal Conductivity K (W/in. · °C)
6061T6 Aluminum	4.25
7075 Aluminum	3.30
Copper	8.80–10.12
Gold	7.56
Silver	10.63
Stainless steels	0.29–0.93
Magnesium alloys	1.05–3.47
Tungsten	4.24
Molybdenum	3.71
Tantalum	1.38
Titanium alloys	0.17–0.51
Beryllium alloys	3.82–5.40
Nickel alloys	0.25–2.18
Silicon	2.13
Nickel superalloys	0.23–1.19
Lead alloys	0.70–0.86
Kovar (Fernico)	0.40
42 Nickel–iron	0.32

tems can be assigned an average value of K for the temperature range of interest without introducing excessive error in the thermal analysis. Tables 12-2 and 12-3 present K values for some materials of interest near room temperature. Ranges of values are due to variability in material purity and preparation.

12.4 THERMAL NETWORK MODEL

The thermal analysis of satellite subsystems is based on a conceptual model of the physical subsystem. This model is used in order to translate the physical data into input data for a digital computer program. The model is called a thermal network model because it consists of discrete segments called nodes, which represent pieces of the actual physical subsystem, and which are connected to one another by thermal resistances. The thermal network is analogous to an electrical network, as shown in Table 12-4. The electrical and thermal resistivities and conductivities in Table 12-4 are intrinsic properties of the material only, while the other parameters depend on the system design. Another thermal parameter that is an intrinsic property of the material only,

TABLE 12.3 Thermal Conductivity Values for Some Nonmetals

Material[a]	Thermal Conductivity K (W/in. · °C)
Acetals	0.0059–0.0095
Alkyds	0.013–0.27
Diallyl phthalate (DAP)	0.015–0.017
Aminos	0.0074–0.018
Cellulosics	0.0040–0.0074
Epoxy molding and coating resins	0.0043–0.032
Fluoroplastics	0.0026–0.0067
Nylons	0.0055–0.0062
Phenolics	0.0075–0.017
Polycarbonates	0.0049–0.0054
Polyesters	0.0040–0.033
Polyimides	0.0081–0.041
Polyethylenes	0.0085–0.013
Polypropylene	0.0030–0.0043
Polyphenylene ether	0.0040–0.0055
Polyphenylene sulfide	0.0073–0.015
Polystyrene	0.0037–0.012
Polyurethane foams, rigid	0.0015–0.0029
Polyvinyl chloride	0.0032–0.0053
Sulfone polymers	0.0034–0.0066
Fiberglass epoxy GF or FR4 printed circuit board	0.0075
Alumina	0.89–0.94
Beryllia	5.5–5.8
Glass	0.015–0.027
Quartz	0.14–0.36
Fused silica	0.037

Source: Ref. 2

[a]Some plastics have a wide range of K values owing to the addition of fillers such as fiberglass and other types of fibers that have different K values than the base material and are used in varying concentrations and fiber lengths.

and which is not included in Table 12-4, is the thermal diffusivity α:

$$\alpha = K/\rho C \quad \text{in.}^2/\text{sec} \cdot {}^\circ\text{C} \tag{12-24}$$

where K = thermal conductivity (Table 12.2 or 12.3)
ρ = material density, g/in.^{3}
C = heat capacity = W · sec/g

Thermal diffusivity is only of importance in time-dependent cases and is not used in this chapter. Symbols used for the thermal parameters referred to in

TABLE 12-4 Electrical / Thermal Network Analogies

Electrical Parameter	Thermal Parameter
Charge (coulomb)	Heat (BTU, W · sec)
Current (coulomb/sec = amp)	Heat flow (BTU/hour, W)
Potential (V)	Temperature (°F, °C)
Resistance (V/amp = ohms)	Thermal resistance (°F · hr/BTU, °C/W)
Resistivity (ohm · cm)	Thermal resistivity (°F · hr · ft/BTU, °C · in./W)
Conductance (amp/V = mho)	Thermal conductance (BTU/hr · °F, W/°C)
Conductivity (mho/cm)	Thermal conductivity (BTU/hr · ft · °F, W/in. · °C)
Capacitance (coulomb/V = farad)	Thermal capacitance (BTU/°F, W · sec/°C)

this chapter are listed here:

$$Q = \text{heat flow}$$
$$T = \text{temperature}$$
$$R = \text{thermal resistance}$$
$$G = 1/R = \text{thermal conductance}$$
$$K = \text{thermal conductivity}$$

When breaking up the physical system into nodes (discrete segments), a compromise must be made between accuracy, which improves with increasing number of nodes, and engineering effort and machine time, which are less with fewer nodes. It is usually important to have a node for each significant heat source, such as semiconductors, electrical, and electromechanical parts, that dissipate power. External heat sources and heat sinks exchanging heat with the subsystem by conduction or radiation may be represented by nodes having specified temperatures. Structural elements of the subsystem may each be represented by a uniform or nonuniform, one-, two-, or three-dimensional grid of nodes where the nodal pattern may be square, rectangular, triangular, or some other shape.

Once the node network has been established, it is necessary to calculate the thermal resistance between adjacent nodes. For two-dimensional plates like printed circuit boards and housing walls, a square grid is common, and each node will have four adjacent nodes, requiring the calculation of approximately $2n$ resistances for a plate having n nodes. As shown in Fig. 12.10, resistances are required only in the nearest-neighbor directions, not along diagonals. For a

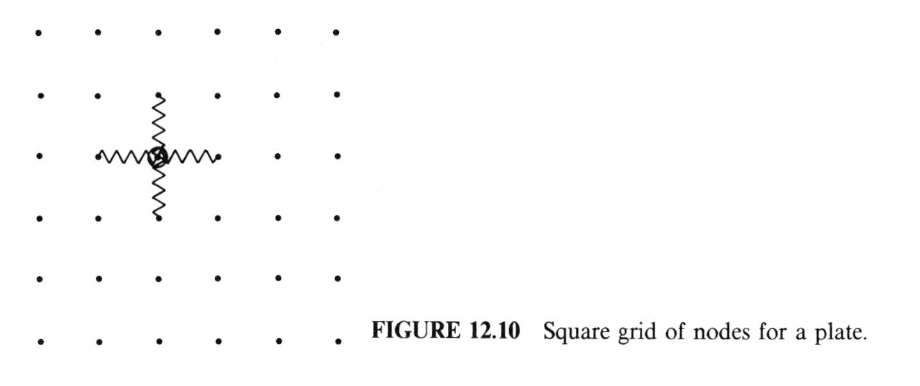

FIGURE 12.10 Square grid of nodes for a plate.

three-dimensional element, a cubic grid may be used where each node will have six nearest-neighbor adjacent nodes, requiring approximately $3n$ resistances for a solid element with n nodes. For a uniform square grid representing a homogeneous plate of constant thickness, the resistance between adjacent nodes is

$$R = 1/Kt \quad {}^\circ\mathrm{C/W} \tag{12-25}$$

where K is the thermal conductivity of the material (W/in.· °C) and t is the thickness of the plate (in.). Note that Eq. (12-25) is independent of the distance between the nodes in the grid. For a uniform cubic grid representing a homogeneous solid, the resistance between adjacent nodes is

$$R = 1/Kl \quad {}^\circ\mathrm{C/W} \tag{12-26}$$

where l is the spacing between nodes (in.).

When two distinct solid elements are in contact with one another there is heat conduction across their interface if the two surfaces are at different temperatures. See Figure 12.11. However, the thermal resistivity across the interface is much greater than the thermal resistivity of the solid material,

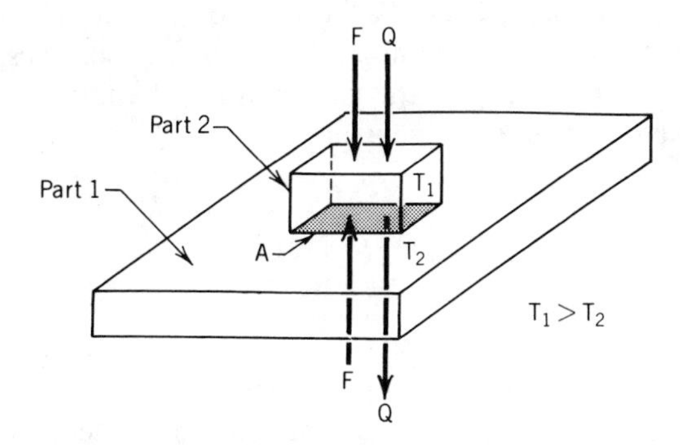

FIGURE 12.11 Contact thermal resistance between adjacent surfaces.

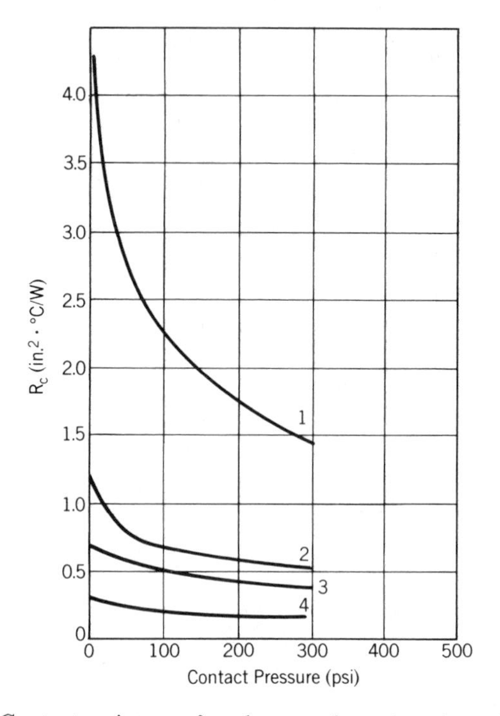

FIGURE 12.12 Contact resistance for clean steel–steel surfaces in air at 1 atm at 200°F, to 300 psi. (1 = 1000 μin. finish; 2 = 125 μin. finish; 3 = 63 μin. finish; 4 = 4 μin. finish.)

resulting in a sharp increase in temperature gradient and a distinct temperature increment across the interface. This effect is called thermal contact resistance, or just contact resistance, and is a function of the two materials that are in contact, the surface condition (e.g., oxide layer), surface finish, and flatness of the two contact surfaces, the pressure forcing together the two contact surfaces, the temperature at the interface, and the characteristics and pressure of the gas (if any) between the two contact surfaces. The better the surface condition, finish, and flatness of the contact surfaces, the larger the contact pressure, and the higher the gas pressure, the less will be the contact resistance. The contact resistance has units of

$$R_c = {}^\circ\mathrm{C} \cdot \mathrm{in.}^2/\mathrm{W} \tag{12-27}$$

Values for R_c for various types of contact interfaces are shown in Figs. 12.12–12.18. Note that the thermal contact resistance defined by Eq. (12-27) has different units than the thermal resistance of Eq. (12-20). The R_c value

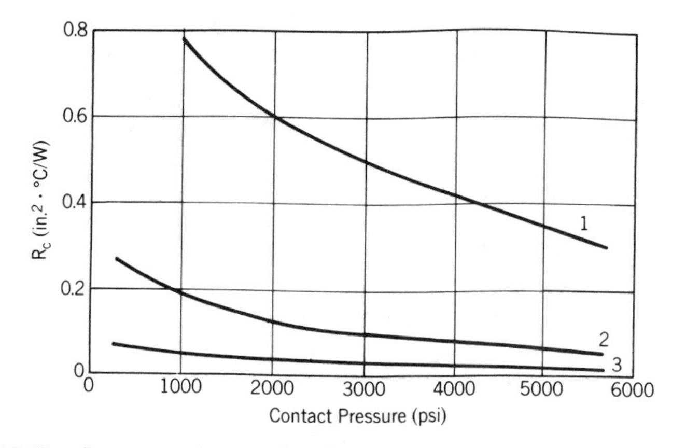

FIGURE 12.13 Contact resistance for clean steel–steel surfaces in air at 1 atm at room temperature, to 6000 psi. (1 = 285 μin. finish; 2 = 219 μin. finish; 3 = 78 μin. finish.)

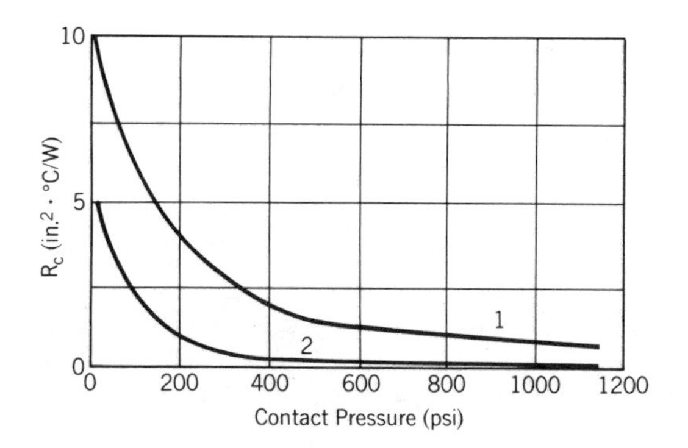

FIGURE 12.14 Contact resistance for clean steel–steel surfaces in air at 10^{-4} torr at room temperature, to 1200 psi. (1 = 42–60 μin finish; 2 = 10–15 μin. finish.)

must be divided by an interface contact area to obtain a thermal resistance:

$$R = R_c/A \quad °\text{C/W} \tag{12-28}$$

where A is the interface contact area between the two surfaces (in.^{2}).

When two adjacent nodes of a thermal network model are separated by a contact interface such as described in the preceding paragraph, the contact resistance of Eq. (12-28) must be added to the resistance between nodes given by Eq. (12-25) or (12-26). The interface pressure, which must be determined in order to use Fig. 12.12–12.18, may be estimated as follows:

$$P = nF/A \text{ psi} \tag{12-29}$$

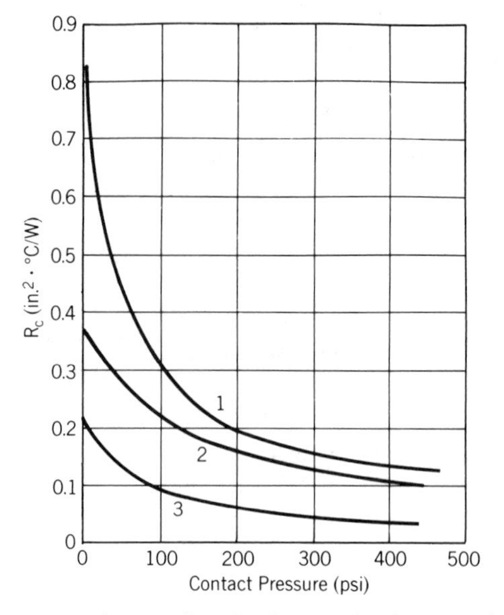

FIGURE 12.15 Contact resistance for aluminum–aluminum surfaces in air at 1 atm at 200–400°F, to 450 psi. (1 = 120 μin. finish; 2 = 65 μin. finish; 3 = 10 μin. finish.)

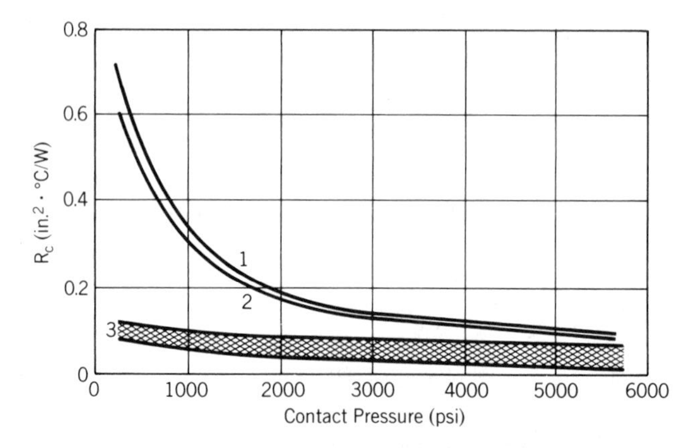

FIGURE 12.16 Contact resistance for aluminum–aluminum surfaces in air at 1 atm at room temperature, to 6000 psi. (1 = 176 μin. finish; 2 = 105 μin. finish; 3 = 52 μin. finish.)

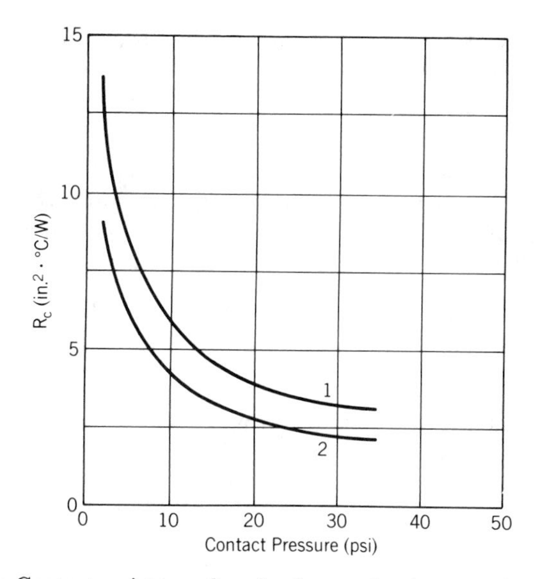

FIGURE 12.17 Contact resistance for aluminum–aluminum surfaces in air at 10^{-4} torr at 100°F, to 35 psi. (1 = 48–65 μin. finish; 2 = 8–18 μin. finish.)

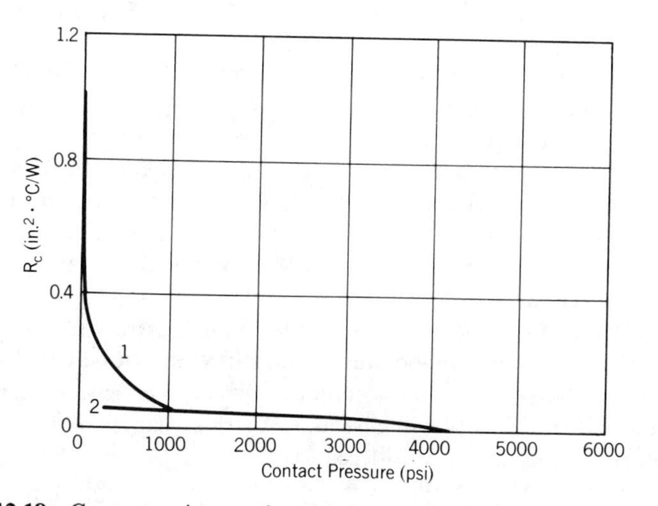

FIGURE 12.18 Contact resistance for aluminum–aluminum surfaces in air at 10^{-4} torr and 4.5 torr at room temperature, to 4500 psi. (1 = 50–60 μin. finish at 10^{-4} torr; 2 = 53 μin. finish at 4.5 torr.)

where n = number of screws or bolts holding the two surfaces together
F = force per screw or bolt (lbf)
A = interface area (in.^{2})

Caution should be used in applying Eq. (12-29). If the screws or bolts are around the perimeter of a large contact area, the central section of the area will not be under as much pressure as the perimeter area. In this case, a better estimate is to redefine A:

$$A = nA_0 \text{ in.}^2 \tag{12-30}$$

where $A_0 = bA_s$ (in.^{2})
A_s = area of screw or bolt head (in.^{2})
b = 1 for very thin plates
b = 10 for thick plates
$nbA_s = nA_0 <$ interface area

If Eq. (12-30) is used, the resulting value of A should be applied to both Eqs. (12-28) and (12-29).

12.5 PRINTED-CIRCUIT-BOARD (PCB) THERMAL ANALYSIS

Semiconductor parts (diodes, transistors, integrated circuits) usually have an upper temperature limit specified for the solid-state junction, which is inside the component case and often is at a significantly higher operating temperature than the case. These parts are mounted on PCBs along with other electrical components such as resistors, capacitors, inductors, and transformers. The semiconductor parts and some of the other electrical components are heat sources. It is sometimes necessary to do a detailed thermal analysis of a PCB and its electrical components to determine if these components will be maintained within their specified temperature limits. In this section, attention is focused on the upper temperature limits in the "worst-case" operating condition when the system or the component or both are generating maximum heat per unit time. As in previous sections of this chapter, only steady-state (equilibrium) conditions are considered.

The thermal path from the PCB-mounted-component heat source to a thermal network node on the PCB is illustrated with the example of an integrated circuit (IC). In Fig. 12.19 the heat source Q is the semiconductor junction at temperature T_J. The heat flows through the thermal resistance between the IC junction and the IC case R_{JC} (usually label θ_{JC} in IC data books). From the IC case the heat will flow through ungrounded IC leads R_{CU} and grounded IC leads R_{CG} to the PCB. The thermal resistance through a grounded IC lead to the PCB is less than that through an ungrounded IC lead to the PCB because the grounded lead is thermally connected to one or more

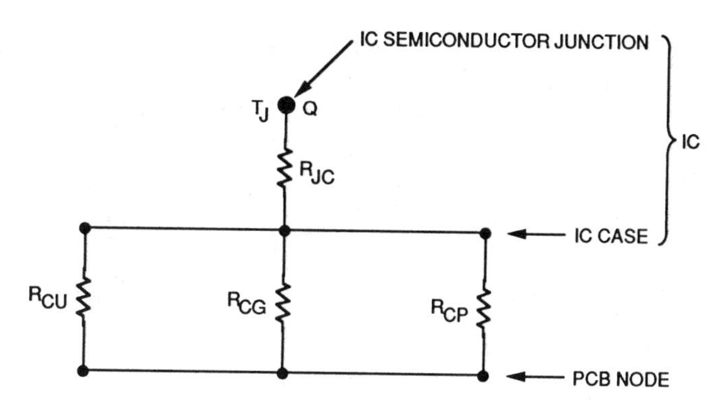

FIGURE 12.19 Thermal path from IC semiconductor junction to node on PCB.

of the copper ground planes that are responsible for most of the PCB thermal conductivity. By contrast, the ungrounded lead is separated from the copper ground planes by a high-resistance thermal path through PCB copper-free board material, for example, fiberglass epoxy. However, there are usually many more ungrounded leads than grounded leads on the IC. If thermal enhancement is used, the IC may be thermally connected directly to the PCB copper ground plane through a higher conductivity material that bonds the IC case to the copper, with a thermal resistance R_{CP}.

The total thermal resistance from the IC junction to the PCB node is

$$R_{JP} = R_{JC} + \left[(R_{CUO}/N_U)^{-1} + (R_{CGO}/N_G)^{-1} + (R_{CP})^{-1} \right]^{-1} \quad °C/W \tag{12-31}$$

where N_U = number of ungrounded IC leads
R_{CUO} = thermal resistance from IC case, through one ungrounded lead, through PCB board material, to PCB copper ground planes
N_G = number of grounded IC leads
R_{CGO} = thermal resistance from IC case, through one grounded lead, to PCB copper ground planes

and

$$R_{CP} = t/KA \quad °C/W \tag{12-32a}$$

or

$$R_{CP} = R_C/A \quad °C/W \tag{12-32b}$$

where A = bond area between IC and PCB (in.^{2})
t = thickness of bond material (in.)
K = thermal conductivity of bond material (W/in.· °C)
R_c = contact resistance of bond material (°C · in.^{2}/W)

The use of Eq. (12-32a) vs (12-32b) depends on whether the manufacturer of the thermal-enhancement bonding material provides property data in terms of thermal conductivity K or contact resistance R_c. These thermal-enhancement materials generally have thermal conductivities in the range from 0.02 to 0.11 W/in.· °C. Equations (12-32) should be used with caution. If there is no copper ground plane beneath the IC, the application of Eqs. (12-32) is not valid. If there is a copper ground plane beneath the IC, there will still be copper-free areas where the ungrounded pins are located, and these areas may introduce additional thermal resistance to heat flow and must be taken into account in calculating the thermal resistance between the IC node on the PCB and its nearest-neighbor nodes on the PCB.

In the following, examples are provided of the calculation of some of the thermal resistances discussed in the preceding paragraphs. Fig. 12.20 shows two rows of pin holes for an 18-pin ceramic DIP IC with Kovar pins soldered into copper-plated holes in a copper-free area of the PCB. Six of the pins are grounded, and twelve are ungrounded. Figure 12-21 shows the pin dimensions. The thermal resistance through a pin, from the IC case to the seating plane (Fig. 12-21) R_1, is the sum of two thermal resistances in series: the upper rectangular part of the pin from the lower edge of the case down to where the tapered section starts; and the lower, tapered section of the pin to the seating plane:

$$R_1 = L_1/KtW_1 + L_2 \ln(W_1/W_2)/Kt(W_1 - W_2) = 323 \text{ °C/W} \quad (12\text{-}33)$$

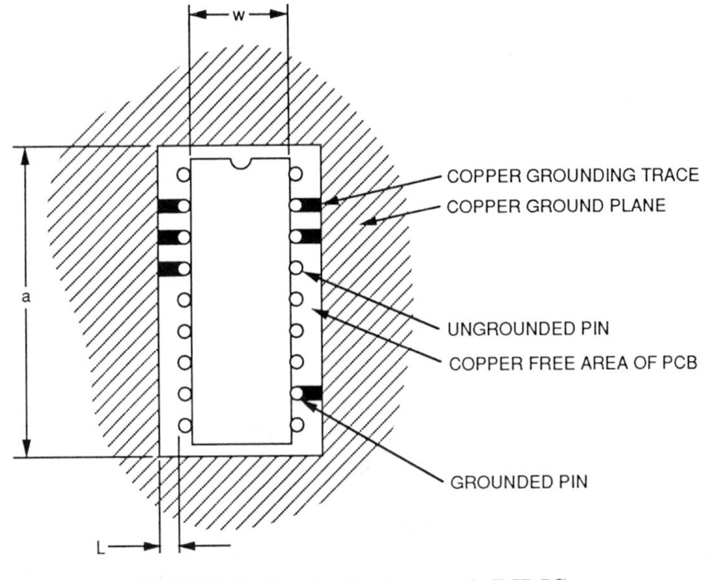

FIGURE 12.20 An 18-pin ceramic DIP IC.

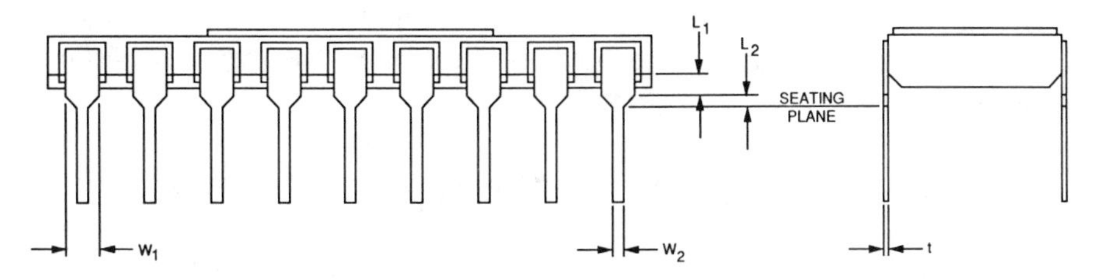

FIGURE 12.21 Pin dimensions for the 18-pin ceramic DIP IC.

where L_1 = length of thermal path for upper, rectangular section = 0.045 in.
K = thermal conductivity of Kovar pin = 0.40 W/in.· °C
t = thickness of pin = 0.011 in.
W_1 = width of upper, rectangular section = 0.055 in.
L_2 = length of thermal path for lower, tapered section = 0.020 in.
W_2 = minor width of lower, tapered section = 0.018 in.

The thermal resistance through the pin soldered into the copper-plated through hole in the PCB consists of three thermal resistances in parallel:

$$R_2 = \left(\sum_{i=1}^{3} \frac{K_i A_i}{h} \right)^{-1} = 9.5 \text{ °C/W} \tag{12-34}$$

where i = 1 for copper plating inside through hole in PCB
i = 2 for Kovar pin
i = 3 for solder
K_1 = 9.4 W/in.· °C for copper
K_2 = 0.40 W/in.· °C for Kovar
K_3 = 1.26 W/in.· °C for solder
$A_1 = (\pi/4)[D^2 - (D - 0.006)^2]$ in.^{2} for 0.003-in. copper plating = 0.000631 in.^{2}
D = diameter of through hole = 0.070 in.
$A_2 = W_2 t = 0.018 \cdot 0.011 = 0.000198$ in.^{2} for Kovar
$A_3 = (\pi/4)(D - 0.006)^2 - A_2 = 0.00302$ in.^{2} for solder
h = PCB thickness = length of hole = 0.093 in.

For grounded pins, the thermal resistance from the pin through the copper grounding trace (Fig. 12.20) to the copper ground plane is

$$R_3 = L/Kwt = 25.4 \text{ °C/W} \tag{12-35}$$

where L = length of ground trace = 0.050 in.
K = 9.4 W/in.· °C for copper
w = width of ground trace = 0.070 in.
t = thickness of ground trace = 0.003 in.

If the PCB has a copper ground plane on both outer surfaces and the grounded pins are grounded to both of them, then the thermal resistance R_{CGO} of Eq. (12-31) is

$$R_{CGO} = R_1 + \left[(R_3)^{-1} + (R_2 + R_3)^{-1}\right]^{-1} = 338\ °C/W \quad (12\text{-}36)$$

For ungrounded pins, the thermal resistance through the copper-free PCB area (Fig. 12.20) is

$$R_4 = 2(L^2 + h^2/4)^{1/2}[\ln(2a/nD)]/Kh(2a/n - D) = 2200\ °C/W \quad (12\text{-}37)$$

where L = path length through copper-free PCB area = 0.050 in.
h = PCB thickness = 0.093 in.
a = length of copper-free PCB area (Fig. 12.20) = 1.00 in.
n = total number of pins = 18
D = diameter of through hole = 0.070 in.
K = thermal conductivity of PCB material = 0.0075 W/in.· °C for epoxy fiberglass

For copper ground planes on both PCB surfaces, the thermal resistance R_{CUO} of Eq. (12-31) is

$$R_{CUO} = R_1 + \tfrac{1}{2}\left(\tfrac{1}{2}R_2 + R_4\right) = 1425\ °C/W \quad (12\text{-}38)$$

With 6 grounded pins, 12 ungrounded pins, and no thermal enhancement ($R_{CP} = 0$), Eq. (12-31) yields

$$R_{JP} = R_{JC} + \left[(1425/12)^{-1} + (338/6)^{-1}\right]^{-1} = (R_{JC} + 38)\ °C/W \quad (12\text{-}39)$$

If the IC is thermally enhanced by bonding the case to the PCB copper ground plane with a material that has a thermal conductivity of $K = 0.031$ W/in.· °C and is applied as a 0.005-in.-thick layer, the thermal resistance by Eq. (12-32a) is $R_{CP} = 0.538$ °C/W, where a bond area of $A = 0.3$ in.^{2} is used. The copper ground plane under the IC is partially separated from the rest of the ground plane by the two copper-free areas around the rows of pins, but is directly connected to the rest of the copper ground plane at each end, as shown in Fig. 12.22. The additional thermal resistance is

$$R'_{CP} = \tfrac{1}{2}\left(\tfrac{1}{2}a\right)/Kwt = 29.6\ °C/W$$

where a = 1.00 in.
K = 9.4 W/in.· °C for copper ground plane
w = 0.3 in.
t = 0.003 in. copper ground plane thickness

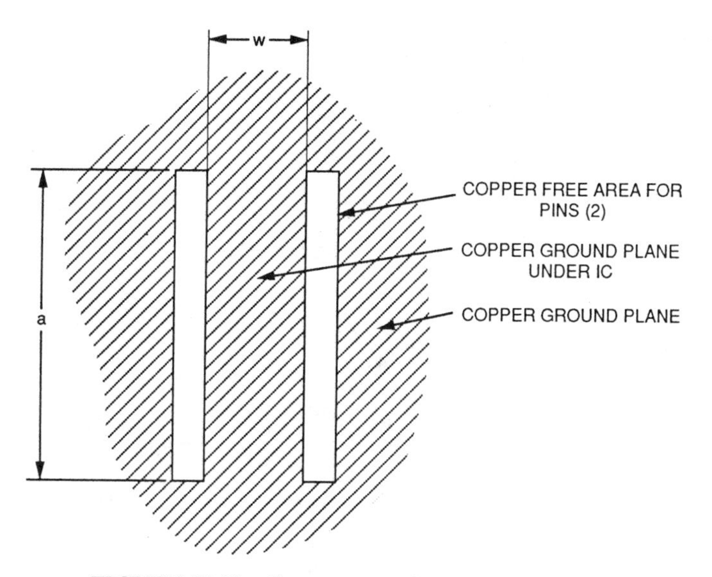

FIGURE 12.22 Copper ground plane under an IC.

Using $(R'_{CP} + R_{CP}) = 30.1$ °C/W for R_{CP} in Eq. (12-31) yields

$$\begin{aligned} R_{JP} &= R_{JC} + \left[(1425/12)^{-1} + (338/6)^{-1} + (30.1)^{-1}\right]^{-1} \\ &= (R_{JC} + 17)\ °\text{C/W} \end{aligned} \tag{12-40}$$

Comparing Eqs. (12-39) and (12-40), it is seen that thermal enhancement reduces the total thermal resistance R_{JP} by 21 °C/W. Additional reduction of R_{JP} may be accomplished by increasing the width or the thickness or both of the copper ground traces connecting the grounded pins to the copper ground plane (Fig. 12.20).

When a number of heat-producing components, such as ICs, are mounted on a PCB, it is often advantageous to use node shapes and sizes that accurately represent the actual geometry and facilitate the type of analysis demonstrated in the preceding paragraphs. This will result in an assembly of nonuniform nodes that are thermally connected to nearest neighbors by PCB-board/copper-ground-plane thermal paths. The ICs are usually mounted so that their long axes are parallel, or sometimes perpendicular, to one another. The thermal resistance from the IC junction to the PCB node R_{JP}, should include all thermal resistances to the sides of the rectangle enclosing the IC and the copper-free areas on the PCB, and in contact with the copper ground plane, as shown in Fig. 12.23. In this case, the thermal paths between nearest-neighbor IC/PCB nodes are trapezoidal sections of PCB/copper laminate, as shown in Fig. 12.24, where the rectangles represent the IC/PCB node boundaries (as shown in Fig. 12.23) and the trapezoids are the thermal

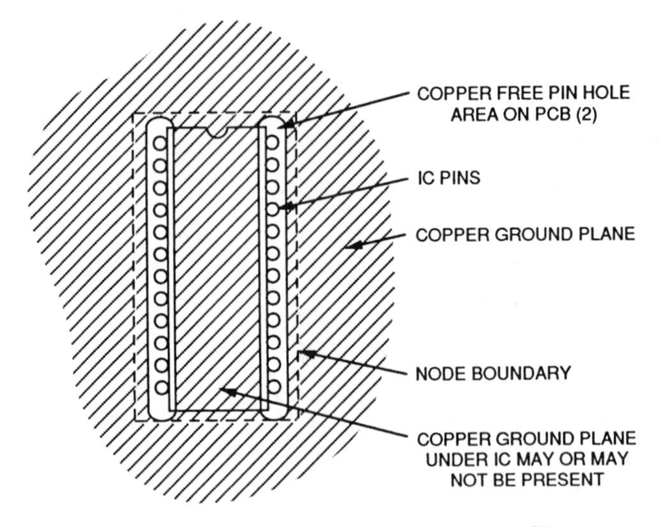

FIGURE 12.23 IC node boundaries on PCB.

paths. The thermal resistances of the thermal paths shown in Fig. 12.24 are calculated using Eq. (12-21), with reference to Fig. 12.8, where Kw is given by

$$Kw = K_1 t_1 + K_2 t_2 \quad \text{W/°C} \tag{12-41}$$

where K_1 = conductivity of copper ground plane (W/in.· °C)
t_1 = total thickness of all copper ground planes and copper power planes in the PCB (in.)
K_2 = conductivity of PCB board material (W/in.· °C)
t_2 = total thickness of PCB board material (in.)

Resistors mounted on PCBs will be heat sources. The thermal resistance through the resistor lead wires, PCB through holes, and copper-free PCB areas to the copper ground plane will cause the resistor temperatures to be higher than the PCB temperatures. Inductors and transformers may also dissipate heat, and the major thermal path will often be through the base in contact with the PCB, and through screws or other mounting hardware that secure the component to the PCB.

In terms of the PCB nodal network, a heat-dissipating mounted component is represented by a defined geometric node area on the PCB, with the edges of the geometric node area in contact with PCB copper ground planes, and a specified power source in the node area. It is usually adequate to assume a uniform power density per unit perimeter of geometric node area, and the geometry and power are independent of the thermal resistances analyzed in the preceding paragraphs. These thermal resistances are only needed to

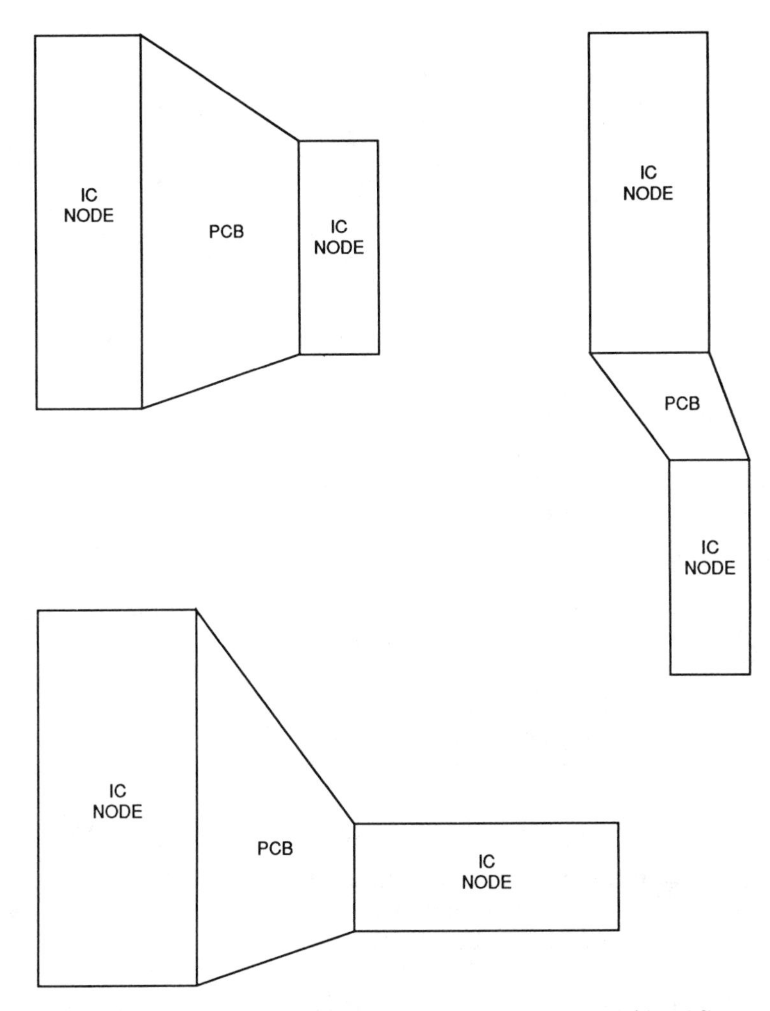

FIGURE 12.24 PCB thermal paths between nearest-neighbor ICs.

calculate the temperature rise between the perimeter of the node area and the heat source in the mounted component, for example, the IC junction.

In addition to the PCB-mounted component, heat source nodes, and internodal thermal resistances, nodes must be defined for the screws, standoffs, card guides, and frames that mechanically and thermally connect the PCB to the subsystem housing. These nodes will be referred to as heat-transfer nodes, and thermal resistances must be calculated between these heat-transfer nodes and nearest-neighbor nodes, which can include both PCB heat-source nodes and other heat-transfer nodes. These thermal resistances will normally include both conductive thermal resistances (e.g., through PCBs, screws, frames) and contact thermal resistances (e.g., between PCBs and screw heads/nuts, be-

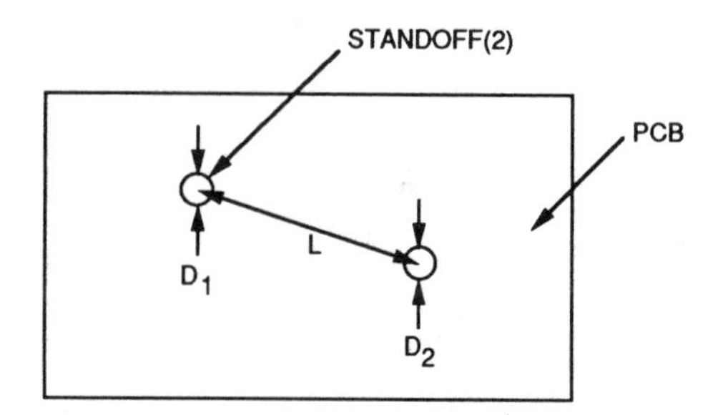

FIGURE 12.25 Thermal resistance through PCB between two standoffs.

tween PCBs and standoffs or frames). By this means the section of the thermal network model representing a PCB and its connection to the housing may be integrated into the subsystem thermal network model, which may include many PCBs as well as the internal structure, the housing, and the connection to the base plate/heat sink. Calculation of conductive thermal resistances is fairly straightforward in most areas using material presented earlier in this chapter. In the following, examples are provided of additional conductive thermal resistances and contact thermal resistances.

PCBs are sometimes supported by standoffs that also conduct heat from the PCB to the frame. If it is necessary to calculate the thermal resistance through the PCB between two standoffs located as shown in Fig. 12.25, the following equation[3] is a fair approximation:

$$R = \cosh^{-1}\left[\left(4L^2 - D_1^2 - D_2^2\right)/2D_1D_2\right]/2\pi Kt \quad {}^\circ\text{C/W} \tag{12-42}$$

where L = center-to-center distance between standoffs (in.)
D_1, D_2 = diameters of standoffs (in.)
Kt = PCB conductivity, Eq. (12-41) (W/°C)

If the two standoffs are both near the perimeter of the PCB, the value of R is twice that found from Eq. (12-42). If the standoffs are square, the $D_1, D_2 \ll L$, Eq. (12-42) may be used with D_1 and D_2, the sides of the square standoffs. In most cases, $D_1 = D_2$.

Other conductive thermal resistances which sometimes occur are:

1. The resistance through a conducting medium of thickness t between a circular surface and a concentric, surrounding square perimeter, Fig. 12.26:

$$R = \ln(1.08S/D)/2\pi Kt \quad {}^\circ\text{C/W} \tag{12-43}$$

where S = side of square (in.)
D = diameter of circle (in.)
t = thickness (in.)
K = conductivity of medium (W/in.· °C)

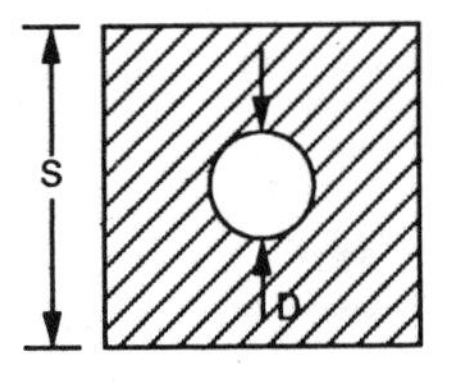

FIGURE 12.26 Thermal resistance between circle and concentric surrounding square.

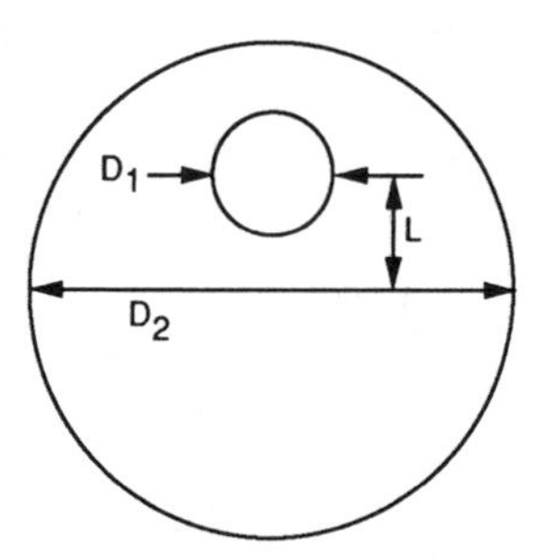

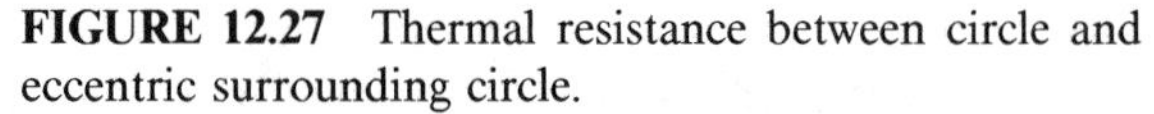

FIGURE 12.27 Thermal resistance between circle and eccentric surrounding circle.

2. The resistance through a conducting medium of thickness t between a circular surface and an eccentric, surrounding circular perimeter, Fig. 12.27:

$$R = \cosh^{-1}\left[\left(D_1^2 + D_2^2 - 4L^2\right)/2D_1D_2\right]/2\pi Kt \quad {}^\circ\mathrm{C/W} \qquad (12\text{-}44)$$

where L = center-to-center distance between circles (in.)
D_1, D_2 = diameters of circles (in.)
t = thickness (in.)
K = conductivity of medium (W/in.· °C)

3. The resistance through a conducting medium of thickness t between a rectangular surface and a surrounding, parallel, rectangular perimeter, Fig. 12.28:

$$R = \left\{\left[x(a_1 - a_2 - x)\ln(b_1/b_2)\right]^{-1} + \left[y(b_1 - b_2 - y)\ln(a_1/a_2)\right]^{-1}\right\}^{-1}/Kt(a_1 - a_2)(b_1 - b_2) \quad {}^\circ\mathrm{C/W} \qquad (12\text{-}45)$$

where a_1, b_1 = sides of larger rectangle (in.)
a_2, b_2 = sides of smaller rectangle (in.)
x, y = coordinates of one corner of smaller rectangle relative to origin at corresponding corner of larger rectangle (in.)
t = thickness (in.)
K = conductivity of medium (W/in.· °C)

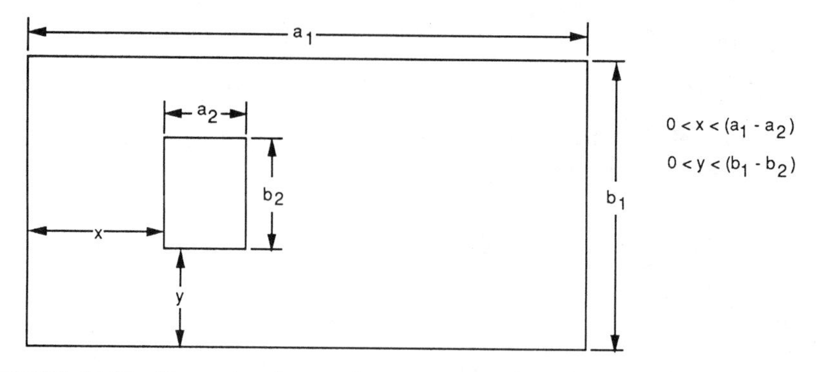

FIGURE 12.28 Thermal resistance between rectangle and surrounding parallel rectangle.

4. The contact resistance between a PCB and the standoff to which it is screwed, where the copper ground plane on the PCB surface is in contact with an aluminum standoff, Fig. 12.29, and Eq. (12-28):

$$R = 2R_c(F/A)/A \quad °\text{C/W} \tag{12-46}$$

where $R_c(F/A)$ may be obtained from Figs. 12.15–12.18 (°C · in.^{2}/W)
F/A = contact pressure (psi)
F = tension force in screw securing PCB to standoff (lbf)
A = contact area of standoff (in.^{2})

Aluminum standoffs should be used rather than steel, since aluminum thermal conductivity is superior. Since steel screws are commonly used, the additional thermal conductivity from the other copper ground plane (the upper one in Fig. 12.29) to the screw head and through the steel screw shank to the standoff is negligible compared to the conductivity through the lower copper ground plane in direct contact with the standoff. The factor of 2 in Eq. (12-46) is due to the approximation that the top and bottom ground planes

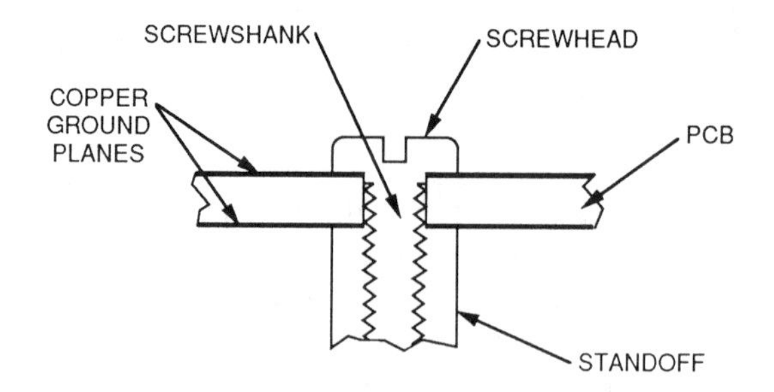

FIGURE 12.29 Thermal contact resistance between PCB and standoff.

each carry half of the heat flow, and the upper ground plane heat flow has much greater thermal resistance to reach the standoff.

An alternative method of modeling the thermal network is to use a fine-meshed square or rectangular nodal grid, where the spacing between nearest-neighbor nodes is less than the width of the copper-free PCB area where the pin holes are located or less than the pin hole diameter. This will require many more nodes than the preceding method and will take much more computer time. If the nodal mesh is not sufficiently fine, significant inaccuracies can occur in the results.

REFERENCES

1 *Heat Transfer Data Book*, Schenectady, NY: General Electric Company, Technology Marketing Operation.

2 *Machine Design*, April 14, 1988, Materials Reference Issue.

3 W. H. Rohsenow and J. P. Hartnett, Eds., *Handbook of Heat Transfer*, McGraw-Hill, New York, 1973.

APPENDIX

ALUMINUM PROPERTIES*

A.1 TYPICAL PROPERTIES

The following typical properties are not guaranteed, since in most cases they are averages for various sizes, product forms, and methods of manufacture and may not be exactly representative of any particular product or size. These data are intended only as a basis for comparing alloys and tempers and should not be specified as engineering requirements or used for design purposes.

*Adapted, with permission, from *Aluminum Standards and Data, 1988*, The Aluminum Association, Incorporated, Washington, D.C. This document is referred to as "the Standard" in the footnotes to the tables in this Appendix.

Typical Mechanical Properties[1,2]

ALLOY AND TEMPER	TENSION				HARDNESS	SHEAR	FATIGUE	MODULUS
	STRENGTH ksi		ELONGATION percent in 2 in.		BRINELL NUMBER	ULTIMATE SHEARING STRENGTH	ENDURANCE[3] LIMIT	MODULUS[4] OF ELASTICITY
	ULTIMATE	YIELD	1/16 in. Thick Specimen	1/2 in. Diameter Specimen	500 kg load 10 mm ball	ksi	ksi	ksi $\times 10^3$
1060-0	10	4	43	..	19	7	3	10.0
1060-H12	12	11	16	..	23	8	4	10.0
1060-H14	14	13	12	..	26	9	5	10.0
1060-H16	16	15	8	..	30	10	6.5	10.0
1060-H18	19	18	6	..	35	11	6.5	10.0
1100-0	13	5	35	45	23	9	5	10.0
1100-H12	16	15	12	25	28	10	6	10.0
1100-H14	18	17	9	20	32	11	7	10.0
1100-H16	21	20	6	17	38	12	9	10.0
1100-H18	24	22	5	15	44	13	9	10.0
1350-0	12	4	..	..[5]	..	8	..	10.0
1350-H12	14	12	..	..	..	9	..	10.0
1350-H14	16	14	..	..	..	10	..	10.0
1350-H16	18	16	..	..	..	11	..	10.0
1350-H19	27	24	..	..[6]	..	15	7	10.0
2011-T3	55	43	..	15	95	32	18	10.2
2011-T8	59	45	..	12	100	35	18	10.2
2014-0	27	14	..	18	45	18	13	10.6
2014-T4, T451	62	42	..	20	105	38	20	10.6
2014-T6, T651	70	60	..	13	135	42	18	10.6
Alclad 2014-0	25	10	21	..	..	18	..	10.5
Alclad 2014-T3	63	40	20	..	..	37	..	10.5
Alclad 2014-T4, T451	61	37	22	..	..	37	..	10.5
Alclad 2014-T6, T651	68	60	10	..	..	41	..	10.5
2017-0	26	10	..	22	45	18	13	10.5
2017-T4, T451	62	40	..	22	105	38	18	10.5
2018-T61	61	46	..	12	120	39	17	10.8
2024-0	27	11	20	22	47	18	13	10.6
2024-T3	70	50	18	..	120	41	20	10.6
2024-T4, T351	68	47	20	19	120	41	20	10.6
2024-T361[7]	72	57	13	..	130	42	18	10.6
Alclad 2024-0	26	11	20	..	..	18	..	10.6
Alclad 2024-T3	65	45	18	..	..	40	..	10.6
Alclad 2024-T4, T351	64	42	19	..	..	40	..	10.6
Alclad 2024-T361[7]	67	53	11	..	..	41	..	10.6
Alclad 2024-T81, T851	65	60	6	..	..	40	..	10.6
Alclad 2024-T861[7]	70	66	6	..	..	42	..	10.6
2025-T6	58	37	..	19	110	35	18	10.4
2036-T4	49	28	24	..	..	..	18[9]	10.3
2117-T4	43	24	..	27	70	28	14	10.3
2124-T851	70	64	..	8	..	..	..	10.6
2218-T72	48	37	..	11	95	30	..	10.8

Typical Mechanical Properties (*continued*)

ALLOY AND TEMPER	TENSION				HARDNESS	SHEAR	FATIGUE	MODULUS
	STRENGTH ksi		ELONGATION percent in 2 in.		BRINELL NUMBER	ULTIMATE SHEARING STRENGTH	ENDURANCE① LIMIT	MODULUS④ OF ELASTICITY
	ULTIMATE	YIELD	1/16 in. Thick Specimen	1/2 in. Diameter Specimen	500 kg load 10 mm ball	ksi	ksi	ksi × 10³
2219-0	25	11	18	...	...	...	...	10.6
2219-T42	52	27	20	...	...	...	...	10.6
2219-T31, T351	52	36	17	...	...	...	...	10.6
2219-T37	57	46	11	...	...	...	...	10.6
2219-T62	60	42	10	...	...	...	15	10.6
2219-T81, T851	66	51	10	...	...	...	15	10.6
2219-T87	69	57	10	...	...	...	15	10.6
2618-T61	64	54	...	10	115	38	18	10.8
3003-0	16	6	30	40	28	11	7	10.0
3003-H12	19	18	10	20	35	12	8	10.0
3003-H14	22	21	8	16	40	14	9	10.0
3003-H16	26	25	5	14	47	15	10	10.0
3003-H18	29	27	4	10	55	16	10	10.0
Alclad 3003-0	16	6	30	40	...	11	...	10.0
Alclad 3003-H12	19	18	10	20	...	12	...	10.0
Alclad 3003-H14	22	21	8	16	...	14	...	10.0
Alclad 3003-H16	26	25	5	14	...	15	...	10.0
Alclad 3003-H18	29	27	4	10	...	16	...	10.0
3004-0	26	10	20	25	45	16	14	10.0
3004-H32	31	25	10	17	52	17	15	10.0
3004-H34	35	29	9	12	63	18	15	10.0
3004-H36	38	33	5	9	70	20	16	10.0
3004-H38	41	36	5	6	77	21	16	10.0
Alclad 3004-0	26	10	20	25	...	16	...	10.0
Alclad 3004-H32	31	25	10	17	...	17	...	10.0
Alclad 3004-H34	35	29	9	12	...	18	...	10.0
Alclad 3004-H36	38	33	5	9	...	20	...	10.0
Alclad 3004-H38	41	36	5	6	...	21	...	10.0
3105-0	17	8	24	...	...	12	...	10.0
3105-H12	22	19	7	...	...	14	...	10.0
3105-H14	25	22	5	...	...	15	...	10.0
3105-H16	28	25	4	...	...	16	...	10.0
3105-H18	31	28	3	...	...	17	...	10.0
3105-H25	26	23	8	...	...	15	...	10.0
4032-T6	55	46	...	9	120	38	16	11.4
5005-0	18	6	25	...	28	11	...	10.0
5005-H12	20	19	10	...	...	14	...	10.0
5005-H14	23	22	6	...	...	14	...	10.0
5005-H16	26	25	5	...	...	15	...	10.0
5005-H18	29	28	4	...	...	16	...	10.0
5005-H32	20	17	11	...	36	14	...	10.0
5005-H34	23	20	8	...	41	14	...	10.0
5005-H36	26	24	6	...	46	15	...	10.0
5005-H38	29	27	5	...	51	16	...	10.0
5050-0	21	8	24	...	36	15	12	10.0
5050-H32	25	21	9	...	46	17	13	10.0
5050-H34	28	24	8	...	53	18	13	10.0
5050-H36	30	26	7	...	58	19	14	10.0
5050-H38	32	29	6	...	63	20	14	10.0

Typical Mechanical Properties (*continued*)

ALLOY AND TEMPER	TENSION				HARDNESS	SHEAR	FATIGUE	MODULUS
	STRENGTH ksi		ELONGATION percent in 2 in.		BRINELL NUMBER	ULTIMATE SHEARING STRENGTH	ENDURANCE[3] LIMIT	MODULUS[4] OF ELASTICITY
	ULTIMATE	YIELD	1/16 in. Thick Specimen	1/2 in. Diameter Specimen	500 kg load 10 mm ball	ksi	ksi	ksi × 10^3
5052-0	28	13	25	30	47	18	16	10.2
5052-H32	33	28	12	18	60	20	17	10.2
5052-H34	38	31	10	14	68	21	18	10.2
5052-H36	40	35	8	10	73	23	19	10.2
5052-H38	42	37	7	8	77	24	20	10.2
5056-0	42	22	..	35	65	26	20	10.3
5056-H18	63	59	..	10	105	34	22	10.3
5056-H38	60	50	..	15	100	32	22	10.3
5083-0	42	21	..	22	..	25	..	10.3
5083-H321, H116	46	33	..	16	..	..	23	10.3
5086-0	38	17	22	..	..	23	..	10.3
5086-H32, H116	42	30	12	..	..	..	..	10.3
5086-H34	47	37	10	..	..	27	..	10.3
5086-H112	39	19	14	..	..	..	..	10.3
5154-0	35	17	27	..	58	22	17	10.2
5154-H32	39	30	15	..	67	22	18	10.2
5154-H34	42	33	13	..	73	24	19	10.2
5154-H36	45	36	12	..	78	26	20	10.2
5154-H38	48	39	10	..	80	28	21	10.2
5154-H112	35	17	25	..	63	..	17	10.2
5252-H25	34	25	11	..	68	21	..	10.0
5252-H38, H28	41	35	5	..	75	23	..	10.0
5254-0	35	17	27	..	58	22	17	10.2
5254-H32	39	30	15	..	67	22	18	10.2
5254-H34	42	33	13	..	73	24	19	10.2
5254-H36	45	36	12	..	78	26	20	10.2
5254-H38	48	39	10	..	80	28	21	10.2
5254-H112	35	17	25	..	63	..	17	10.2
5454-0	36	17	22	..	62	23	..	10.2
5454-H32	40	30	10	..	73	24	..	10.2
5454-H34	44	35	10	..	81	26	..	10.2
5454-H111	38	26	14	..	70	23	..	10.2
5454-H112	36	18	18	..	62	23	..	10.2
5456-0	45	23	..	24	..	..	..	10.3
5456-H112	45	24	..	22	..	..	..	10.3
5456-H321, H116	51	37	..	16	90	30	..	10.3
5457-0	19	7	22	..	32	12	..	10.0
5457-H25	26	23	12	..	48	16	..	10.0
5457-H38, H28	30	27	6	..	55	18	..	10.0
5652-0	28	13	25	30	47	18	16	10.2
5652-H32	33	28	12	18	60	20	17	10.2
5652-H34	38	31	10	14	68	21	18	10.2
5652-H36	40	35	8	10	73	23	19	10.2
5652-H38	42	37	7	8	77	24	20	10.2
5657-H25	23	20	12	..	40	14	..	10.0
5657-H38, H28	28	24	7	..	50	15	..	10.0

Typical Mechanical Properties (*concluded*)

ALLOY AND TEMPER	TENSION				HARDNESS	SHEAR	FATIGUE	MODULUS
	STRENGTH ksi		ELONGATION percent in 2 in.		BRINELL NUMBER	ULTIMATE SHEARING STRENGTH	ENDURANCE[3] LIMIT	MODULUS[4] OF ELASTICITY
	ULTIMATE	YIELD	1/16 in. Thick Specimen	1/2 in. Diameter Specimen	500 kg load 10 mm ball	ksi	ksi	ksi × 10³
6061-0	18	8	25	30	30	12	9	10.0
6061-T4, T451	35	21	22	25	65	24	14	10.0
6061-T6, T651	45	40	12	17	95	30	14	10.0
Alclad 6061-0	17	7	25	..	..	11	..	10.0
Alclad 6061-T4, T451	33	19	22	..	..	22	..	10.0
Alclad 6061-T6, T651	42	37	12	..	..	27	..	10.0
6063-0	13	7	..	..	25	10	8	10.0
6063-T1	22	13	20	..	42	14	9	10.0
6063-T4	25	13	22	..	..	..	..	10.0
6063-T5	27	21	12	..	60	17	10	10.0
6063-T6	35	31	12	..	73	22	10	10.0
6063-T83	37	35	9	..	82	22	..	10.0
6063-T831	30	27	10	..	70	18	..	10.0
6063-T832	42	39	12	..	95	27	..	10.0
6066-0	22	12	..	18	43	14	..	10.0
6066-T4, T451	52	30	..	18	90	29	..	10.0
6066-T6, T651	57	52	..	12	120	34	16	10.0
6070-T6	55	51	10	..	..	34	14	10.0
6101-H111	14	11	..	..	..	..	..	10.0
6101-T6	32	28	15	..	71	20	..	10.0
6262-T9	58	55	..	10	120	35	13	10.0
6351-T4	36	22	20	..	..	..	..	10.0
6351-T6	45	41	14	..	95	29	13	10.0
6463-T1	22	13	20	..	42	14	10	10.0
6463-T5	27	21	12	..	60	17	10	10.0
6463-T6	35	31	12	..	74	22	10	10.0
7049-T73	75	65	..	12	135	44	..	10.4
7049-T7352	75	63	..	11	135	43	..	10.4
7050-T73510, T73511	72	63	..	12	..	..	..	10.4
7050-T7451[10]	76	68	..	11	..	44	..	10.4
7050-T7651	80	71	..	11	..	47	..	10.4
7075-0	33	15	17	16	60	22	..	10.4
7075-T6, T651	83	73	11	11	150	48	23	10.4
Alclad 7075-0	32	14	17	..	..	22	..	10.4
Alclad 7075-T6, T651	76	67	11	..	..	46	..	10.4
7178-0	33	15	15	16	..	..	..	10.4
7178-T6, T651	88	78	10	11	..	..	..	10.4
7178-T76, T7651	83	73	..	11	..	..	..	10.3
Alclad 7178-0	32	14	16	..	..	..	..	10.4
Alclad 7178-T6, T651	81	71	10	..	..	..	..	10.4
8176-H24	17	14	15	..	..	10	..	10.0

[1]The mechanical property limits are listed by major product in subsequent tables.

[2]The indicated typical mechanical properties for all except 0 temper material are higher than the specified minimum properties. For 0 temper products typical ultimate and yield values are slightly lower than specified (maximum) values.

[3]Based on 500,000,000 cycles of completely reversed stress using the R. R. Moore type of machine and specimen.

[4]Average of tension and compression moduli. Compression modulus is about 2% greater than tension modulus.

[5]1350-0 wire will have an elongation of approximately 23% in 10 in.

[6]1350-H19 wire will have an elongation of approximately $1\frac{1}{2}$% in 10 in.

[7]Tempers T361 and T861 were formerly designated T36 and T86, respectively.

[8]Based on $\frac{1}{4}$-in.-thick specimen.

[9]Based on 10^7 cycles using flexural type testing of sheet specimens.

[10]T7451 although not previously registered has appeared in literature and in some specifications as T73651.

Typical Physical Properties

ALLOY	AVERAGE① COEFFICIENT OF THERMAL EXPANSION	MELTING RANGE②③ APPROX.	TEMPER	THERMAL CONDUCTIVITY AT 77°F	ELECTRICAL CONDUCTIVITY AT 68°F Percent of International Annealed Copper Standard		ELECTRICAL RESISTIVITY AT 68°F
	68° to 212°F per °F	°F		English Units④	Equal Volume	Equal Weight	Ohm—Cir. Mil/Foot
1060	13.1	1195–1215	0	1625	62	204	17
			H18	1600	61	201	17
1100	13.1	1190–1215	0	1540	59	194	18
			H18	1510	57	187	18
1350	13.2	1195–1215	All	1625	62	204	17
2011	12.7	1005–1190⑥	T3	1050	39	123	27
			T8	1190	45	142	23
2014	12.8	945–1180⑤	0	1340	50	159	21
			T4	930	34	108	31
			T6	1070	40	127	26
2017	13.1	955–1185⑤	0	1340	50	159	21
			T4	930	34	108	31
2018	12.4	945–1180⑥	T61	1070	40	127	26
2024	12.9	935–1180⑤	0	1340	50	160	21
			T3,T4,T361	840	30	96	35
			T6,T81,T861	1050	38	122	27
2025	12.6	970–1185⑤	T6	1070	40	128	26
2036	13.0	1030–1200⑥	T4	1100	41	135	25
2117	13.2	1030–1200⑥	T4	1070	40	130	26
2124	12.7	935–1180⑤	T851	1055	38	122	27
2218	12.4	940–1175⑤	T72	1070	40	126	26
2219	12.4	1010–1190⑤	0	1190	44	138	24
			T31,T37	780	28	88	37
			T6,T81,T87	840	30	94	35
2618	12.4	1020–1180	T6	1020	37	120	28
3003	12.9	1190–1210	0	1340	50	163	21
			H12	1130	42	137	25
			H14	1100	41	134	25
			H18	1070	40	130	26
3004	13.3	1165–1210	All	1130	42	137	25
3105	13.1	1175–1210	All	1190	45	148	23
4032	10.8	990–1060⑤	0	1070	40	132	26
			T6	960	35	116	30
4043	12.3	1065–1170	0	1130	42	140	25
4045	11.7	1065–1110	All	1190	45	151	23
4343	12.0	1070–1135	All	1250	47	158	25
5005	13.2	1170–1210	All	1390	52	172	20
5050	13.2	1155–1205	All	1340	50	165	21
5052	13.2	1125–1200	All	960	35	116	30
5056	13.4	1055–1180	0	810	29	98	36
			H38	750	27	91	38
5083	13.2	1095–1180	0	810	29	98	36
5086	13.2	1085–1185	All	870	31	104	33
5154	13.3	1100–1190	All	870	32	107	32
5252	13.2	1125–1200	All	960	35	116	30
5254	13.3	1100–1190	All	870	32	107	32
5356	13.4	1060–1175	0	810	29	98	36
5454	13.1	1115–1195	0	930	34	113	31
			H38	930	34	113	31
5456	13.3	1055–1180	0	810	29	98	36
5457	13.2	1165–1210	All	1220	46	153	23
5652	13.2	1125–1200	All	960	35	116	30
5657	13.2	1180–1215	All	1420	54	180	19

Typical Physical Properties (*concluded*)

ALLOY	AVERAGE[1] COEFFICIENT OF THERMAL EXPANSION	MELTING RANGE[2] [3] APPROX.	TEMPER	THERMAL CONDUCTIVITY AT 77°F	ELECTRICAL CONDUCTIVITY AT 68°F Percent of International Annealed Copper Standard		ELECTRICAL RESISTIVITY AT 68°F
	68° to 212°F per °F	°F		English Units[4]	Equal Volume	Equal Weight	Ohm—Cir. Mil/Foot
6005	13.0	1125–1210[6]	T1	1250	47	155	22
			T5	1310	49	161	21
6053	12.8	1070–1205[6]	0	1190	45	148	23
			T4	1070	40	132	26
			T6	1130	42	139	25
6061	13.1	1080–1205[6]	0	1250	47	155	22
			T4	1070	40	132	26
			T6	1160	43	142	24
6063	13.0	1140–1210	0	1510	58	191	18
			T1	1340	50	165	21
			T5	1450	55	181	19
			T6,T83	1390	53	175	20
6066	12.9	1045–1195[5]	0	1070	40	132	26
			T6	1020	37	122	28
6070	. .	1050–1200[5]	T6	1190	44	145	24
6101	13.0	1150–1210	T6	1510	57	188	18
			T61	1540	59	194	18
			T63	1510	58	191	18
			T64	1570	60	198	17
			T65	1510	58	191	18
6105	13.0	1110–1200[6]	T1	1220	46	151	23
			T5	1340	50	165	21
6151	12.9	1090–1200[6]	0	1420	54	178	19
			T4	1130	42	138	25
			T6	1190	45	148	23
6201	13.0	1125–1210[6]	T81	1420	54	180	19
6253	. .	1100–1205	. .	. .	. .	. .	. .
6262	13.0	1080–1205[6]	T9	1190	44	145	24
6351	13.0	1030–1200	T6	1220	46	151	23
6463	13.0	1140–1210	T1	1340	50	165	21
			T5	1450	55	181	19
			T6	1390	53	175	20
6951	13.0	1140–1210	0	1480	56	186	19
			T6	1370	52	172	20
7049	13.0	890–1175	T73	1070	40	132	26
7050	13.4	910–1165	T74[8]	1090	41	135	25
7072	13.1	1185–1215	0	1540	59	193	18
7075	13.1	890–1175[7]	T6	900	33	105	31
7178	13.0	890–1165[7]	T6	870	31	98	33
8017	13.1	1190–1215	H12,H22	. .	59	193	18
			H212	. .	61	200	17
8030	13.1	1190–1215	H221	1600	61	201	17
8176	13.1	1190–1215	H24	1600	61	201	17

[1]Coefficient to be multiplied by 10^{-6}. Example: $12.2 \times 10^{-6} = 0.0000122$.

[2]Melting ranges shown apply to wrought products of $\frac{1}{4}$-in. thickness or greater.

[3]Based on typical composition of the indicated alloys.

[4]English units = BTU-in./ft^2 hr°F

[5]Eutectic melting is not eliminated by homogenization.

[6]Eutectic melting can be completely eliminated by homogenization.

[7]Homogenization may raise eutectic melting temperature 20–40°F but usually does not eliminate eutectic melting.

[8]Although not formerly registered, the literature and some specifications have used T736 as the designation for this temper.

A.2 SHEET AND PLATE

Mechanical Property Limits—Nonheat-Treatable Alloys[1, 10]

ALLOY AND TEMPER	SPECIFIED THICKNESS[2] IN.	TENSILE STRENGTH—ksi				ELONGATION PERCENT MIN IN 2 IN. OR 4D[5]
		ULTIMATE		YIELD		
		min	max	min	max	
		1060				
1060-0	0.006–0.019	8.0	14.0	2.5	..	15
	0.020–0.050	8.0	14.0	2.5	..	22
	0.051–3.000	8.0	14.0	2.5	..	25
1060-H12[3]	0.017–0.050	11.0	16.0	9.0	..	6
	0.051–2.000	11.0	16.0	9.0	..	12
1060-H14[3]	0.009–0.019	12.0	17.0	10.0	..	1
	0.020–0.050	12.0	17.0	10.0	..	5
	0.051–1.000	12.0	17.0	10.0	..	10
1060-H16[3]	0.006–0.019	14.0	19.0	11.0	..	1
	0.020–0.050	14.0	19.0	11.0	..	4
	0.051–0.162	14.0	19.0	11.0	..	5
1060-H18[3]	0.006–0.019	16.0	..	12.0	..	1
	0.020–0.050	16.0	..	12.0	..	3
	0.051–0.128	16.0	..	12.0	..	4
1060-H112	0.250–0.499	11.0	..	..	..	10
	0.500–1.000	10.0	..	..	..	20
	1.001–3.000	9.0	..	..	..	25
		1100				
1100-0	0.006–0.019	11.0	15.5	3.5	..	15
	0.020–0.031	11.0	15.5	3.5	..	20
	0.032–0.050	11.0	15.5	3.5	..	25
	0.051–0.249	11.0	15.5	3.5	..	30
	0.250–3.000	11.0	15.5	3.5	..	28
1100-H12[3]	0.017–0.019	14.0	19.0	11.0	..	3
	0.020–0.031	14.0	19.0	11.0	..	4
	0.032–0.050	14.0	19.0	11.0	..	6
	0.051–0.113	14.0	19.0	11.0	..	8
	0.114–0.499	14.0	19.0	11.0	..	9
	0.500–2.000	14.0	19.0	11.0	..	12
1100-H14[3]	0.009–0.012	16.0	21.0	14.0	..	1
	0.013–0.019	16.0	21.0	14.0	..	2
	0.020–0.031	16.0	21.0	14.0	..	3
	0.032–0.050	16.0	21.0	14.0	..	4
	0.051–0.113	16.0	21.0	14.0	..	5
	0.114–0.499	16.0	21.0	14.0	..	6
	0.500–1.000	16.0	21.0	14.0	..	10
1100-H16[3]	0.006–0.019	19.0	24.0	17.0	..	1
	0.020–0.031	19.0	24.0	17.0	..	2
	0.032–0.050	19.0	24.0	17.0	..	3
	0.051–0.162	19.0	24.0	17.0	..	4
1100-H18	0.006–0.019	22.0	..	..	..	1
	0.020–0.031	22.0	..	..	..	2
	0.032–0.050	22.0	..	..	..	3
	0.051–0.128	22.0	..	..	..	4
1100-H19	0.006–0.063	24.0	..	..	..	1
1100-H112	0.250–0.499	13.0	..	7.0	..	9
	0.500–2.000	12.0	..	5.0	..	14
	2.001–3.000	11.5	..	4.0	..	20

Mechanical Property Limits—Nonheat-Treatable Alloys (*continued*)

ALLOY AND TEMPER	SPECIFIED THICKNESS② IN.	TENSILE STRENGTH—ksi				ELONGATION PERCENT MIN IN 2 IN. OR 4D⑤
		ULTIMATE		YIELD		
		min	max	min	max	
			1350			
1350-0	0.006–0.019	8.0	14.0	..	..	15
	0.020–0.031	8.0	14.0	..	..	20
	0.032–0.050	8.0	14.0	..	..	25
	0.051–0.249	8.0	14.0	..	..	30
	0.250–3.000	8.0	14.0	..	..	28
1350-H12	0.017–0.019	12.0	17.0	..	..	3
	0.020–0.031	12.0	17.0	..	..	4
	0.032–0.050	12.0	17.0	..	..	6
	0.051–0.113	12.0	17.0	..	..	8
	0.114–0.499	12.0	17.0	..	..	9
	0.500–2.000	12.0	17.0	..	..	12
1350-H14	0.009–0.012	14.0	19.0	..	..	1
	0.013–0.019	14.0	19.0	..	..	2
	0.020–0.031	14.0	19.0	..	..	3
	0.032–0.050	14.0	19.0	..	..	4
	0.051–0.113	14.0	19.0	..	..	5
	0.114–0.499	14.0	19.0	..	..	6
	0.500–1.000	14.0	19.0	..	..	10
1350-H16	0.006–0.019	16.0	21.0	..	..	1
	0.020–0.031	16.0	21.0	..	..	2
	0.032–0.050	16.0	21.0	..	..	3
	0.051–0.162	16.0	21.0	..	..	4
1350-H18	0.006–0.019	18.0	..	..	..	1
	0.020–0.031	18.0	..	..	..	2
	0.032–0.050	18.0	..	..	..	3
	0.051–0.128	18.0	..	..	..	4
1350-H112	0.250–0.499	11.0	..	..	..	10
	0.500–1.000	10.0	..	..	..	16
	1.001–1.500	9.0	..	..	..	22
			3003			
3003-0	0.006–0.007	14.0	19.0	5.0	..	14
	0.008–0.012	14.0	19.0	5.0	..	18
	0.013–0.031	14.0	19.0	5.0	..	20
	0.032–0.050	14.0	19.0	5.0	..	23
	0.051–0.249	14.0	19.0	5.0	..	25
	0.250–3.000	14.0	19.0	5.0	..	23
3003-H12③	0.017–0.019	17.0	23.0	12.0	..	3
	0.020–0.031	17.0	23.0	12.0	..	4
	0.032–0.050	17.0	23.0	12.0	..	5
	0.051–0.113	17.0	23.0	12.0	..	6
	0.114–0.161	17.0	23.0	12.0	..	7
	0.162–0.249	17.0	23.0	12.0	..	8
	0.250–0.499	17.0	23.0	12.0	..	9
	0.500–2.000	17.0	23.0	12.0	..	10
3003-H14③	0.009–0.012	20.0	26.0	17.0	..	1
	0.013–0.019	20.0	26.0	17.0	..	2
	0.020–0.031	20.0	26.0	17.0	..	3
	0.032–0.050	20.0	26.0	17.0	..	4
	0.051–0.113	20.0	26.0	17.0	..	5
	0.114–0.161	20.0	26.0	17.0	..	6
	0.162–0.249	20.0	26.0	17.0	..	7
	0.250–0.499	20.0	26.0	17.0	..	8
	0.500–1.000	20.0	26.0	17.0	..	10
3003-H16③	0.006–0.019	24.0	30.0	21.0	..	1
	0.020–0.031	24.0	30.0	21.0	..	2
	0.032–0.050	24.0	30.0	21.0	..	3
	0.051–0.162	24.0	30.0	21.0	..	4

Mechanical Property Limits—Nonheat-Treatable Alloys (*continued*)

ALLOY AND TEMPER	SPECIFIED THICKNESS② IN.	TENSILE STRENGTH—ksi				ELONGATION PERCENT MIN IN 2 IN. OR 4D⑤
		ULTIMATE		YIELD		
		min	max	min	max	
		3003 (Continued)				
3003-H18③	0.006–0.019	27.0	. .	24.0	. .	1
	0.020–0.031	27.0	. .	24.0	. .	2
	0.032–0.050	27.0	. .	24.0	. .	3
	0.051–0.128	27.0	. .	24.0	. .	4
3003-H19	0.006–0.063	29.0	. .	. .	. .	1
3003-H112	0.250–0.499	17.0	. .	10.0	. .	8
	0.500–2.000	15.0	. .	6.0	. .	12
	2.001–3.000	14.5	. .	6.0	. .	18
		ALCLAD 3003⑧				
Alclad 3003-0	0.006–0.007	13.0	18.0	4.5	. .	14
	0.008–0.012	13.0	18.0	4.5	. .	18
	0.013–0.031	13.0	18.0	4.5	. .	20
	0.032–0.050	13.0	18.0	4.5	. .	23
	0.051–0.249	13.0	18.0	4.5	. .	25
	0.250–0.499	13.0	18.0	4.5	. .	23
	0.500–3.000	14.0④	19.0④	5.0④	. .	23
Alclad 3003-H12③	0.017–0.031	16.0	22.0	11.0	. .	4
	0.032–0.050	16.0	22.0	11.0	. .	5
	0.051–0.113	16.0	22.0	11.0	. .	6
	0.114–0.161	16.0	22.0	11.0	. .	7
	0.162–0.249	16.0	22.0	11.0	. .	8
	0.250–0.499	16.0	22.0	11.0	. .	9
	0.500–2.000	17.0④	23.0④	12.0④	. .	10
Alclad 3003-H14③	0.009–0.012	19.0	25.0	16.0	. .	1
	0.013–0.019	19.0	25.0	16.0	. .	2
	0.020–0.031	19.0	25.0	16.0	. .	3
	0.032–0.050	19.0	25.0	16.0	. .	4
	0.051–0.113	19.0	25.0	16.0	. .	5
	0.114–0.161	19.0	25.0	16.0	. .	6
	0.162–0.249	19.0	25.0	16.0	. .	7
	0.250–0.499	19.0	25.0	16.0	. .	8
	0.500–1.000	20.0④	26.0④	17.0④	. .	10
Alclad 3003-H16③	0.006–0.019	23.0	29.0	20.0	. .	1
	0.020–0.031	23.0	29.0	20.0	. .	2
	0.032–0.050	23.0	29.0	20.0	. .	3
	0.051–0.162	23.0	29.0	20.0	. .	4
Alclad 3003-H18	0.006–0.019	26.0	. .	. .	. .	1
	0.020–0.031	26.0	. .	. .	. .	2
	0.032–0.050	26.0	. .	. .	. .	3
	0.051–0.128	26.0	. .	. .	. .	4
Alclad 3003-H112	0.250–0.499	16.0	. .	9.0	. .	8
	0.500–2.000	15.0④	. .	6.0④	. .	12
	2.001–3.000	14.5④	. .	6.0④	. .	18
		3004				
3004-0	0.006–0.007	22.0	29.0	8.5	. .	. .
	0.008–0.019	22.0	29.0	8.5	. .	10
	0.020–0.031	22.0	29.0	8.5	. .	14
	0.032–0.050	22.0	29.0	8.5	. .	16
	0.051–0.249	22.0	29.0	8.5	. .	18
	0.250–3.000	22.0	29.0	8.5	. .	16
3004-H32③	0.017–0.019	28.0	35.0	21.0	. .	1
	0.020–0.031	28.0	35.0	21.0	. .	3
	0.032–0.050	28.0	35.0	21.0	. .	4
	0.051–0.113	28.0	35.0	21.0	. .	5
	0.114–2.000	28.0	35.0	21.0	. .	6

Mechanical Property Limits—Nonheat-Treatable Alloys (*continued*)

ALLOY AND TEMPER	SPECIFIED THICKNESS② IN.	TENSILE STRENGTH—ksi ULTIMATE min	TENSILE STRENGTH—ksi ULTIMATE max	TENSILE STRENGTH—ksi YIELD min	TENSILE STRENGTH—ksi YIELD max	ELONGATION PERCENT MIN IN 2 IN. OR 4D⑤
			3004 (Continued)			
3004-H34③	0.009–0.019	32.0	38.0	25.0	..	1
	0.020–0.050	32.0	38.0	25.0	..	3
	0.051–0.113	32.0	38.0	25.0	..	4
	0.114–1.000	32.0	38.0	25.0	..	5
3004-H36③	0.006–0.007	35.0	41.0	28.0	..	..
	0.008–0.019	35.0	41.0	28.0	..	1
	0.020–0.031	35.0	41.0	28.0	..	2
	0.032–0.050	35.0	41.0	28.0	..	3
	0.051–0.162	35.0	41.0	28.0	..	4
3004-H38③	0.006–0.007	38.0	..	31.0	..	..
	0.008–0.019	38.0	..	31.0	..	1
	0.020–0.031	38.0	..	31.0	..	2
	0.032–0.050	38.0	..	31.0	..	3
	0.051–0.128	38.0	..	31.0	..	4
3004-H112	0.250–3.000	23.0	..	9.0	..	7
			ALCLAD 3004⑧			
Alclad 3004-0	0.006–0.007	21.0	28.0	8.0	..	..
	0.008–0.019	21.0	28.0	8.0	..	10
	0.020–0.031	21.0	28.0	8.0	..	14
	0.032–0.050	21.0	28.0	8.0	..	16
	0.051–0.249	21.0	28.0	8.0	..	18
	0.250–0.499	21.0	28.0	8.0	..	16
	0.500–3.000	22.0④	29.0④	8.5④	..	16
Alclad 3004-H32③	0.017–0.019	27.0	34.0	20.0	..	1
	0.020–0.031	27.0	34.0	20.0	..	3
	0.032–0.050	27.0	34.0	20.0	..	4
	0.051–0.113	27.0	34.0	20.0	..	5
	0.114–0.249	27.0	34.0	20.0	..	6
	0.250–0.499	27.0	34.0	20.0	..	6
	0.500–2.000	28.0④	35.0④	21.0④	..	6
Alclad 3004-H34③	0.009–0.019	31.0	37.0	24.0	..	1
	0.020–0.050	31.0	37.0	24.0	..	3
	0.051–0.113	31.0	37.0	24.0	..	4
	0.114–0.249	31.0	37.0	24.0	..	5
	0.250–0.499	31.0	37.0	24.0	..	5
	0.500–1.000	32.0④	38.0④	25.0④	..	5
Alclad 3004-H36③	0.006–0.007	34.0	40.0	27.0	..	..
	0.008–0.019	34.0	40.0	27.0	..	1
	0.020–0.031	34.0	40.0	27.0	..	2
	0.032–0.050	34.0	40.0	27.0	..	3
	0.051–0.162	34.0	40.0	27.0	..	4
Alclad 3004-H38	0.006–0.007	37.0	..	..	..	..
	0.008–0.019	37.0	..	..	..	1
	0.020–0.031	37.0	..	..	..	2
	0.032–0.050	37.0	..	..	..	3
	0.051–0.128	37.0	..	..	..	4
Alclad 3004-H112	0.250–0.499	22.0	..	8.5	..	7
	0.500–3.000	23.0④	..	9.0④	..	7
			3005			
3005-0	0.006–0.007	17.0	24.0	6.5	..	10
	0.008–0.012	17.0	24.0	6.5	..	12
	0.013–0.019	17.0	24.0	6.5	..	14
	0.020–0.031	17.0	24.0	6.5	..	16
	0.032–0.050	17.0	24.0	6.5	..	18
	0.051–0.249	17.0	24.0	6.5	..	20

Mechanical Property Limits—Nonheat-Treatable Alloys (*continued*)

ALLOY AND TEMPER	SPECIFIED THICKNESS② IN.	TENSILE STRENGTH—ksi				ELONGATION PERCENT MIN IN 2 IN. OR 4D⑤
		ULTIMATE		YIELD		
		min	max	min	max	
3005 (Continued)						
3005-H12	0.017–0.019	20.0	27.0	17.0	..	1
	0.020–0.050	20.0	27.0	17.0	..	2
	0.051–0.113	20.0	27.0	17.0	..	3
	0.114–0.161	20.0	27.0	17.0	..	4
	0.162–0.249	20.0	27.0	17.0	..	5
3005-H14	0.009–0.031	24.0	31.0	21.0	..	1
	0.032–0.050	24.0	31.0	21.0	..	2
	0.051–0.113	24.0	31.0	21.0	..	3
	0.114–0.249	24.0	31.0	21.0	..	4
3005-H16	0.006–0.031	28.0	35.0	25.0	..	1
	0.032–0.113	28.0	35.0	25.0	..	2
	0.114–0.162	28.0	35.0	25.0	..	3
3005-H18	0.006–0.031	32.0	..	29.0	..	1
	0.032–0.128	32.0	..	29.0	..	2
3005-H19	0.006–0.012	34.0	..	..	..	..
	0.013–0.063	34.0	..	..	..	1
3005-H25	0.006–0.019	26.0	34.0	22.0	..	1
	0.020–0.031	26.0	34.0	22.0	..	2
	0.032–0.050	26.0	34.0	22.0	..	3
	0.051–0.080	26.0	34.0	22.0	..	4
3005-H26	0.006–0.019	28.0	36.0	24.0	..	1
	0.020–0.031	28.0	36.0	24.0	..	2
	0.032–0.050	28.0	36.0	24.0	..	3
	0.051–0.080	28.0	36.0	24.0	..	4
3005-H27	0.006–0.019	29.5	37.5	26.0	..	1
	0.020–0.031	29.5	37.5	26.0	..	2
	0.032–0.050	29.5	37.5	26.0	..	3
	0.051–0.080	29.5	37.5	26.0	..	4
3005-H28	0.006–0.019	31.0	..	27.0	..	1
	0.020–0.031	31.0	..	27.0	..	2
	0.032–0.050	31.0	..	27.0	..	3
	0.051–0.080	31.0	..	27.0	..	4
3105						
3105-0	0.013–0.019	14.0	21.0	5.0	..	16
	0.020–0.031	14.0	21.0	5.0	..	18
	0.032–0.080	14.0	21.0	5.0	..	20
3105-H12	0.017–0.019	19.0	26.0	15.0	..	1
	0.020–0.031	19.0	26.0	15.0	..	1
	0.032–0.050	19.0	26.0	15.0	..	2
	0.051–0.080	19.0	26.0	15.0	..	3
3105-H14	0.013–0.019	22.0	29.0	18.0	..	1
	0.020–0.031	22.0	29.0	18.0	..	1
	0.032–0.050	22.0	29.0	18.0	..	2
	0.051–0.080	22.0	29.0	18.0	..	2
3105-H16	0.013–0.031	25.0	32.0	21.0	..	1
	0.032–0.050	25.0	32.0	21.0	..	2
	0.051–0.080	25.0	32.0	21.0	..	2
3105-H18	0.013–0.031	28.0	..	24.0	..	1
	0.032–0.050	28.0	..	24.0	..	1
	0.051–0.080	28.0	..	24.0	..	2
3105-H25	0.013–0.019	23.0	..	19.0	..	2
	0.020–0.031	23.0	..	19.0	..	3
	0.032–0.050	23.0	..	19.0	..	4
	0.051–0.080	23.0	..	19.0	..	6

Mechanical Property Limits—Nonheat-Treatable Alloys (*continued*)

ALLOY AND TEMPER	SPECIFIED THICKNESS② IN.	TENSILE STRENGTH—ksi				ELONGATION PERCENT MIN IN 2 IN. OR 4D⑤
		ULTIMATE		YIELD		
		min	max	min	max	
			5005			
5005-0	0.006–0.007	15.0	21.0	5.0	. .	12
	0.008–0.012	15.0	21.0	5.0	. .	14
	0.013–0.019	15.0	21.0	5.0	. .	16
	0.020–0.031	15.0	21.0	5.0	. .	18
	0.032–0.050	15.0	21.0	5.0	. .	20
	0.051–0.113	15.0	21.0	5.0	. .	21
	0.114–0.249	15.0	21.0	5.0	. .	22
	0.250–3.000	15.0	21.0	5.0	. .	22
5005-H12	0.017–0.019	18.0	24.0	14.0	. .	2
	0.020–0.031	18.0	24.0	14.0	. .	3
	0.032–0.050	18.0	24.0	14.0	. .	4
	0.051–0.113	18.0	24.0	14.0	. .	6
	0.114–0.161	18.0	24.0	14.0	. .	7
	0.162–0.249	18.0	24.0	14.0	. .	8
	0.250–0.499	18.0	24.0	14.0	. .	9
	0.500–2.000	18.0	24.0	14.0	. .	10
5005-H14	0.009–0.031	21.0	27.0	17.0	. .	1
	0.032–0.050	21.0	27.0	17.0	. .	2
	0.051–0.113	21.0	27.0	17.0	. .	3
	0.114–0.161	21.0	27.0	17.0	. .	5
	0.162–0.249	21.0	27.0	17.0	. .	6
	0.250–0.499	21.0	27.0	17.0	. .	8
	0.500–1.000	21.0	27.0	17.0	. .	10
5005-H16	0.006–0.031	24.0	30.0	20.0	. .	1
	0.032–0.050	24.0	30.0	20.0	. .	2
	0.051–0.162	24.0	30.0	20.0	. .	3
5005-H18	0.006–0.031	27.0	. .	. .	. .	1
	0.032–0.050	27.0	. .	. .	. .	2
	0.051–0.128	27.0	. .	. .	. .	3
5005-H32③	0.017–0.019	17.0	23.0	12.0	. .	3
	0.020–0.031	17.0	23.0	12.0	. .	4
	0.032–0.050	17.0	23.0	12.0	. .	5
	0.051–0.113	17.0	23.0	12.0	. .	7
	0.114–0.161	17.0	23.0	12.0	. .	8
	0.162–0.249	17.0	23.0	12.0	. .	9
	0.250–2.000	17.0	23.0	12.0	. .	10
5005-H34③	0.009–0.012	20.0	26.0	15.0	. .	2
	0.013–0.031	20.0	26.0	15.0	. .	3
	0.032–0.050	20.0	26.0	15.0	. .	4
	0.051–0.113	20.0	26.0	15.0	. .	5
	0.114–0.161	20.0	26.0	15.0	. .	6
	0.162–0.249	20.0	26.0	15.0	. .	7
	0.250–0.499	20.0	26.0	15.0	. .	8
	0.500–1.000	20.0	26.0	15.0	. .	10
5005-H36③	0.006–0.007	23.0	29.0	18.0	. .	1
	0.008–0.019	23.0	29.0	18.0	. .	2
	0.020–0.031	23.0	29.0	18.0	. .	3
	0.032–0.162	23.0	29.0	18.0	. .	4
5005-H38	0.006–0.012	26.0	. .	. .	. .	1
	0.013–0.019	26.0	. .	. .	. .	2
	0.020–0.031	26.0	. .	. .	. .	3
	0.032–0.128	26.0	. .	. .	. .	4
5005-H39	0.006–0.063	28.0	. .	. .	. .	1
5005-H112	0.250–0.499	17.0	. .	. .	. .	8
	0.500–2.000	15.0	. .	. .	. .	12
	2.001–3.000	14.5	. .	. .	. .	18

Mechanical Property Limits—Nonheat-Treatable Alloys (*continued*)

ALLOY AND TEMPER	SPECIFIED THICKNESS② IN.	TENSILE STRENGTH—ksi				ELONGATION PERCENT MIN IN 2 IN. OR 4D⑤
		ULTIMATE		YIELD		
		min	max	min	max	
			5050			
5050-0	0.006–0.007	18.0	24.0	6.0	. .	. .
	0.008–0.019	18.0	24.0	6.0	. .	16
	0.020–0.031	18.0	24.0	6.0	. .	18
	0.032–0.113	18.0	24.0	6.0	. .	20
	0.114–0.249	18.0	24.0	6.0	. .	22
	0.250–3.000	18.0	24.0	6.0	. .	20
5050-H32③	0.017–0.050	22.0	28.0	16.0	. .	4
	0.051–0.249	22.0	28.0	16.0	. .	6
5050-H34③	0.009–0.031	25.0	31.0	20.0	. .	3
	0.032–0.050	25.0	31.0	20.0	. .	4
	0.051–0.249	25.0	31.0	20.0	. .	5
5050-H36③	0.006–0.019	27.0	33.0	22.0	. .	2
	0.020–0.050	27.0	33.0	22.0	. .	3
	0.051–0.162	27.0	33.0	22.0	. .	4
5050-H38	0.006–0.007	29.0	. .	. .	. .	. .
	0.008–0.031	29.0	. .	. .	. .	2
	0.032–0.050	29.0	. .	. .	. .	3
	0.051–0.128	29.0	. .	. .	. .	4
5050-H39	0.006–0.063	31.0	. .	. .	. .	1
5050-H112	0.250–3.000	20.0	. .	8.0	. .	12
			5052			
5052-0	0.006–0.007	25.0	31.0	9.5	. .	. .
	0.008–0.012	25.0	31.0	9.5	. .	14
	0.013–0.019	25.0	31.0	9.5	. .	15
	0.020–0.031	25.0	31.0	9.5	. .	16
	0.032–0.050	25.0	31.0	9.5	. .	18
	0.051–0.113	25.0	31.0	9.5	. .	19
	0.114–0.249	25.0	31.0	9.5	. .	20
	0.250–3.000	25.0	31.0	9.5	. .	18
5052-H32③	0.017–0.019	31.0	38.0	23.0	. .	4
	0.020–0.050	31.0	38.0	23.0	. .	5
	0.051–0.113	31.0	38.0	23.0	. .	7
	0.114–0.249	31.0	38.0	23.0	. .	9
	0.250–0.499	31.0	38.0	23.0	. .	11
	0.500–2.000	31.0	38.0	23.0	. .	12
5052-H34③	0.009–0.019	34.0	41.0	26.0	. .	3
	0.020–0.050	34.0	41.0	26.0	. .	4
	0.051–0.113	34.0	41.0	26.0	. .	6
	0.114–0.249	34.0	41.0	26.0	. .	7
	0.250–1.000	34.0	41.0	26.0	. .	10
5052-H36③	0.006–0.007	37.0	44.0	29.0	. .	2
	0.008–0.031	37.0	44.0	29.0	. .	3
	0.032–0.162	37.0	44.0	29.0	. .	4
5052-H38③	0.006–0.007	39.0	. .	32.0	. .	2
	0.008–0.031	39.0	. .	32.0	. .	3
	0.032–0.128	39.0	. .	32.0	. .	4
5052-H39	0.006–0.063	41.0	. .	. .	. .	1
5052-H112	0.250–0.499	28.0	. .	16.0	. .	7
	0.500–2.000	25.0	. .	9.5	. .	12
	2.001–3.000	25.0	. .	9.5	. .	16
5052-H391	0.008–0.125	42.0	. .	35.0	. .	3
			5083			
5083-0	0.051–1.500	40.0	51.0	18.0	29.0	16
	1.501–3.000	39.0	50.0	17.0	29.0	16
	3.001–4.000	38.0	. .	16.0	. .	16
	4.001–5.000	38.0	. .	16.0	. .	14
	5.001–7.000	37.0	. .	15.0	. .	14
	7.001–8.000	36.0	. .	14.0	. .	12

Mechanical Property Limits—Nonheat-Treatable Alloys (*continued*)

ALLOY AND TEMPER	SPECIFIED THICKNESS② IN.	TENSILE STRENGTH—ksi ULTIMATE min	ULTIMATE max	YIELD min	YIELD max	ELONGATION PERCENT MIN IN 2 IN. OR 4D⑤
			5083 (Continued)			
5083-H112	0.250–1.500	40.0	..	18.0	..	12
	1.501–3.000	39.0	..	17.0	..	12
5083-H116⑧⑨	0.063–0.499	44.0	..	31.0	..	10
	0.500–1.250	44.0	..	31.0	..	12
	1.251–1.500	44.0	..	31.0	..	12
	1.501–3.000	41.0	..	29.0	..	12
5083-H321	0.188–1.500	44.0	56.0	31.0	43.0	12
	1.501–3.000	41.0	56.0	29.0	43.0	12
			5086			
5086-0	0.020–0.050	35.0	44.0	14.0	..	15
	0.051–0.249	35.0	44.0	14.0	..	18
	0.250–2.000	35.0	44.0	14.0	..	16
5086-H32③	0.020–0.050	40.0	47.0	28.0	..	6
	0.051–0.249	40.0	47.0	28.0	..	8
	0.250–2.000	40.0	47.0	28.0	..	12
5086-H34③	0.009–0.019	44.0	51.0	34.0	..	4
	0.020–0.050	44.0	51.0	34.0	..	5
	0.051–0.249	44.0	51.0	34.0	..	6
	0.250–1.000	44.0	51.0	34.0	..	10
5086-H36③	0.006–0.019	47.0	54.0	38.0	..	3
	0.020–0.050	47.0	54.0	38.0	..	4
	0.051–0.162	47.0	54.0	38.0	..	6
5086-H38③	0.006–0.020	50.0	..	41.0	..	3
5086-H112	0.188–0.499	36.0	..	18.0	..	8
	0.500–1.000	35.0	..	16.0	..	10
	1.001–2.000	35.0	..	14.0	..	14
	2.001–3.000	34.0	..	14.0	..	14
5086-H116⑧⑨	0.063–0.249	40.0	..	28.0	..	8
	0.250–0.499	40.0	..	28.0	..	10
	0.500–1.250	40.0	..	28.0	..	10
	1.251–2.000	40.0	..	28.0	..	10
			5154			
5154-0	0.020–0.031	30.0	41.0	11.0	..	12
	0.032–0.050	30.0	41.0	11.0	..	14
	0.051–0.113	30.0	41.0	11.0	..	16
	0.114–3.000	30.0	41.0	11.0	..	18
5154-H32③	0.020–0.050	36.0	43.0	26.0	..	5
	0.051–0.249	36.0	43.0	26.0	..	8
	0.250–2.000	36.0	43.0	26.0	..	12
5154-H34③	0.009–0.050	39.0	46.0	29.0	..	4
	0.051–0.161	39.0	46.0	29.0	..	6
	0.162–0.249	39.0	46.0	29.0	..	7
	0.250–1.000	39.0	46.0	29.0	..	10
5154-H36③	0.006–0.050	42.0	49.0	32.0	..	3
	0.051–0.113	42.0	49.0	32.0	..	4
	0.114–0.162	42.0	49.0	32.0	..	5
5154-H38③	0.006–0.050	45.0	..	35.0	..	3
	0.051–0.113	45.0	..	35.0	..	4
	0.114–0.128	45.0	..	35.0	..	5
5154-H112	0.250–0.499	32.0	..	18.0	..	8
	0.500–2.000	30.0	..	11.0	..	11
	2.001–3.000	30.0	..	11.0	..	15

Mechanical Property Limits—Nonheat-Treatable Alloys (*continued*)

ALLOY AND TEMPER	SPECIFIED THICKNESS② IN.	TENSILE STRENGTH—ksi ULTIMATE min	TENSILE STRENGTH—ksi ULTIMATE max	TENSILE STRENGTH—ksi YIELD min	TENSILE STRENGTH—ksi YIELD max	ELONGATION PERCENT MIN IN 2 IN. OR 4D⑤
			5252			
5252-H24	0.030–0.090	30.0	38.0	..	..	10
5252-H25	0.030–0.090	31.0	39.0	..	..	9
5252-H28	0.030–0.090	38.0	..	..	..	3
			5254			
5254-0	0.020–0.031	30.0	41.0	11.0	..	12
	0.032–0.050	30.0	41.0	11.0	..	14
	0.051–0.113	30.0	41.0	11.0	..	16
	0.114–3.000	30.0	41.0	11.0	..	18
5254-H32③	0.020–0.050	36.0	43.0	26.0	..	5
	0.051–0.249	36.0	43.0	26.0	..	8
	0.250–2.000	36.0	43.0	26.0	..	12
5254-H34③	0.009–0.050	39.0	46.0	29.0	..	4
	0.051–0.161	39.0	46.0	29.0	..	6
	0.162–0.249	39.0	46.0	29.0	..	7
	0.250–1.000	39.0	46.0	29.0	..	10
5254-H36③	0.006–0.050	42.0	49.0	32.0	..	3
	0.051–0.113	42.0	49.0	32.0	..	4
	0.114–0.162	42.0	49.0	32.0	..	5
5254-H38③	0.006–0.050	45.0	..	35.0	..	3
	0.051–0.113	45.0	..	35.0	..	4
	0.114–0.128	45.0	..	35.0	..	5
5254-H112	0.250–0.499	32.0	..	18.0	..	8
	0.500–2.000	30.0	..	11.0	..	11
	2.001–3.000	30.0	..	11.0	..	15
			5454			
5454-0	0.020–0.031	31.0	41.0	12.0	..	12
	0.032–0.050	31.0	41.0	12.0	..	14
	0.051–0.113	31.0	41.0	12.0	..	16
	0.114–3.000	31.0	41.0	12.0	..	18
5454-H32③	0.020–0.050	36.0	44.0	26.0	..	5
	0.051–0.249	36.0	44.0	26.0	..	8
	0.250–2.000	36.0	44.0	26.0	..	12
5454-H34③	0.020–0.050	39.0	47.0	29.0	..	4
	0.051–0.161	39.0	47.0	29.0	..	6
	0.162–0.249	39.0	47.0	29.0	..	7
	0.250–1.000	39.0	47.0	29.0	..	10
5454-H112	0.250–0.499	32.0	..	18.0	..	8
	0.500–2.000	31.0	..	12.0	..	11
	2.001–3.000	31.0	..	12.0	..	15
			5456			
5456-0	0.051–1.500	42.0	53.0	19.0	30.0	16
	1.501–3.000	41.0	52.0	18.0	30.0	16
	3.001–5.000	40.0	..	17.0	..	14
	5.001–7.000	39.0	..	16.0	..	14
	7.001–8.000	38.0	..	15.0	..	12
5456-H112	0.250–1.500	42.0	..	19.0	..	12
	1.501–3.000	41.0	..	18.0	..	12
5456-H116⑥⑨	0.063–0.499	46.0	..	33.0	..	10
	0.500–1.250	46.0	..	33.0	..	12
	1.251–1.500	44.0	..	31.0	..	12
	1.501–3.000	41.0	..	29.0	..	12
	3.001–4.000	40.0	..	25.0	..	12
5456-H321	0.188–0.499	46.0	59.0	33.0	46.0	12
	0.500–1.500	44.0	56.0	31.0	44.0	12
	1.501–3.000	41.0	54.0	29.0	43.0	12

Mechanical Property Limits—Nonheat-Treatable Alloys (*concluded*)

ALLOY AND TEMPER	SPECIFIED THICKNESS[2] IN.	TENSILE STRENGTH—ksi				ELONGATION PERCENT MIN IN 2 IN. OR 4D[5]
		ULTIMATE		YIELD		
		min	max	min	max	
			5457			
5457-0	0.030–0.090	16.0	22.0	. .	. .	20
			5652			
5652-0	0.006–0.007	25.0	31.0	9.5	. .	. .
	0.008–0.012	25.0	31.0	9.5	. .	14
	0.013–0.019	25.0	31.0	9.5	. .	15
	0.020–0.031	25.0	31.0	9.5	. .	16
	0.032–0.050	25.0	31.0	9.5	. .	18
	0.051–0.113	25.0	31.0	9.5	. .	19
	0.114–0.249	25.0	31.0	9.5	. .	20
	0.250–3.000	25.0	31.0	9.5	. .	18
5652-H32[3]	0.017–0.019	31.0	38.0	23.0	. .	4
	0.020–0.050	31.0	38.0	23.0	. .	5
	0.051–0.113	31.0	38.0	23.0	. .	7
	0.114–0.249	31.0	38.0	23.0	. .	9
	0.250–0.499	31.0	38.0	23.0	. .	11
	0.500–2.000	31.0	38.0	23.0	. .	12
5652-H34[3]	0.009–0.019	34.0	41.0	26.0	. .	3
	0.020–0.050	34.0	41.0	26.0	. .	4
	0.051–0.113	34.0	41.0	26.0	. .	6
	0.114–0.249	34.0	41.0	26.0	. .	7
	0.250–1.000	34.0	41.0	26.0	. .	10
5652-H36[3]	0.006–0.007	37.0	44.0	29.0	. .	2
	0.008–0.031	37.0	44.0	29.0	. .	3
	0.032–0.162	37.0	44.0	29.0	. .	4
5652-H38[3]	0.006–0.007	39.0	. .	32.0	. .	2
	0.008–0.031	39.0	. .	32.0	. .	3
	0.032–0.128	39.0	. .	32.0	. .	4
5652-H112	0.250–0.499	28.0	. .	16.0	. .	7
	0.500–2.000	25.0	. .	9.5	. .	12
	2.001–3.000	25.0	. .	9.5	. .	16
			5657			
5657-H241[7]	0.030–0.090	18.0	26.0	. .	. .	13
5657-H25	0.030–0.090	20.0	28.0	. .	. .	8
5657-H26	0.030–0.090	22.0	30.0	. .	. .	7
5657-H28	0.030–0.090	25.0	. .	. .	. .	5

[1]The data base and criteria upon which these mechanical property limits are established are outlined in the Standard.

[2]Type of test specimen used depends on thickness of material; see "Sampling and Testing," pages 60–63, in the Standard.

[3]For the corresponding H2 temper, limits for maximum ultimate tensile strength and minimum yield strength do not apply.

[4]This table specifies the properties applicable to the test specimens, and since for plate 0.500 in. or over in thickness the cladding material is removed during preparation of the test specimens the listed properties are applicable to the core material only. Tensile and yield strengths of the composite plate are slightly lower depending on the thickness of cladding.

[5]D represents specimen diameter.

[6]Also applies to material previously designated H117.

[7]This material is subject to some recrystallization and the attendant loss of brightness.

[8]See page 96 of the Standard for specific cladding thicknesses.

[9]Material in this temper when tested upon receipt by the purchaser is required to pass the exfoliation corrosion resistance test (ASSET Method) indicated on page 60 of the Standard. The improved resistance to exfoliation corrosion of individual lots is determined by microscopic examination to assure a microstructure which is predominately free of a continuous grain boundary network of aluminum–magnesium precipitate. The microstructure is compared to that in a previously established acceptable reference photomicrograph.

[10]Processes such as flattening, levelling, or straightening coiled products subsequent to shipment by the producer may alter the mechanical properties of the metal (refer to Certification, Section 4 of the Standard).

Mechanical Property Limits—Heat-Treatable Alloys[1,14]

ALLOY AND TEMPER	SPECIFIED THICKNESS② IN.	TENSILE STRENGTH—ksi				ELONGATION PERCENT MIN IN 2 IN. OR 4D③
		ULTIMATE		YIELD		
		min	max	min	max	
		2014				
2014-0 Sheet and plate	0.020–0.499 0.500–1.000	. . . .	32.0 32.0	. . . .	16.0 . .	16 10
2014-T3 Flat sheet	0.020–0.039 0.040–0.249	59.0 59.0	. . . .	35.0 36.0	. . . .	14 14
2014-T4 Coiled sheet	0.020–0.249	59.0	. .	35.0	. .	14
2014-T451⑦⑧ Plate	0.250–0.499 0.500–1.000 1.001–2.000 2.001–3.000	58.0 58.0 58.0 57.0	. . . .	36.0 36.0 36.0 36.0	. . . .	14 14 12 8
2014-T42④⑩ Sheet and plate	0.020–1.000	58.0	. .	34.0	. .	14
2014-T6 and T62④⑩ Sheet	0.020–0.039 0.040–0.249	64.0 66.0	. . . .	57.0 58.0	. . . .	6 7
2014-T62④⑩ and T651⑦ Plate	0.250–0.499 0.500–1.000 1.001–2.000 2.001–2.500 2.501–3.000 3.001–4.000	67.0 67.0 67.0 65.0 63.0 59.0	. . . .	59.0 59.0 59.0 58.0 57.0 55.0	. . . .	7 6 4 2 2 1
		ALCLAD 2014⑧				
Alclad 2014-0 Sheet and plate	0.020–0.499 0.500–1.000	. . . .	30.0 32.0⑤	. . . .	14.0 . .	16 10
Alclad 2014-T3 Flat sheet	0.020–0.024 0.025–0.039 0.040–0.249	54.0 55.0 57.0	. . . . . .	33.0 34.0 35.0	. . . . . .	14 14 15
Alclad 2014-T4 Coiled sheet	0.020–0.024 0.025–0.039 0.040–0.128 0.129–0.249	54.0 55.0 57.0 57.0	. . . .	31.0 32.0 34.0 34.0	. . . .	14 14 15 15
Alclad 2014-T451⑦⑧ Plate	0.250–0.499 0.500–1.000 1.001–2.000 2.001–3.000	57.0 58.0⑤ 58.0⑤ 57.0⑤	. . . .	36.0 36.0⑤ 36.0⑤ 36.0⑤	. . . .	15 14 12 8
Alclad 2014-T42④⑩ Sheet and plate	0.020–0.024 0.025–0.039 0.040–0.128 0.129–0.249 0.250–0.499 0.500–1.000	54.0 55.0 57.0 57.0 57.0 58.0⑤	. . . .	31.0 32.0 34.0 34.0 34.0 34.0⑤	. . . .	14 14 15 15 15 14
Alclad 2014-T6 and T62④⑩ Sheet	0.020–0.024 0.025–0.039 0.040–0.249	62.0 63.0 64.0	. . . . . .	54.0 55.0 57.0	. . . . . .	7 7 8
Alclad 2014-T62④⑩ and T651⑦ Plate	0.250–0.499 0.500–1.000 1.001–2.000 2.001–2.500 2.501–3.000 3.001–4.000	64.0 67.0⑤ 67.0⑤ 65.0⑤ 63.0⑤ 59.0⑤	. . . .	57.0 59.0⑤ 59.0⑤ 58.0⑤ 57.0⑤ 55.0⑤	. . . .	8 6 4 2 2 1
		2024				
2024-0 Sheet and plate	0.010–0.499 0.500–1.750	. . . .	32.0 32.0	. . . .	14.0	12 12
2024-T3⑧ Flat sheet	0.008–0.009 0.010–0.020 0.021–0.128 0.129–0.249	63.0 63.0 63.0 64.0	. . . .	42.0 42.0 42.0 42.0	. . . .	10 12 15 15

Mechanical Property Limits—Heat-Treatable Alloys (*continued*)

ALLOY AND TEMPER	SPECIFIED THICKNESS② IN.	TENSILE STRENGTH—ksi				ELONGATION PERCENT MIN IN 2 IN. OR 4D③
		ULTIMATE		YIELD		
		min	max	min	max	
		2024 (Continued)				
2024-T351⑦⑧ Plate	0.250–0.499	64.0	. .	42.0	. .	12
	0.500–1.000	63.0	. .	42.0	. .	8
	1.001–1.500	62.0	. .	42.0	. .	7
	1.501–2.000	62.0	. .	42.0	. .	6
	2.001–3.000	60.0	. .	42.0	. .	4
	3.001–4.000	57.0	. .	41.0	. .	4
2024-T361⑧⑪ Flat sheet and plate	0.020–0.062	67.0	. .	50.0	. .	8
	0.063–0.249	68.0	. .	51.0	. .	9
	0.250–0.499	66.0	. .	49.0	. .	9
	0.500	66.0	. .	49.0	. .	10
2024-T4 Coiled sheet	0.010–0.020	62.0	. .	40.0	. .	12
	0.021–0.249	62.0	. .	40.0	. .	15
2024-T42④⑩ Sheet and plate	0.010–0.020	62.0	. .	38.0	. .	12
	0.021–0.249	62.0	. .	38.0	. .	15
	0.250–0.499	62.0	. .	38.0	. .	12
	0.500–1.000	61.0	. .	38.0	. .	8
	1.001–1.500	60.0	. .	38.0	. .	7
	1.501–2.000	60.0	. .	38.0	. .	6
	2.001–3.000	58.0	. .	38.0	. .	4
2024-T62④⑩ Sheet and plate	0.010–0.499	64.0	. .	50.0	. .	5
	0.500–3.000	63.0	. .	50.0	. .	5
2024-T72④⑩ Sheet	0.010–0.249	60.0	. .	46.0	. .	5
2024-T81 Flat sheet	0.010–0.249	67.0	. .	58.0	. .	5
2024-T851⑦ Plate	0.250–0.499	67.0	. .	58.0	. .	5
	0.500–1.000	66.0	. .	58.0	. .	5
	1.001–1.499	66.0	. .	57.0	. .	5
2024-T861⑪ Flat sheet and plate	0.020–0.062	70.0	. .	62.0	. .	3
	0.063–0.249	71.0	. .	66.0	. .	4
	0.250–0.499	70.0	. .	64.0	. .	4
	0.500	70.0	. .	64.0	. .	4
		ALCLAD 2024⑥				
Alclad 2024-0 Sheet and plate	0.008–0.009	. .	30.0	. .	14.0	10
	0.010–0.032	. .	30.0	. .	14.0	12
	0.033–0.062	. .	30.0	. .	14.0	12
	0.063–0.187	. .	32.0	. .	14.0	12
	0.188–0.499	. .	32.0	. .	14.0	12
	0.500–1.750	. .	32.0⑤	. .	. .	12
Alclad 2024-T3⑧ Flat sheet	0.008–0.009	58.0	. .	39.0	. .	10
	0.010–0.020	59.0	. .	39.0	. .	12
	0.021–0.062	59.0	. .	39.0	. .	15
	0.063–0.128	61.0	. .	40.0	. .	15
	0.129–0.249	62.0	. .	40.0	. .	15
Alclad 2024-T351⑦⑧ Plate	0.250–0.499	62.0	. .	40.0	. .	12
	0.500–1.000	63.0⑤	. .	42.0⑤	. .	8
	1.001–1.500	62.0⑤	. .	42.0⑤	. .	7
	1.501–2.000	62.0⑤	. .	42.0⑤	. .	6
	2.001–3.000	60.0⑤	. .	42.0⑤	. .	4
	3.001–4.000	57.0⑤	. .	41.0⑤	. .	4
Alcald 2024-T361⑧⑪ Flat sheet and plate	0.020–0.062	61.0	. .	47.0	. .	8
	0.063–0.187	64.0	. .	48.0	. .	9
	0.188–0.249	64.0	. .	48.0	. .	9
	0.250–0.499	64.0	. .	48.0	. .	9
	0.500	66.0⑤	. .	49.0⑤	. .	10
Alclad 2024-T4 Coiled sheet	0.010–0.020	58.0	. .	36.0	. .	12
	0.021–0.062	58.0	. .	36.0	. .	15
	0.063–0.128	61.0	. .	38.0	. .	15

Mechanical Property Limits—Heat-Treatable Alloys (*continued*)

ALLOY AND TEMPER	SPECIFIED THICKNESS② IN.	TENSILE STRENGTH—ksi				ELONGATION PERCENT MIN IN 2 IN. OR 4D③
		ULTIMATE		YIELD		
		min	max	min	max	
		ALCLAD 2024 (Continued)				
Alclad 2024-T42④⑩ Sheet and plate	0.008–0.009	55.0	..	34.0	..	10
	0.010–0.020	57.0	..	34.0	..	12
	0.021–0.062	57.0	..	34.0	..	15
	0.063–0.187	60.0	..	36.0	..	15
	0.188–0.249	60.0	..	36.0	..	15
	0.250–0.499	60.0	..	36.0	..	12
	0.500–1.000	61.0⑤	..	38.0⑤	..	8
	1.001–1.500	60.0⑤	..	38.0⑤	..	7
	1.501–2.000	60.0⑤	..	38.0⑤	..	6
	2.001–3.000	58.0⑤	..	38.0⑤	..	4
Alclad 2024-T62④⑩ Sheet and plate	0.010–0.062	60.0	..	47.0	..	5
	0.063–0.187	62.0	..	49.0	..	5
	0.188–0.499	62.0	..	49.0	..	5
Alclad 2024-T72④⑩ Sheet	0.010–0.062	56.0	..	43.0	..	5
	0.063–0.187	58.0	..	45.0	..	5
	0.188–0.249	58.0	..	45.0	..	5
Alclad 2024-T81 Flat sheet	0.010–0.062	62.0	..	54.0	..	5
	0.063–0.187	65.0	..	56.0	..	5
	0.188–0.249	65.0	..	56.0	..	5
Alclad 2024-T851⑦ Plate	0.250–0.499	65.0	..	56.0	..	5
	0.500–1.000	66.0⑤	..	58.0⑤	..	5
Alclad 2024-T861⑪ Flat sheet and plate	0.020–0.062	64.0	..	58.0	..	3
	0.063–0.187	69.0	..	64.0	..	4
	0.188–0.249	69.0	..	64.0	..	4
	0.250–0.499	68.0	..	62.0	..	4
	0.500	70.0⑤	..	64.0⑤	..	4
		1½% ALCLAD 2024⑥				
1½% Alclad 2024-0 Sheet and plate	0.188–0.499	..	32.0	..	14.0	12
	0.500–1.750	..	32.0⑤	..	..	12
1½% Alclad 2024-T3⑧ Flat sheet	0.188–0.249	63.0	..	41.0	..	15
1½% Alclad 2024-T351⑦⑧ Plate	0.250–0.499	63.0	..	41.0	..	12
	0.500–1.000	63.0⑤	..	42.0⑤	..	8
	1.001–1.500	62.0⑤	..	42.0⑤	..	7
	1.501–2.000	62.0⑤	..	42.0⑤	..	6
	2.001–3.000	60.0⑤	..	42.0⑤	..	4
	3.001–4.000	57.0⑤	..	41.0⑤	..	4
1½% Alclad 2024-T361⑧⑪ Flat sheet and plate	0.188–0.249	65.0	..	49.0	..	9
	0.250–0.499	65.0	..	48.0	..	9
	0.500	66.0⑤	..	49.0⑤	..	10
1½% Alclad 2024-T42 Sheet and plate④⑩	0.188–0.249	61.0	..	37.0	..	15
	0.250–0.499	61.0	..	37.0	..	12
	0.500–1.000	61.0⑤	..	38.0⑤	..	8
	1.001–1.500	60.0⑤	..	38.0⑤	..	7
	1.501–2.000	60.0⑤	..	38.0⑤	..	6
	2.001–3.000	58.0⑤	..	38.0⑤	..	4
1½% Alclad 2024-T62④⑩ Sheet and plate	0.188–0.499	62.0	..	49.0	..	5
1½% Alclad 2024-T72④⑩ Sheet	0.188–0.249	59.0	..	45.0	..	5
1½% Alclad 2024-T81 Flat sheet	0.188–0.249	66.0	..	57.0	..	5
1½% Alclad 2024-T851⑦ Plate	0.250–0.499	66.0	..	57.0	..	5
	0.500–1.000	66.0⑤	..	58.0⑤	..	5
1½% Alclad 2024-T861⑪ Flat sheet and plate	0.188–0.249	70.0	..	65.0	..	4
	0.250–0.499	69.0	..	63.0	..	4
	0.500	70.0⑤	..	64.0⑤	..	4

Mechanical Property Limits—Heat-Treatable Alloys (*continued*)

ALLOY AND TEMPER	SPECIFIED THICKNESS② IN.	TENSILE STRENGTH—ksi				ELONGATION PERCENT MIN IN 2 IN. OR 4D③
		ULTIMATE		YIELD		
		min	max	min	max	
		ALCLAD ONE SIDE 2024⑥				
Alclad One Side 2024-0 Sheet and plate	0.008–0.009 0.010–0.062 0.063–0.499	. . . . . .	31.0 31.0 32.0	. . . . . .	14.0 14.0 14.0	10 12 12
Alclad One Side 2024-T3⑧ Flat sheet	0.010–0.020 0.021–0.062 0.063–0.128 0.129–0.249	61.0 61.0 62.0 63.0	. . . .	40.0 40.0 41.0 41.0	. . . .	12 15 15 15
Alclad One Side 2024-T351⑦⑧ Plate	0.250–0.499	63.0	. .	41.0	. .	12
Alclad One Side 2024-T361⑧⑪ Sheet and plate	0.020–0.062 0.063–0.249 0.250–0.499	64.0 66.0 65.0	. . . . . .	48.0 49.0 48.0	. . . . . .	8 9 9
Alclad One Side 2024-T42④⑩ Sheet and plate	0.010–0.020 0.021–0.062 0.063–0.249 0.250–0.499	59.0 59.0 61.0 61.0	. . . .	35.0 36.0 37.0 37.0	. . . .	12 15 15 12
Alclad One Side 2024-T62④⑩ Sheet and plate	0.010–0.062 0.063–0.499	62.0 63.0	. . . .	48.0 49.0	. . . .	5 5
Alclad One Side 2024-T72④⑩ Flat sheet	0.010–0.062 0.063–0.249	58.0 59.0	. . . .	44.0 45.0	. . . .	5 5
Alclad One Side 2024-T81 Flat sheet	0.010–0.062 0.063–0.249	64.0 66.0	. . . .	56.0 57.0	. . . .	5 5
Alclad One Side 2024-T851⑦ Plate	0.250–0.499	66.0	. .	57.0	. .	5
Alclad One Side 2024-T861⑪ Sheet and plate	0.020–0.062 0.063–0.249 0.250–0.499	67.0 70.0 69.0	. . . . . .	60.0 65.0 63.0	. . . . . .	3 4 4
		1½% ALCLAD ONE SIDE 2024⑥				
1½% Alclad One Side 2024-0 Sheet and plate	0.188–0.499		32.0	. .	14.0	12
1½% Alclad One Side 2024-T3⑧ Flat sheet	0.188–0.249	63.0	. .	41.0	. .	15
1½% Alclad One Side 2024-T351⑦⑧ Plate	0.250–0.499	63.0	. .	41.0	. .	12
1½% Alclad One Side 2024-T361⑧⑪ Sheet and plate	0.188–0.249 0.250–0.499	66.0 65.0	. . . .	49.0 48.0	. . . .	9 9
1½% Alclad One Side 2024-T42④⑩ Sheet and plate	0.188–0.249 0.250–0.499	61.0 61.0	. . . .	37.0 37.0	. . . .	15 12
1½% Alclad One Side 2024-T62④⑩ Sheet and plate	0.188–0.499	63.0	. .	49.0	. .	5
1½% Alclad One Side 2024-T72④⑩ Flat sheet	0.188–0.249	59.0	. .	45.0	. .	5
1½% Alclad One Side 2024-T81 Flat sheet	0.188–0.249	66.0	. .	57.0	. .	5
1½% Alclad One Side 2024-T851⑦ Plate	0.250–0.499	66.0	. .	57.0	. .	5
1½% Alclad One Side 2024-T861⑪ Sheet and plate	0.188–0.249 0.250–0.499	70.0 69.0	. . . .	65.0 63.0	. . . .	4 4
		2036				
2036-T4 Flat sheet	0.025–0.125	42.0	. .	23.0	. .	20

Mechanical Property Limits—Heat-Treatable Alloys (*continued*)

ALLOY AND TEMPER	SPECIFIED THICKNESS[2] IN.	AXIS OF TEST SPECIMEN	TENSILE STRENGTH—ksi				ELONGATION PERCENT MIN IN. 2 IN. OR 4 D[3]
			ULTIMATE		YIELD		
			min	max	min	max	
			2124				
2124-T351[7] Plate	1.001–1.499	Longitudinal	61.0	..	45.0	..	14
		Long Transverse	62.0	..	42.0	..	13
	1.500–2.000	Longitudinal	61.0	..	45.0	..	14
		Long Transverse	62.0	..	42.0	..	13
		Short Transverse	58.0	..	38.0	..	4.5
	2.001–3.000	Longitudinal	61.0	..	45.0	..	14
		Long Transverse	62.0	..	41.0	..	10
		Short Transverse	58.0	..	38.0	..	4
2124-T851[7][16] Plate	1.001–2.000	Longitudinal	66.0	..	57.0	..	6
		Long Transverse	66.0	..	57.0	..	5
		Short Transverse	64.0[15]	..	55.0[15]	..	1.5[15]
	2.001–3.000	Longitudinal	65.0	..	57.0	..	6
		Long Transverse	65.0	..	57.0	..	4
		Short Transverse	63.0	..	55.0	..	1.5
	3.001–4.000	Longitudinal	65.0	..	56.0	..	5
		Long Transverse	65.0	..	56.0	..	4
		Short Transverse	62.0	..	54.0	..	1.5
	4.001–5.000	Longitudinal	64.0	..	55.0	..	5
		Long Transverse	64.0	..	55.0	..	4
		Short Transverse	61.0	..	53.0	..	1.5
	5.001–6.000	Longitudinal	63.0	..	54.0	..	5
		Long Transverse	63.0	..	54.0	..	4
		Short Transverse	58.0	..	51.0	..	1.5

ALLOY AND TEMPER	SPECIFIED THICKNESS[2] IN.	TENSILE STRENGTH—ksi				ELONGATION PERCENT MIN IN 2 IN. OR 4D[3]
		ULTIMATE		YIELD		
		min	max	min	max	
		2219				
2219-0 Sheet and plate	0.020–2.000	..	32.0	..	16.0	12
2219-T31[8] Flat sheet	0.020–0.039	46.0	..	29.0	..	8
	0.040–0.249	46.0	..	28.0	..	10
2219-T351[7][8] Plate	0.250–2.000	46.0	..	28.0	..	10
	2.001–3.000	44.0	..	28.0	..	10
	3.001–4.000	42.0	..	27.0	..	9
	4.001–5.000	40.0	..	26.0	..	9
	5.001–6.000	39.0	..	25.0	..	8
2219-T37[8] Flat sheet and plate	0.020–0.039	49.0	..	38.0	..	6
	0.040–2.000	49.0	..	37.0	..	6
	2.001–2.500	49.0	..	37.0	..	6
	2.501–3.000	47.0	..	36.0	..	6
	3.001–4.000	45.0	..	35.0	..	5
	4.001–5.000	43.0	..	34.0	..	4
2219-T62[4][10] Sheet and plate	0.020–0.039	54.0	..	36.0	..	6
	0.040–0.249	54.0	..	36.0	..	7
	0.250–1.000	54.0	..	36.0	..	8
	1.001–2.000	54.0	..	36.0	..	7
2219-T81 Flat sheet	0.020–0.039	62.0	..	46.0	..	6
	0.040–0.249	62.0	..	46.0	..	7
2219-T851[7] Plate	0.250–1.000	62.0	..	46.0	..	8
	1.001–2.000	62.0	..	46.0	..	7
	2.001–3.000	62.0	..	45.0	..	6
	3.001–4.000	60.0	..	44.0	..	5
	4.001–5.000	59.0	..	43.0	..	5
	5.001–6.000	57.0	..	42.0	..	4

[1] The data base and criteria upon which these mechanical property limits are established are outlined on page 93 of the Standard under "Mechanical Properties."

[2] Type of specimen used depends on thickness of material; see "Sampling and Testing," pages 60–63 of the Standard.

[3] D represents specimen diameter.

Mechanical Property Limits—Heat-Treatable Alloys (*continued*)

ALLOY AND TEMPER	SPECIFIED THICKNESS② IN.	TENSILE STRENGTH—ksi				ELONGATION PERCENT MIN IN 2 IN. OR 4D③
		ULTIMATE		YIELD		
		min	max	min	max	
2219 (Continued)						
2219-T87	0.020–0.039	64.0	..	52.0	..	5
Flat sheet and plate	0.040–0.249	64.0	..	52.0	..	6
	0.250–1.000	64.0	..	51.0	..	7
	1.001–2.000	64.0	..	51.0	..	6
	2.001–3.000	64.0	..	51.0	..	6
	3.001–4.000	62.0	..	50.0	..	4
	4.001–5.000	61.0	..	49.0	..	3
ALCLAD 2219 ⑥						
Alclad 2219-0	0.020–0.499	..	32.0	..	16.0	12
Sheet and plate	0.500–2.000	..	32.0⑤	..	16.0⑤	..
Alclad 2219-T31⑧	0.040–0.099	42.0	..	25.0	..	10
Flat sheet	0.100–0.249	44.0	..	26.0	..	10
Alclad 2219-T351⑦⑧ Plate	0.250–0.499	44.0	..	26.0	..	10
Alclad 2219-TT37⑧	0.040–0.099	45.0	..	34.0	..	6
Flat sheet and plate	0.100–0.249	47.0	..	35.0	..	6
	0.250–0.499	47.0	..	35.0	..	6
Alclad 2219-T62④⑩	0.020–0.039	44.0	..	29.0	..	6
Sheet and plate	0.040–0.099	49.0	..	32.0	..	7
	0.100–0.249	51.0	..	34.0	..	7
	0.250–0.499	51.0	..	34.0	..	8
	0.500–1.000	54.0⑤	..	36.0⑤	..	8
	1.001–2.000	54.0⑤	..	36.0⑤	..	7
Alclad 2219-T81	0.020–0.039	49.0	..	37.0	..	6
Flat sheet	0.040–0.099	55.0	..	41.0	..	7
	0.100–0.249	58.0	..	43.0	..	7
Alclad 2219-T851⑦ Plate	0.250–0.499	58.0	..	42.0	..	8
Alclad 2219-T87	0.040–0.099	57.0	..	46.0	..	6
Flat sheet and plate	0.100–0.249	60.0	..	48.0	..	6
	0.250–0.499	60.0	..	48.0	..	7
6061						
6061-0	0.006–0.007	..	22.0	..	12.0	10
Sheet and plate	0.008–0.009	..	22.0	..	12.0	12
	0.010–0.020	..	22.0	..	12.0	14
	0.021–0.128	..	22.0	..	12.0	16
	0.129–0.499	..	22.0	..	12.0	18
	0.500–1.000	..	22.0	..		18
	1.001–3.000	..	22.0	..		16
6061-T4	0.006–0.007	30.0	..	16.0	..	10
Sheet	0.008–0.009	30.0	..	16.0	..	12
	0.010–0.020	30.0	..	16.0	..	14
	0.021–0.249	30.0	..	16.0	..	16
6061-T451⑦⑧	0.250–1.000	30.0	..	16.0	..	18
Plate	1.001–3.000	30.0	..	16.0	..	16
6061-T42④⑩	0.006–0.007	30.0	..	14.0	..	10
Sheet and plate	0.008–0.009	30.0	..	14.0	..	12
	0.010–0.020	30.0	..	14.0	..	14
	0.021–0.249	30.0	..	14.0	..	16
	0.250–1.000	30.0	..	14.0	..	18
	1.001–3.000	30.0	..	14.0	..	16
6061-T6 and T62④⑩	0.006–0.007	42.0	..	35.0	..	4
Sheet	0.008–0.009	42.0	..	35.0	..	6
	0.010–0.020	42.0	..	35.0	..	8
	0.021–0.249	42.0	..	35.0	..	10
6061-T62④⑩ and	0.250–0.499	42.0	..	35.0	..	10
T651⑦ Plate	0.500–1.000	42.0	..	35.0	..	9
	1.001–2.000	42.0	..	35.0	..	8
	2.001–4.000	42.0	..	35.0	..	6
	4.001–6.000⑨	40.0	..	35.0	..	6

[4] These properties can usually be obtained by the user, when the material is properly solution heat treated or solution and precipitation heat treated from the O (annealed) or F (as fabricated) temper. These properties also apply to samples of material in the O or F tempers, which are solution heat treated or solution and precipitation treated by the producer to determine that the material will respond to proper heat treatment. Properties attained by the user, however, may be lower than those listed if the material has been formed or otherwise cold or hot worked,

Mechanical Property Limits—Heat-Treatable Alloys (*continued*)

ALLOY AND TEMPER	SPECIFIED THICKNESS② IN.	TENSILE STRENGTH—ksi ULTIMATE		TENSILE STRENGTH—ksi YIELD		ELONGATION PERCENT MIN IN 2 IN. OR 4D③
		min	max	min	max	
		ALCLAD 6061⑥				
Alclad 6061-0 Sheet and plate	0.010–0.020	..	20.0	..	12.0	14
	0.021–0.128	..	20.0	..	12.0	16
	0.129–0.499	..	20.0	..	12.0	18
	0.500–1.000	..	22.0⑤	..	..	18
	1.001–3.000	..	22.0⑤	..	..	16
Alclad 6061-T4 Sheet	0.010–0.020	27.0	..	14.0	..	14
	0.021–0.249	27.0	..	14.0	..	16
Alclad 6061-T451⑦⑧ Plate	0.250–0.499	27.0	..	14.0	..	18
	0.500–1.000	30.0⑤	..	16.0⑤	..	18
	1.001–3.000	30.0⑤	..	16.0⑤	..	16
Alclad 6061-T42④⑩ Sheet and plate	0.010–0.020	27.0	..	12.0	..	14
	0.021–0.249	27.0	..	12.0	..	16
	0.250–0.499	27.0	..	12.0	..	18
	0.500–1.000	30.0⑤	..	14.0⑤	..	18
	1.001–3.000	30.0⑤	..	14.0⑤	..	16
Alclad 6061-T6 and T62④⑩ Sheet	0.010–0.020	38.0	..	32.0	..	8
	0.021–0.249	38.0	..	32.0	..	10
Alclad 6061-T62④⑩ and T651⑦ Plate	0.250–0.499	38.0	..	32.0	..	10
	0.500–1.000	42.0⑤	..	35.0⑤	..	9
	1.001–2.000	42.0⑤	..	35.0⑤	..	8
	2.001–3.000	42.0⑤	..	35.0⑤	..	6
	3.001–4.000	42.0⑤	..	35.0⑤	..	6
	4.001–5.000⑨	40.0⑤	..	35.0⑤	..	6

ALLOY AND TEMPER	SPECIFIED THICKNESS② IN.	AXIS OF TEST SPECIMEN	TENSILE STRENGTH—ksi ULTIMATE		TENSILE STRENGTH—ksi YIELD		ELONGATION PERCENT MIN IN 2 IN. OR 4 D③
			min	max	min	max	
			7050				
7050-T7451⑯⑰⑱⑳ Plate	0.250–2.000	Longitudinal	74.0	..	64.0	..	10
		Long Transverse	74.0	..	64.0	..	9
	2.001–3.000	Longitudinal	73.0	..	63.0	..	9
		Long Transverse	73.0	..	63.0	..	8
		Short Transverse	68.0	..	59.0	..	2
	3.001–4.000	Longitudinal	72.0	..	62.0	..	9
		Long Transverse	72.0	..	62.0	..	6
		Short Transverse	68.0	..	58.0	..	2
	4.001–5.000	Longitudinal	71.0	..	61.0	..	9
		Long Transverse	71.0	..	61.0	..	5
		Short Transverse	67.0	..	57.0	..	2
	5.001–6.000	Longitudinal	70.0	..	60.0	..	8
		Long Transverse	70.0	..	60.0	..	4
		Short Transverse	67.0	..	57.0	..	2
7050-T7651⑯⑲㉑ Plate	0.250–1.000	Longitudinal	76.0	..	66.0	..	9
		Long Transverse	76.0	..	66.0	..	8
	1.001–1.500	Longitudinal	77.0	..	67.0	..	9
		Long Transverse	77.0	..	67.0	..	8
	1.501–2.000	Longitudinal	76.0	..	66.0	..	9
		Long Transverse	76.0	..	66.0	..	8
	2.001–3.000	Longitudinal	76.0	..	66.0	..	8
		Long Transverse	76.0	..	66.0	..	7
		Short Transverse	70.0	..	60.0	..	1.5

particularly in the annealed temper, prior to solution heat treatment.

[5]This table specifies the properties applicable to the test specimens, and since for plate 0.500 in. or over in thickness the cladding material is removed during preparation of the test specimens the listed properties are applicable to the core material only. Tensile and yield strengths of the composite plate are slightly lower depending on the thickness of cladding.

[6]See page 96 of the Standard for specific cladding thicknesses.

[7]For stress-relieved tempers the characteristics and properties other than those specified may differ somewhat from the corresponding characteristics and properties of material in the basic temper.

Mechanical Property Limits—Heat-Treatable Alloys (*continued*)

ALLOY AND TEMPER	SPECIFIED THICKNESS② IN.	TENSILE STRENGTH—ksi ULTIMATE min	TENSILE STRENGTH—ksi ULTIMATE max	TENSILE STRENGTH—ksi YIELD min	TENSILE STRENGTH—ksi YIELD max	ELONGATION PERCENT MIN IN 2 IN. OR 4D③
7075						
7075-0	0.015–0.499	..	40.0	..	21.0	10
Sheet and plate	0.500–2.000	..	40.0	..	..	10
7075-T6 and T62④⑩	0.008–0.011	74.0	..	63.0	..	5
Sheet	0.012–0.039	76.0	..	67.0	..	8
	0.040–0.062	78.0	..	68.0	..	9
	0.063–0.125	78.0	..	68.0	..	9
	0.126–0.187	79.0	..	69.0	..	9
	0.188–0.249	80.0	..	69.0	..	9
7075-T62④⑩ and T651⑦	0.250–0.499	78.0	..	67.0	..	9
Plate	0.500–1.000	78.0	..	68.0	..	7
	1.001–2.000	77.0	..	67.0	..	6
	2.001–2.500	76.0	..	64.0	..	5
	2.501–3.000	72.0	..	61.0	..	5
	3.001–3.500	71.0	..	58.0	..	5
	3.501–4.000	67.0	..	54.0	..	3
7075-T73 Sheet	0.040–0.249	67.0	..	56.0	..	8
7075-T7351⑦⑫	0.250–1.000	69.0	..	57.0	..	7
Plate	1.001–2.000	69.0	..	57.0	..	6
	2.001–2.500	66.0	..	52.0	..	6
	2.501–3.000	64.0	..	49.0	..	6
	3.001–3.500	63.0	..	49.0	..	6
	3.501–4.000	61.0	..	48.0	..	6
7075-T76⑬ Sheet	0.125–0.249	73.0	..	62.0	..	8
7075-T7651⑦⑬	0.250–0.499	72.0	..	61.0	..	8
Plate	0.500–1.000	71.0	..	60.0	..	6
ALCLAD 7075⑥						
Alclad 7075-0	0.008–0.014	..	36.0	..	20.0	9
Sheet and plate	0.015–0.062	..	36.0	..	20.0	10
	0.063–0.187	..	38.0	..	20.0	10
	0.188–0.499	..	39.0	..	21.0	10
	0.500–1.000	..	40.0⑤	..	..	10
Alclad 7075-T6 and T62④⑩	0.008–0.011	68.0	..	58.0	..	5
Sheet	0.012–0.039	71.0	..	61.0	..	8
	0.040–0.062	72.0	..	62.0	..	9
	0.063–0.125	74.0	..	64.0	..	9
	0.126–0.187	74.0	..	64.0	..	9
	0.188–0.249	76.0	..	65.0	..	9
Alclad 7075-T62④⑩	0.250–0.499	75.0	..	65.0	..	9
and T651⑦	0.500–1.000	78.0⑤	..	68.0⑤	..	7
Plate	1.001–2.000	77.0⑤	..	67.0⑤	..	6
	2.001–2.500	76.0⑤	..	64.0⑤	..	5
	2.501–3.000	72.0⑤	..	61.0⑤	..	5
	3.001–3.500	71.0⑤	..	58.0⑤	..	5
	3.501–4.000	67.0⑤	..	54.0⑤	..	3
Alclad 7075-T73	0.040–0.062	63.0	..	51.0	..	8
Sheet	0.063–0.187	64.0	..	52.0	..	8
	0.188–0.249	66.0	..	54.0	..	8
Alclad 7075-T7351⑦⑫	0.250–0.499	66.0	..	54.0	..	8
Plate	0.500–1.000	69.0⑤	..	57.0⑤	..	7
Alclad 7075-T76⑬	0.125–0.187	68.0	..	57.0	..	8
Sheet	0.188–0.249	70.0	..	59.0	..	8
Alclad 7075-T7651⑦⑬	0.250–0.499	69.0	..	58.0	..	8
Plate	0.500–1.000	71.0⑤	..	60.0⑤	..	6
2½% ALCLAD 7075⑥						
2½% Alclad 7075-0	0.188–0.499	..	39.0	..	20.0	10
Sheet and plate	0.500–1.000	..	40.0⑤	..	..	10
2½% Alclad 7075-T6 and T62④⑩ Sheet	0.188–0.249	74.0	..	64.0	..	8

[8]Upon artificial aging, T3 and T31, T37, T351, T361 and T451 temper material shall be capable of developing the mechanical properties applicable to the T81, T87, T851, T861 and T651 tempers, respectively.

[9]The properties for this thickness apply only to the T651 temper.

[10]This temper is not available from the material producer.

[11]Tempers T361 and T861 formerly designated T36 and T86, respectively.

[12]Material in this temper, 0.750 in. and thicker, when tested in accordance with ASTM G47 in the short transverse direction at a stress level of 75% of the specified minimum yield strength will exhibit no evidence of stress corrosion cracking. Capability of individual lots to resist stress

Mechanical Property Limits—Heat-Treatable Alloys (*continued*)

ALLOY AND TEMPER	SPECIFIED THICKNESS② IN.	TENSILE STRENGTH—ksi				ELONGATION PERCENT MIN IN 2 IN. OR 4D③
		ULTIMATE		YIELD		
		min	max	min	max	
2½% ALCLAD 7075⑥ (Continued)						
2½% Alclad 7075-T62④⑩	0.250–0.499	74.0	. .	64.0	. .	9
and T651⑦	0.500–1.000	78.0⑤	. .	68.0⑤	. .	7
Plate	1.001–2.000	77.0⑤	. .	67.0⑤	. .	6
	2.001–2.500	76.0⑤	. .	64.0⑤	. .	5
	2.501–3.000	72.0⑤	. .	61.0⑤	. .	5
	3.001–3.500	71.0⑤	. .	58.0⑤	. .	5
	3.501–4.000	67.0⑤	. .	54.0⑤	. .	3
2½% Alclad 7075-T73 Sheet	0.188–0.249	64.0	. .	53.0	. .	8
2½% Alclad 7075-T7351⑦⑫	0.250–0.499	65.0	. .	54.0	. .	8
Plate	0.500–1.000	69.0⑤	. .	57.0⑤	. .	7
2½% Alclad 7075-T76⑬ Sheet	0.188–0.249	69.0	. .	59.0	. .	8
2½% Alclad 7075-T7651⑦⑩	0.250–0.499	68.0	. .	58.0	. .	8
Plate	0.500–1.000	71.0⑤	. .	60.0⑤	. .	6
ALCLAD ONE SIDE 7075⑥						
Alclad One Side 7075-0	0.015–0.062	. .	38.0	. .	21.0	10
Sheet and plate	0.063–0.187	. .	39.0	. .	21.0	10
	0.188–0.499	. .	39.0	. .	21.0	10
	0.500–1.000	. .	40.0⑤	. .	. .	10
Alclad One Side	0.008–0.011	71.0	. .	60.0	. .	5
7075-T6 and T62④⑩ Sheet	0.012–0.039	74.0	. .	64.0	. .	8
	0.040–0.062	75.0	. .	65.0	. .	9
	0.063–0.125	76.0	. .	66.0	. .	9
	0.126–0.187	77.0	. .	66.0	. .	9
	0.188–0.249	78.0	. .	67.0	. .	9
Alclad One Side 7075	0.250–0.499	76.0	. .	66.0	. .	9
T62④⑩ and T651⑦ Plate	0.500–1.000	78.0⑤	. .	68.0⑤	. .	7
	1.001–2.000	77.0⑤	. .	67.0⑤	. .	6
2½% ALCLAD ONE SIDE 7075⑥						
2½% Alclad One Side 7075-0	0.188–0.499	. .	39.0	. .	21.0	10
Sheet and plate	0.500–1.000	. .	40.0⑤	. .	. .	10
2½% Alclad One Side 7075-T6 and T62④⑩ Sheet	0.188–0.249	76.0	. .	65.0	. .	8
2½% Alclad One Side 7075	0.250–0.499	76.0	. .	65.0	. .	9
T62④⑩ and T651⑦	0.500–1.000	78.0⑤	. .	68.0⑤	. .	7
Plate	1.001–2.000	77.0⑤	. .	67.0⑤	. .	6
7008 ALCLAD 7075⑥						
7008 Alclad 7075-0	0.015–0.062	. .	40.0	. .	21.0	10
Sheet and plate	0.063–0.187	. .	40.0	. .	21.0	10
	0.188–0.499	. .	40.0	. .	21.0	10
	0.500–2.000	. .	40.0⑤	. .	. .	10
7008 Alclad 7075-T6 and	0.015–0.039	73.0	. .	63.0	. .	7
T62④⑩ Sheet	0.040–0.062	75.0	. .	65.0	. .	8
	0.063–0.187	75.0	. .	65.0	. .	8
	0.188–0.249	76.0	. .	66.0	. .	8
7008 Alclad 7075-T62④⑩	0.250–0.499	76.0	. .	66.0	. .	9
and T651⑦ Plate	0.500–1.000	78.0⑤	. .	68.0⑤	. .	7
	1.001–2.000	77.0⑤	. .	67.0⑤	. .	6
	2.001–2.500	76.0⑤	. .	64.0⑤	. .	5
	2.501–3.000	72.0⑤	. .	61.0⑤	. .	5
	3.001–3.500	71.0⑤	. .	58.0⑤	. .	5
	3.501–4.000	67.0⑤	. .	54.0⑤	. .	3
7008 Alclad 7075-T76⑬	0.040–0.062	70.0	. .	59.0	. .	8
Sheet	0.063–0.187	71.0	. .	60.0	. .	8
	0.188–0.249	72.0	. .	61.0	. .	8
7008 Alclad 7075-T7651⑦⑬	0.250–0.499	71.0	. .	60.0	. .	8
Plate	0.500–1.000	71.0⑤	. .	60.0⑤	. .	6

corrosion is determined by testing the previously selected tensile test sample in accordance with the applicable lot acceptance criteria outlined on page 100 of the Standard.

[13]Material in this temper when tested in accordance with ASTM G34-72 will exhibit exfoliation less than that shown in Category B, Figure 2 of ASTM G34-72. Also, material, 0.750 in. and thicker, when tested in accordance with ASTM G47 in the short transverse direction at a stress level of 25 ksi will exhibit no evidence of stress corrosion cracking. Capability of individual lots to resist exfoliation corrosion and stress corrosion cracking is determined by testing the previously selected tensile test sample in accordance with the applicable lot acceptance criteria outlined on page 100 of the Standard.

Mechanical Property Limits—Heat-Treatable Alloys (*concluded*)

ALLOY AND TEMPER	SPECIFIED THICKNESS② IN.	TENSILE STRENGTH—ksi				ELONGATION PERCENT MIN IN 2 IN. OR 4D③
		ULTIMATE		YIELD		
		min	max	min	max	
			7178			
7178-0 Sheet and plate	0.015–0.499 0.500	. . . .	40.0 40.0	. . . .	21.0 . .	10 10
7178-T6 and T62④⑩ Sheet	0.015–0.044 0.045–0.249	83.0 84.0	. . . .	72.0 73.0	. . . .	7 8
7178-T62④⑩ and T651⑦ Plate	0.250–0.499 0.500–1.000 1.001–1.500 1.501–2.000	84.0 84.0 84.0 80.0	. . . .	73.0 73.0 73.0 70.0	. . . .	8 6 4 3
7178-T76⑬ Sheet	0.045–0.249	75.0	. .	64.0	. .	8
7178-T7651⑦⑬ Plate	0.250–0.499 0.500–1.000	74.0 73.0	. . . .	63.0 62.0	. . . .	8 6
			ALCLAD 7178⑥			
Alclad 7178-0 Sheet and plate	0.015–0.062 0.063–0.187 0.188–0.499 0.500	. . . .	36.0 38.0 40.0 40.0	. . . .	20.0 20.0 21.0 . .	10 10 10 10
Alclad 7178-T6 and T62④⑩ Sheet	0.015–0.044 0.045–0.062 0.063–0.187 0.188–0.249	76.0 78.0 80.0 82.0	. . . .	66.0 68.0 70.0 71.0	. . . .	7 8 8 8
Alclad 7178-T62④⑩ and T651⑦ Plate	0.250–0.499 0.500–1.000 1.001–1.500 1.501–2.000	82.0 84.0⑤ 84.0⑤ 80.0⑤	. . . .	71.0 73.0⑤ 73.0⑤ 70.0⑤	. . . .	8 6 4 3
Alclad 7178-T76⑬ Sheet	0.045–0.062 0.063–0.187 0.188–0.249	71.0 71.0 73.0	. . . . . .	60.0 60.0 61.0	. . . . . .	8 8 8
Alclad 7178-T7651⑦⑬ Plate	0.250–0.499 0.500–1.000	72.0 73.0⑤	. . . .	60.0 62.0⑤	. . . .	8 6

[14]Processes such as flattening, levelling, or straightening coiled products subsequent to shipment by the producer may alter the mechanical properties of the metal (refer to Certification, Section 4 of the Standard).

[15]Applicable only to 1.500 in. and greater thickness.

[16]See Table 6.6 of the Standard for fracture toughness limits.

[17]T7451 temper although not previously registered has appeared in the literature and in some specifications as T73651.

[18]Material in this temper, 0.750 in. and thicker, when tested in accordance with ASTM G47 in the short transverse direction at a stress level of 35 ksi will exhibit no evidence of stress corrosion cracking. Capability of individual lots to resist stress corrosion is determined by testing the previously selected tensile test sample in accordance with the applicable lot acceptance criteria outlined on page 100 of the Standard.

[19]Material in this temper, 0.750 in. and thicker, when tested in accordance with ASTM G47 in the short transverse direction at a stress level of 25 ksi will exhibit no evidence of stress corrosion cracking. Capability of individual lots to resist stress corrosion is determined by testing the previously selected tensile test sample in accordance with the applicable lot acceptance criteria outlined on page 100 of the Standard.

[20]Material in this temper when tested at any plane in accordance with ASTM G34-72 will exhibit exfoliation less than that shown in Category B, Figure 2 of ASTM G34-72. Capability of individual lots to resist exfoliation corrosion and stress corrosion cracking is determined by testing the previously selected tensile test sample in accordance with the applicable lot acceptance criteria outlined on page 100 of the Standard.

[21]Material in this temper when tested at t/10 plane in accordance with ASTM G34-72 will exhibit exfoliation less than that shown in Category B, Figure 2 of ASTM G34-72. Capability of individual lots to resist exfoliation corrosion and stress corrosion cracking is determined by testing the previously selected tensile test sample in accordance with the applicable lot acceptance criteria outlined on page 100 of the Standard.

A.3 WIRE, ROD, AND BAR — ROLLED OR COLD-FINISHED

Mechanical Property Limits—Wire, Rod, and Bar—Rolled or Cold-Finished—Nonheat-Treatable Alloys[1,16]

ALLOY AND TEMPER	SPECIFIED DIAMETER OR THICKNESS in.	TENSILE STRENGTH—ksi			ELONGATION② percent min in 2 in. or 4D③
		ULTIMATE		YIELD②⑪ min	
		min	max		
1100					
1100–0	All	11.0	15.5	3.0	25
1100–H112	All	11.0	. .	3.0	. .
1100–H12	Up thru 0.374	14.0	. .	. .	. .
1100–H14	Up thru 0.374	16.0	. .	. .	. .
1100–H16	Up thru 0.374	19.0	. .	. .	. .
1100–H18	Up thru 0.374	22.0	. .	. .	. .
1100–F⑫	0.375 and over	. .	. .	. .	. .
1345					
1345–0	Up thru 0.374	. .	14.0	. .	25
1345–H12	Up thru 0.374	13.0	. .	. .	. .
1345–H14	Up thru 0.314	15.0	. .	. .	. .
1345–H16	Up thru 0.314	17.0	. .	. .	. .
1345–H18	Up thru 0.314	19.0	. .	. .	. .
1345–H19	Up thru 0.204	21.0	. .	. .	. .
3003					
3003–0	All	14.0	19.0	5.0	25
3003–H112	All	14.0	. .	5.0	. .
3003–H12	Up thru 0.374	17.0	. .	. .	. .
3003–H14	Up thru 0.374	20.0	. .	. .	. .
3003–H16	Up thru 0.374	24.0	. .	. .	. .
3003–H18	Up thru 0.374	27.0	. .	. .	. .
3003–F⑫	0.375 and over	. .	. .	. .	. .
5050					
5050–0	All	18.0	26.0	. .	25
5050–H32	Up thru 0.374	22.0	. .	. .	. .
5050–H34	Up thru 0.374	25.0	. .	. .	. .
5050–H36	Up thru 0.374	27.0	. .	. .	. .
5050–H38	Up thru 0.374	29.0	. .	. .	. .
5050–F⑫	0.375 and over	. .	. .	. .	. .
5052					
5052–0	All	25.0	32.0	9.5	25
5052–H32	Up thru 0.374	31.0	. .	23.0	. .
5052–H34	Up thru 0.374	34.0	. .	26.0	. .
5052–H36	Up thru 0.374	37.0	. .	29.0	. .
5052–H38	Up thru 0.374	39.0	. .	. .	. .
5052–F⑫	0.375 and over	. .	. .	. .	. .

[1]Mechanical test specimens are taken as detailed under "Sampling and Testing," pages 60–63. The data base and criteria upon which these mechanical property limits are established are outlined on page 93 under "Mechanical Properties," all in the Standard.

[2]The measurement of elongation and yield strength is not required for wire less than 0.125 in. in thickness or diameter.

[3]D represents specimen diameter.

[4]Properties listed for this full size increment are applicable to rod. Properties listed are only applicable for square, rectangular, hexagonal, or octagonal bar having a maximum thickness of 4 in. and a maximum cross-sectional area of 36 $in.^2$

[5]Properties listed for this full size increment are applicable to rod. Properties listed are only applicable for square, hexagonal or octagonal bar having a maximum thickness of $3\frac{1}{2}$ in; for rectangular bar having a maximum thickness of 3 in. with corresponding maximum width of 6 in. For rectangular bar less than 3 in. in thickness, maximum width is 10 in.

[6]For bar maximum cross-sectional area is 50 $in.^2$

[7]Rivet and cold heading wire and rod, and the fasteners produced from it, shall upon proper heat treatment (T4 temper) or heat treatment and aging (T6, T61, T7 and T73 tempers) be capable of

Mechanical Property Limits—Wire, Rod, and Bar—Rolled or Cold-Finished—Nonheat-Treatable Alloys (*concluded*)

ALLOY AND TEMPER	SPECIFIED DIAMETER OR THICKNESS in.	TENSILE STRENGTH—ksi			ELONGATION[2] percent min in 2 in. or 4D[3]
		ULTIMATE		YIELD[2][11] min	
		min	max		
5056					
5056–0	All	. .	46.0	. .	20
5056–H111	Up thru 0.374	44.0	. .	. .	. .
5056–H12	Up thru 0.374	46.0	. .	. .	. .
5056–H32	Up thru 0.374	44.0	. .	. .	. .
5056–H14	Up thru 0.374	52.0	. .	. .	. .
5056–H34	Up thru 0.374	50.0	. .	. .	. .
5056–H18	Up thru 0.374	58.0	. .	. .	. .
5056–H38	Up thru 0.374	55.0	. .	. .	. .
5056–H192	Up thru 0.374	60.0	. .	. .	. .
5056–H392	Up thru 0.374	58.0	. .	. .	. .
5056–F[12]	0.375 and over	. .	. .	. .	. .
ALCLAD 5056					
Alclad 5056–H192	Up thru 0.374	52.0	. .	. .	. .
Alclad 5056–H392	Up thru 0.374	50.0	. .	. .	. .
Alclad 5056–H393	Up thru 0.192	54.0	. .	47.0	. .
5154					
5154–0	All	30.0	41.0	11.0	25
5154–H112	All	30.0	41.0	11.0	. .
5154–H32	Up thru 0.374	36.0	. .	. .	. .
5154–H34	Up thru 0.374	39.0	. .	. .	. .
5154–H36	Up thru 0.374	42.0	. .	. .	. .
5154–H38	Up thru 0.374	45.0	. .	. .	. .
5154–F[12]	0.375 and over	. .	. .	. .	. .

developing the properties presented in Table 7, following. Tensile tests are preferred for the rivet and cold heading wire and rod, and shear tests for the fasteners made from it.

[8]For stress-relieved tempers the characteristics and properties other than those specified may differ somewhat from the corresponding characteristics and properties of material in the basic temper.

[9]Material in this temper, 0.750 in. and thicker, when tested in accordance with ASTM G47 in the short transverse direction at a stress level of 75% of the specified minimum yield strength will exhibit no evidence of stress corrosion cracking. Capability of individual lots to resist stress corrosion is determined by testing the previously selected tensile test sample in accordance with the applicable lot acceptance criteria outlined on page 100 of the Standard.

[10]These properties can usually be obtained by the user when the material is properly solution heat treated or solution and precipitation heat treated from the 0 (annealed) or F (as fabricated) temper. These properties also apply to samples of material in the 0 or F tempers which are solution heat treated or solution and precipitation treated by the producer to determine that the material will respond to proper heat treatment. Properties attained by the user, however, may be lower than those listed if the material has been formed or otherwise cold or hot worked, particularly in the annealed temper, prior to solution heat treatment.

[11]These yield strengths determined only when specifically requested.

[12]Except in the annealed (0 temper) condition, the temper of nonheat-treatable alloy rod and bar cannot be closely controlled and will vary according to size.

[13]Minimum yield strength of coiled 2024-T4 wire and rod is 40.0 ksi.

[14]Applicable to rod only.

[15]This temper is not available from the material producer.

[16]Processes such as flattening, levelling, or straightening coiled products subsequent to shipment by the producer may alter the mechanical properties of the metal (refer to Certification, Section 4 of the Standard).

Mechanical Property Limits—Wire, Rod, and Bar—Rolled or Cold-Finished—Heat-Treatable Alloys

ALLOY AND TEMPER	SPECIFIED DIAMETER OR THICKNESS in.	TENSILE STRENGTH—ksi			ELONGA-TION② percent min in 2 in. or 4D③
		ULTIMATE		YIELD② min	
		min	max		
		2011			
2011–T3	0.125–1.500	45.0	. .	38.0	10
	1.501–2.000	43.0	. .	34.0	12
	2.001–3.250	42.0	. .	30.0	12
2011–T4 and T451⑧	0.375–8.000	40.0	. .	18.0	16
2011–T8	0.125–3.250	54.0	. .	40.0	10
		2014			
2014–0	Up thru 8.000	. .	35.0	. .	12
2014–T4, T42⑩⑮ and T451⑧	Up thru 8.000④	55.0	. .	32.0	16
2014–T6, T62⑩⑮ and T651⑧	Up thru 8.000④	65.0	. .	55.0	8
		2017			
2017–0	Up thru 8.000	. .	35.0	. .	16
2017–T4, T42⑩⑮ and T451⑧	Up thru 8.000⑥	55.0	. .	32.0	12
		2024			
2024–0	Up thru 8.000	. .	35.0	. .	16
2024–T36	Up thru 0.375	69.0	. .	52.0	10
2024–T4	Up thru 0.499	62.0	. .	45.0⑬	10
	0.500–4.500④	62.0	. .	42.0⑬	10
	4.501–6.500⑭	62.0	. .	40.0	10
	6.501–8.000⑭	58.0	. .	38.0	10
2024–T42⑩⑮	Up thru 6.500④	62.0	. .	40.0	10
2024–T351	0.500–6.500④	62.0	. .	45.0	10
2024–T6	Up thru 6.500④	62.0	. .	50.0	5
2024–T62⑩⑮	Up thru 6.500④	60.0	. .	46.0	5
2024–T851	0.500–6.500④	66.0	. .	58.0	5
		2219			
2219–T851	0.500–2.000	58.0	. .	40.0	4
	2.001–4.000	57.0	. .	39.0	4
		6061			
6061–0	Up thru 8.000	. .	22.0	. .	18
6061–T4 and T451⑧	Up thru 8.000⑥	30.0	. .	16.0	18
6061–T42⑩⑮	Up thru 8.000⑥	30.0	. .	14.0	18
6061–T6, T62⑩⑮ and T651⑧	Up thru 8.000⑥	42.0	. .	35.0	10
6061–T89	Up thru 0.374	54.0	. .	47.0	. .
6061–T913	Up thru 0.374	63.0	. .	. .	. .
6061–T94	Up thru 0.374	54.0	. .	47.0	. .
		6262			
6262–T6 and T651⑧	Up thru 8.000④	42.0	. .	35.0	10
6262–T9	0.125–2.000	52.0	. .	48.0	5
	2.001–3.000	50.0	. .	46.0	5
		7075			
7075–0	Up thru 8.000	. .	40.0	. .	10
7075–T6, T62⑩⑮ and T651⑧	Up thru 4.000⑤	77.0	. .	66.0	7
7075–T73⑨ and T7351⑧⑨	Up thru 3.000	68.0	. .	56.0	10

See footnotes on pp. 331–332 to Table 5.

Mechanical Property Limits—Wire, Rod, and Bar—Rolled or Cold-Finished—Heat-Treatable Alloys

ALLOY AND TEMPER	SPECIFIED DIAMETER in.	TENSILE STRENGTH ksi min		ELONGATION② percent min in. 2 in. or 4D③	ULTIMATE SHEARING STRENGTH ksi min
		ULTIMATE	YIELD②		
2017-T4	0.063–1.000	55.0	32.0	12	33.0
2024-T4	0.063–1.000	62.0	40.0	10	37.0
2117-T4	0.063–1.000	38.0	18.0	18	26.0
2219-T6	0.063–1.000	55.0	35.0	6	30.0
6053-T61	0.063–1.000	30.0	20.0	14	20.0
6061-T6	0.063–1.000	42.0	35.0	10	25.0
7050-T7	0.063–1.000	70.0	58.0	10	39.0
7075-T6	0.063–1.000	77.0	66.0	7	42.0
7075-T73	0.063–1.000	68.0	56.0	10	41.0
7178-T6	0.063–1.000	84.0	73.0	5	46.0

See footnote on pp. 331–332 to Table 5.

A.4 WIRE, ROD, BAR, AND SHAPES — EXTRUDED

Mechanical Property Limits—Extruded Wire, Rod, Bar, and Shapes[6]

ALLOY AND TEMPER	SPECIFIED DIAMETER OR THICKNESS[1] in.	AREA sq in.	TENSILE STRENGTH—ksi				ELONGATION[2] percent min in 2 in. or 4D[3]
			ULTIMATE		YIELD		
			min	max	min	max	
		1100					
1100-0	All	All	11.0	15.5	3.0	..	25
1100-H112	All	All	11.0	..	3.0	..	..
		2014					
2014-0	All	All	..	30.0	..	18.0	12
2014-T4, T4510[5][7] and T4511[5][7]	All	All	50.0	..	35.0	..	12
2014-T42[4][8]	All	All	50.0	..	29.0	..	12
2014-T6, T6510[5] and T6511[5]	Up thru 0.499	All	60.0	..	53.0	..	7
	0.500–0.749	All	64.0	..	58.0	..	7
	0.750 and over	Up thru 25	68.0	..	60.0	..	7
	0.750 and over	Over 25 thru 32	68.0	..	58.0	..	6
2014-T62[4][8]	Up thru 0.749	All	60.0	..	53.0	..	7
	0.750 and over	Up thru 25	60.0	..	53.0	..	7
	0.750 and over	Over 25 thru 32	60.0	..	53.0	..	6
		2024					
2024-0	All	All	..	35.0	..	19.0	12
2024-T3, T3510[5][7] and T3511[5][7]	Up thru 0.249	All	57.0	..	42.0	..	12
	0.250–0.749	All	60.0	..	44.0	..	12
	0.750–1.499	All	65.0	..	46.0	..	10
	1.500 and over	Up thru 25	70.0	..	52.0	..	10
	1.500 and over	Over 25 thru 32	68.0	..	48.0	..	8
2024-T42[4][8]	Up thru 0.749	All	57.0	..	38.0	..	12
	0.750–1.499	All	57.0	..	38.0	..	10
	1.500 and over	Up thru 25	57.0	..	38.0	..	10
	1.500 and over	Over 25 thru 32	57.0	..	38.0	..	8
2024-T81, T8510[5] and T8511[5]	0.050–0.249	All	64.0	..	56.0	..	4
	0.250–1.499	All	66.0	..	58.0	..	5
	1.500 and over	Up thru 32	66.0	..	58.0	..	5

[1] The thickness of the cross-section from which the tension test specimen is taken determines the applicable mechanical properties. The data base and criteria upon which these mechanical property limits are established are outlined on page 93 of the Standard under "Mechanical Properties."

[2] For material of such dimensions that a standard test specimen cannot be taken, or for shapes thinner than 0.062 in., the test for elongation is not required.

[3] D represents specimen diameter.

[4] These properties can usually be obtained by the user when the material is properly solution heat treated or solution and precipitation heat treated from the O (annealed) or F (as fabricated) temper. These properties also apply to samples of material in the O or F tempers which are solution heat treated or solution and precipitation treated by the producer to determine that the material will respond to proper heat treatment. Properties attained by the user, however, may be lower than those listed if the material has been formed or otherwise cold or hot worked, particularly in the annealed temper, prior to solution heat treatment.

[5] For stress-relieved tempers the characteristics and properties other than those specified may differ somewhat from the corresponding characteristics and properties of material in the basic temper.

[6] Processes such as flattening, levelling, or straightening coiled products subsequent to shipment by the producer may alter the mechanical properties of the metal (refer to Certification, Section 4 of the Standard).

[7] Upon artificial aging, T3, T31, T3510, T3511, T4, T4510 and T4511 temper material shall be capable of developing the mechanical properties applicable to the T81, T8510, T8511, T6, T6510 and T6511 tempers, respectively.

[8] This temper is not available from the material producer.

Mechanical Property Limits—Extruded Wire, Rod, Bar, and Shapes (*continued*)

ALLOY AND TEMPER	SPECIFIED DIAMETER OR THICKNESS① in.	AREA sq in.	TENSILE STRENGTH—ksi				ELONGATION② percent min in 2 in. or 4D③
			ULTIMATE		YIELD		
			min	max	min	max	
		2219					
2219-0	All	All	. .	32.0	. .	18.0	12
2219-T31, T3510⑤⑦ and T3511⑤⑦	Up thru 0.499	Up thru 25	42.0	. .	26.0	. .	14
	0.500–2.999	Up thru 25	45.0	. .	27.0	. .	14
2219-T62④⑧	Up thru 0.999	Up thru 25	54.0	. .	36.0	. .	6
	1.000 and over	Up thru 32	54.0	. .	36.0	. .	6
2219-T81, T8510⑤ and T8511⑤	Up thru 2.999	Up thru 25	58.0	. .	42.0	. .	6
		3003					
3003-0	All	All	14.0	19.0	5.0	. .	25
3003-H112	All	All	14.0	. .	5.0	. .	. .
		5083					
5083-0	Up thru 5.000	Up thru 32	39.0	51.0	16.0	. .	14
5083-H111	Up thru 5.000	Up thru 32	40.0	. .	24.0	. .	12
5083-H112	Up thru 5.000	Up thru 32	39.0	. .	16.0	. .	12
		5086					
5086-0	Up thru 5.000	Up thru 32	35.0	46.0	14.0	. .	14
5086-H111	Up thru 5.000	Up thru 32	36.0	. .	21.0	. .	12
5086-H112	Up thru 5.000	Up thru 32	35.0	. .	14.0	. .	12
		5154					
5154-0	All	All	30.0	41.0	11.0	. .	. .
5154-H112	All	All	30.0	. .	11.0	. .	. .
		5454					
5454-0	Up thru 5.000	Up thru 32	31.0	41.0	12.0	. .	14
5454-H111	Up thru 5.000	Up thru 32	33.0	. .	19.0	. .	12
5454-H112	Up thru 5.000	Up thru 32	31.0	. .	12.0	. .	12
		6005					
6005-T1	Up thru 0.500	All	25.0	. .	15.0	. .	16
6005-T5	Up thru 0.124	All	38.0	. .	35.0	. .	8
	0.125–1.000	all	38.0	. .	35.0	. .	10
		6061					
6061-0	All	All	. .	22.0	. .	16.0	16
6061-T1	Up thru 0.625	All	26.0	. .	14.0	. .	16
6061-T4, T4510⑤⑦ and T4511⑤⑦	All	All	26.0	. .	16.0	. .	16
6061-T42④⑧	All	All	26.0	. .	12.0	. .	16
6061-T51	Up thru 0.625	All	35.0	. .	30.0	. .	8
6061-T6, T62④⑧, T6510⑤ and T6511⑤	Up thru 0.249	All	38.0	. .	35.0	. .	8
	0.250 and over	All	38.0	. .	35.0	. .	10
		6063					
6063-0	All	All	. .	19.0	. .	. .	18
6063-T1	Up thru 0.500	All	17.0	. .	9.0	. .	12
	0.501–1.000	All	16.0	. .	8.0	. .	12
6063-T4 and T42④⑧	Up thru 0.500	All	19.0	. .	10.0	. .	14
	0.501–1.000	All	18.0	. .	9.0	. .	14
6063-T5	Up thru 0.500	All	22.0	. .	16.0	. .	8
	0.501–1.000	All	21.0	. .	15.0	. .	8

[9]Material in this temper, 0.750 in. and thicker, when tested in accordance with ASTM G47 in the short transverse direction at a stress level of 75% of the specified minimum yield strength will exhibit no evidence of stress corrosion cracking. Capability of individual lots to resist stress corrosion is determined by testing the previously selected tensile test sample in accordance with the applicable lot acceptance criteria outlined on page 100 of the Standard.

[10]Material in this temper when tested in accordance with ASTM G34-72 will exhibit exfoliation less than that shown in Category B, Figure 2 of ASTM G34-72. Also, material, 0.750 in. and thicker, when tested in accordance with ASTM G47 in the short transverse direction at a stress level of 25 ksi will exhibit no evidence of stress corrosion cracking. Capability of individual lots to resist exfoliation corrosion and stress corrosion cracking is determined by testing the previously selected tensile test sample in accordance with the applicable lot acceptance criteria outlined on page 100 of the Standard.

Mechanical Property Limits—Extruded Wire, Rod, Bar, and Shapes (*continued*)

ALLOY AND TEMPER	SPECIFIED DIAMETER OR THICKNESS① in.	AREA sq in.	TENSILE STRENGTH—ksi				ELONGATION② percent min in 2 in. or 4D③
			ULTIMATE		YIELD		
			min	max	min	max	
6063 (Continued)							
6063-T52	Up thru 1.000	All	22.0	30.0	16.0	25.0	8
6063-T6 and T62④⑧	Up thru 0.124	All	30.0	. .	25.0	. .	8
	0.125–1.000	All	30.0	. .	25.0	. .	10
6066							
6066-O	All	All	. .	29.0	. .	18.0	16
6066-T4, T4510⑤⑦ and T4511⑤⑦	All	All	40.0	. .	25.0	. .	14
6066-T42④⑧	All	All	40.0	. .	24.0	. .	14
6066-T6, T6510⑤ and T6511⑤	All	All	50.0	. .	45.0	. .	8
6066-T62④⑧	All	All	50.0	. .	42.0	. .	8
6070							
6070-T6 and T62④⑧	Up thru 2.999	Up thru 32	48.0	. .	45.0	. .	6
6105							
6105-T1	Up thru 0.500	All	25.0	. .	15.0	. .	16
6105-T5	Up thru 0.500	All	38.0	. .	35.0	. .	8
6162							
6162-T5, T5510⑤ and T5511⑤	Up thru 1.000	All	37.0	. .	34.0	. .	7
6162-T6, T6510⑤ and T6511⑤	Up thru 0.249	All	38.0	. .	35.0	. .	8
	0.250–0.499	All	38.0	. .	35.0	. .	10
6262							
6262-T6, T62④⑧, T6510⑤ and T6511⑤	All	All	38.0	. .	35.0	. .	10
6351							
6351-T1	Up thru 0.499	Up thru 20	26.0	. .	13.0	. .	15
6351-T4	Up thru 0.749	All	32.0	. .	19.0	. .	16
6351-T5	Up thru 0.249	All	38.0	. .	35.0	. .	8
	0.250–1.000	All	38.0	. .	35.0	. .	10
6351–T51	0.125–1.000	Up thru 20	36.0	. .	33.0	. .	10
6351-T54	Up thru 0.500	Up thru 20	30.0	. .	20.0	. .	10
6351-T6	Up thru 0.124	All	42.0	. .	37.0	. .	8
	0.125–0.749	All	42.0	. .	37.0	. .	10
6463							
6463-T1	Up thru 0.500	Up thru 20	17.0	. .	9.0	. .	12
6463-T5	Up thru 0.500	Up thru 20	22.0	. .	16.0	. .	8
6463-T6 and T62④⑧	Up thru 0.124	Up thru 20	30.0	. .	25.0	. .	8
	0.125–0.500	Up thru 20	30.0	. .	25.0	. .	10
7005							
7005-T53	Up thru 0.750	All	50.0	. .	44.0	. .	10
7050							
7050-T73510⑨ and T73511⑨	Up thru 5.000	Up thru 32	70.0	. .	60.0	. .	8
7050-T4510⑪⑬ and T74511⑪⑬	Up thru 5.000	Up thru 32	73.0	. .	63.0	. .	7
7050-T76510⑫ and T76511⑫	Up thru 0.499	Up thru 32	77.0	. .	69.0	. .	7
	0.500–5.000	Up thru 32	79.0	. .	69.0	. .	7
7075							
7075-O	All	All	. .	40.0	. .	24.0	10
7075-T6, T62④⑧, T6510⑤ and T6511⑤	Up thru 0.249	All	78.0	. .	70.0	. .	7
	0.250–0.499	All	81.0	. .	73.0	. .	7
	0.500–1.499	All	81.0	. .	72.0	. .	7
	1.500–2.999	All	81.0	. .	72.0	. .	7
	3.000–4.499	Up thru 20	81.0	. .	71.0	. .	7
	3.000–4.499	Over 20 thru 32	78.0	. .	70.0	. .	6
	4.500–5.000	Up thru 32	78.0	. .	68.0	. .	6

[11]Material in this temper when tested at the t/10 plane in accordance with ASTM G34-72 will exhibit exfoliation less than that shown in Category B, Figure 2 of ASTM G34-72. Also, material, 0.750 in. and thicker, when tested in accordance with ASTM G47 in the short transverse direction at a stress level of 35 ksi will exhibit no evidence of stress corrosion cracking. Capability of individual lots to resist exfoliation corrosion and stress corrosion cracking is determined by testing

Mechanical Property Limits—Extruded Wire, Rod, Bar, and Shapes (*concluded*)

ALLOY AND TEMPER	SPECIFIED DIAMETER OR THICKNESS① in.	AREA sq in.	TENSILE STRENGTH—ksi				ELONGATION② percent min in 2 in. or 4D③
			ULTIMATE		YIELD		
			min	max	min	max	
		7075 (Continued)					
7075-T73⑨, T73510⑤⑨ and T73511⑤⑨	0.062–0.249	Up thru 20	68.0	..	58.0	..	7
	0.250–1.499	Up thru 25	70.0	..	61.0	..	8
	1.500–2.999	Up thru 25	69.0	..	59.0	..	8
	3.000–4.499	Up thru 20	68.0	..	57.0	..	7
	3.000–4.499	Over 20 thru 32	65.0	..	55.0	..	7
7075-T76⑩, T76510⑤⑩ and T76511⑤⑩	Up thru 0.124	All	72.0	..	62.0	..	7
	0.125–0.249	Up thru 20	74.0	..	64.0	..	7
	0.250–0.499	Up thru 20	75.0	..	65.0	..	7
	0.500–1.000	Up thru 20	75.0	..	65.0	..	7
		7178					
7178-0	All	Up thru 32	..	40.0	..	24.0	10
7178-T6, T6510⑤ and T6511⑤	Up thru 0.061	All	82.0	..	76.0	..	..
	0.062–0.249	Up thru 20	84.0	..	76.0	..	5
	0.250–1.499	Up thru 25	87.0	..	78.0	..	5
	1.500–2.499	Up thru 25	86.0	..	77.0	..	5
	1.500–2.499	Over 25 thru 32	84.0	..	75.0	..	5
	2.500–2.999	Up thru 32	82.0	..	71.0	..	5
7178-T62④⑧	Up thru 0.061	All	79.0	..	73.0	..	..
	0.062–0.249	Up thru 20	82.0	..	74.0	..	5
	0.250–1.499	Up thru 25	86.0	..	77.0	..	5
	1.500–2.499	Up thru 25	86.0	..	77.0	..	5
	1.500–2.499	Over 25 thru 32	84.0	..	75.0	..	5
	2.500–2.999	Up thru 32	82.0	..	71.0	..	5
7178-T76⑩, T76510⑤⑩ and T76511⑤⑩	0.125–0.249	Up thru 20	76.0	..	66.0	..	7
	0.250–0.499	Up thru 20	77.0	..	67.0	..	7
	0.500–1.000	Up thru 20	77.0	..	67.0	..	7

the previously selected tensile test sample in accordance with the applicable lot acceptance criteria outlined on page 100 of the Standard.

[12]Material in this temper when tested at the t/10 plane in accordance with ASTM G34-72 will exhibit exfoliation less than that shown in Category B, Figure 2 of ASTM G34-72. Also, material, 0.750 in. and thicker, when tested in accordance with ASTM G47 in the short transverse direction at a stress level of 17 ksi will exhibit no evidence of stress corrosion cracking. Capability of individual lots to resist exfoliation corrosion and stress corrosion cracking is determined by testing the previously selected tensile test sample in accordance with the applicable lot acceptance criteria outlined on page 100 of the Standard.

[13]T74 type tempers although not previously registered have appeared in the literature and in some specifications as T736 type tempers.

A.5 TUBE AND PIPE

Mechanical Property Limits—Extruded Tube

ALLOY AND TEMPER	SPECIFIED WALL THICKNESS① in.	AREA sq in.	TENSILE STRENGTH—ksi				ELONGATION percent min in 2 in. or 4D②
			ULTIMATE		YIELD		
			min	max	min	max	
1060							
1060-0	All	All	8.5	14.0	2.5	. .	25
1060-H112	All	All	8.5	. .	2.5	. .	25
1100							
1100-0	All	All	11.0	15.5	3.0	. .	25
1100-H112	All	All	11.0	. .	3.0	. .	25
2014							
2014-0	All	All	. .	30.0	. .	18.0	12
2014-T4, T4510④ and T4511④	All	All	50.0	. .	35.0	. .	12
2014-T42③⑤	All	All	50.0	. .	29.0	. .	12
2014-T6, T6510④ and T6511④	Up thru 0.499	All	60.0	. .	53.0	. .	7
	0.500–0.749	All	64.0	. .	58.0	. .	7
	0.750 and over	Up thru 25	68.0	. .	60.0	. .	7
	0.750 and over	Over 25 thru 32	68.0	. .	58.0	. .	6
2014-T62③⑤	Up thru 0.749	All	60.0	. .	53.0	. .	7
	0.750 and over	Up thru 25	60.0	. .	53.0	. .	7
	0.750 and over	Over 25 thru 32	60.0	. .	53.0	. .	6
2024							
2024-0	All	All	. .	35.0	. .	19.0	12
2024-T3, T3510④ and T3511④	Up thru 0.249	All	57.0	. .	42.0	. .	10
	0.250–0.749	All	60.0	. .	44.0	. .	10
	0.750–1.499	All	65.0	. .	46.0	. .	10
	1.500 and over	Up thru 25	70.0	. .	48.0	. .	10
	1.500 and over	Over 25 thru 32	68.0	. .	46.0	. .	8
2024-T42③⑤	Up thru 0.749	All	57.0	. .	38.0	. .	12
	0.750–1.499	All	57.0	. .	38.0	. .	10
	1.500 and over	Up thru 25	57.0	. .	38.0	. .	10
	1.500 and over	Over 25 thru 32	57.0	. .	38.0	. .	8
2024-T81, T8510④ and T8511④	0.050–0.249	All	64.0	. .	56.0	. .	4
	0.250–1.499	All	66.0	. .	58.0	. .	5
	1.500 and over	Up thru 32	66.0	. .	58.0	. .	5
2219							
2219-0	All	All	. .	32.0	. .	18.0	12
2219-T31, T3510④ and T3511④	Up thru 0.499	Up thru 25	42.0	. .	26.0	. .	14
	0.500–2.999	Up thru 25	45.0	. .	27.0	. .	14
2219-T62③⑤	Up thru 0.999	Up thru 25	54.0	. .	36.0	. .	6
	1.000 and over	Up thru 32	54.0	. .	36.0	. .	6
2219-T81, T8510④, and T8511④	Up thru 2.999	Up thru 25	58.0	. .	42.0	. .	6

Mechanical Property Limits—Extruded Tube (*continued*)

ALLOY AND TEMPER	SPECIFIED WALL THICKNESS① in.	AREA sq in.	TENSILE STRENGTH—ksi ULTIMATE min	ULTIMATE max	YIELD min	YIELD max	ELONGATION percent min in 2 in. or 4D②
		3003					
3003-0	All	All	14.0	19.0	5.0	..	25
3003-H112	All	All	14.0	..	5.0	..	25
		ALCLAD 3003					
Alclad 3003-0	All	All	13.0	18.0	4.5	..	25
Alclad 3003-H112	All	All	13.0	..	4.5	..	25
		3004					
3004-0	All	All	23.0	29.0	8.5	..	..
		5083					
5083-0	All	Up thru 32	39.0	51.0	16.0	..	14
5083-H111	All	Up thru 32	40.0	..	24.0	..	12
5083-H112	All	Up thru 32	39.0	..	16.0	..	12
		5086					
5086-0	All	Up thru 32	35.0	46.0	14.0	..	14
5086-H111	All	Up thru 32	36.0	..	21.0	..	12
5086-H112	All	Up thru 32	35.0	..	14.0	..	12
		5154					
5154-0	All	All	30.0	41.0	11.0	..	..
5154-H112	All	All	30.0	..	11.0	..	..
		5454					
5454-0	All	Up thru 32	31.0	41.0	12.0	..	14
5454-H111	All	Up thru 32	33.0	..	19.0	..	12
5454-H112	All	Up thru 32	31.0	..	12.0	..	12
		6005					
6005-T1	Up thru 0.500	All	25.0	..	15.0	..	16
6005-T5	Up thru 0.124	All	38.0	..	35.0	..	8
	0.125–1.000	All	38.0	..	35.0	..	10
		6061					
6061-0	All	All	..	22.0	..	16.0	16
6061-T1	Up thru 0.625	All	26.0	..	14.0	..	16
6061-T4, T4510④ and T4511④	All	All	26.0	..	16.0	..	16
6061-T42③⑤	All	All	26.0	..	12.0	..	16
6061-T51	Up thru 0.625	All	35.0	..	30.0	..	8
6061-T6, T62③⑤, T6510④	Up thru 0.249	All	38.0	..	35.0	..	8
and T6511④	0.250 and over	All	38.0	..	35.0	..	10
		6063					
6063-0	All	All	..	19.0	..	..	18
6063-T1	Up thru 0.500	All	17.0	..	9.0	..	12
	0.501–1.000	All	16.0	..	8.0	..	12
6063-T4 and T42③⑤	Up thru 0.500	All	19.0	..	10.0	..	14
	0.501–1.000	All	18.0	..	9.0	..	14
6063-T5	Up thru 0.500	All	22.0	..	16.0	..	8
	0.501–1.000	All	21.0	..	15.0	..	8
6063-T52	Up thru 1.000	All	22.0	30.0	16.0	25.0	8
6063-T6 and T62③⑤	Up thru 0.124	All	30.0	..	25.0	..	8
	0.125–1.000	All	30.0	..	25.0	..	10

[1]The thickness of the cross-section from which the tension test specimen is taken determines the applicable mechanical properties. The data base and criteria upon which these mechanical property limits are established are outlined on page 93 of the Standard under "Mechanical Properties."

[2]D represents specimen diameter.

Mechanical Property Limits—Extruded Tube (*concluded*)

ALLOY AND TEMPER	SPECIFIED WALL THICKNESS① in.	AREA sq in.	TENSILE STRENGTH—ksi				ELONGATION percent min in 2 in. or 4D②
			ULTIMATE		YIELD		
			min	max	min	max	
		6066					
6066-0	All	All	. .	29.0	. .	18.0	16
6066-T4, T4510④ and T4511④	All	All	40.0	. .	25.0	. .	14
6066-T42③⑤	All	All	40.0	. .	24.0	. .	14
6066-T6, T6510④ and T6511④	All	All	50.0	. .	45.0	. .	8
6066-T62③⑤	All	All	50.0	. .	42.0	. .	8
		6070					
6070-T6 and T62③⑤	Up thru 2.999	Up thru 32	48.0	. .	45.0	. .	6
		6105					
6105-T1	Up thru 0.500	All	25.0	. .	15.0	. .	16
6105-T5	Up thru 0.500	All	38.0	. .	35.0	. .	8
		6162					
6162-T5, T5510④ and T5511④	Up thru 1.000	All	37.0	. .	34.0	. .	7
6162-T6, T6510④ and T6511④	Up thru 0.249	All	38.0	. .	35.0	. .	8
	0.250–0.499	All	38.0	. .	35.0	. .	10
		6262					
6262-T6, T62③⑤, T6510④ and T6511④	All	All	38.0	. .	35.0	. .	10
		6351					
6351-T4	Up thru 0.749	All	32.0	. .	19.0	. .	16
6351-T6	Up thru 0.124	All	42.0	. .	37.0	. .	8
	0.125–0.749	All	42.0	. .	37.0	. .	10
		7001					
7001-0	All	All	. .	42.0	. .	26.0	10
7001-T6, T62③⑤, T6510④ and T6511④	Up thru 0.249	All	89.0	. .	82.0	. .	5
	0.250–0.499	All	92.0	. .	84.0	. .	5
	0.500–1.999	All	94.0	. .	88.0	. .	5
	2.000–2.999	All	90.0	. .	84.0	. .	5
		7075					
7075-0	All	All	. .	40.0	. .	24.0	10
7075-T6, T62③⑤, T6510④ and T6511④	Up thru 0.249	All	78.0	. .	70.0	. .	7
	0.250–0.499	All	81.0	. .	73.0	. .	7
	0.500–1.499	All	81.0	. .	72.0	. .	7
	1.500–2.999	All	81.0	. .	72.0	. .	7
7075-T73⑥, T73510④⑥ and T73511④⑥	0.062–0.249	All	68.0	. .	58.0	. .	7
	0.250–1.499	Up thru 25	70.0	. .	61.0	. .	8
	1.500–2.999	Up thru 25	69.0	. .	59.0	. .	8

[3]These properties can usually be obtained by the user when the material is properly solution heat treated or solution and precipitation heat treated from the O (annealed) or F (as fabricated) temper. These properties also apply to samples of material in the O or F tempers which are solution heat treated and precipitation treated by the producer to determine that the material will respond to proper heat treatment. Properties attained by the user, however, may be lower than those listed if the material has been formed or otherwise cold or hot worked, particularly in the annealed temper, prior to solution heat treatment.

[4]For stress-relieved tempers the characteristics and properties other than those specified may differ somewhat form the corresponding characteristics and properties of material in the basic temper.

[5]This temper is not available from the material producer.

[6]Material in this temper, 0.750 in. and thicker, when tested in accordance with ASTM G47 in the short transverse direction at a stress level of 75% of the specified minimum yield strength will exhibit no evidence of stress corrosion cracking. Capability of individual lots to resist stress corrosion is determined by testing the previously selected tensile test sample in accordance with the applicable lot acceptance criteria outlined on page 100 of the Standard.

Mechanical Property Limits[1,7]—Drawn Tube

ALLOY AND TEMPER	SPECIFIED WALL THICKNESS in.	TENSILE STRENGTH—ksi				ELONGATION percent min in 2 in. or 4D②	
		ULTIMATE		YIELD		FULL-SECTION SPECIMEN③	CUT-OUT SPECIMEN④
		min	max	min	max		
				1060⑤			
1060–0	0.010–0.500	8.5	13.5	2.5	..	..	..
1060–H12	0.010–0.500	10.0	..	4.0	..	..	..
1060–H14	0.010–0.500	12.0	..	10.0	..	..	..
1060–H18	0.010–0.500	16.0	..	13.0	..	..	..
1060–H113⑪	0.010–0.500	8.5	..	2.5	..	..	..
				1100⑤			
1100–0	0.014–0.500	11.0	15.5	3.5	..	..	..
1100–H12	0.014–0.500	14.0	..	11.0	..	..	..
1100–H14	0.014–0.500	16.0	..	14.0	..	..	..
1100–H16	0.014–0.500	19.0	..	17.0	..	..	..
1100–H18	0.014–0.500	22.0	..	20.0	..	..	..
1100–H113⑪	0.014–0.500	11.0	..	3.5	..	..	..
				2011			
2011-T3	0.018–0.049	47.0	..	40.0	..	..	..
	0.050–0.500	47.0	..	40.0	..	10	8
2011-T4511	0.018–0.049	44.0	..	25.0	..	..	..
	0.050–0.259	44.0	..	25.0	..	20	18
	0.260–0.500	44.0	..	25.0	..	20	20
2011–T8	0.018–0.500	58.0	..	46.0	..	10	8
				2014			
2014–0	0.018–0.500	..	32.0	..	16.0	..	..
2014-T4 and T42⑥⑧	0.018–0.024	54.0	..	30.0	..	10	..
	0.025–0.049	54.0	..	30.0	..	12	10
	0.050–0.259	54.0	..	30.0	..	14	10
	0.260–0.500	54.0	..	30.0	..	16	12
2014-T6 and T62⑥⑧	0.018–0.024	65.0	..	55.0	..	7	..
	0.025–0.049	65.0	..	55.0	..	7	6
	0.050–0.259	65.0	..	55.0	..	8	7
	0.260–0.500	65.0	..	55.0	..	9	8
				2024			
2024–0	0.018–0.500	..	32.0	..	15.0	..	..
2024-T3	0.018–0.024	64.0	..	42.0	..	10	..
	0.025–0.049	64.0	..	42.0	..	12	10
	0.500–0.259	64.0	..	42.0	..	14	10
	0.260–0.500	64.0	..	42.0	..	16	12
2024-T42⑥⑧	0.018–0.024	64.0	..	40.0	..	10	..
	0.025–0.049	64.0	..	40.0	..	12	10
	0.050–0.259	64.0	..	40.0	..	14	10
	0.260–0.500	64.0	..	40.0	..	16	12
				3003⑤			
3003-0	0.010–0.024	14.0	19.0	5.0	..	..	..
	0.025–0.049	14.0	19.0	5.0	..	30	20
	0.050–0.259	14.0	19.0	5.0	..	35	25
	0.260–0.500	14.0	19.0	5.0	..	..	30
3003–H12	0.010–0.500	17.0	..	12.0	..	..	..
3003-H14	0.010–0.024	20.0	..	17.0	..	3	..
	0.025–0.049	20.0	..	17.0	..	5	3
	0.050–0.259	20.0	..	17.0	..	8	4
	0.260–0.500	20.0	..	17.0	..	..	..
3003-H16	0.010–0.024	24.0	..	21.0	..	..	..
	0.025–0.049	24.0	..	21.0	..	3	2
	0.050–0.259	24.0	..	21.0	..	5	4
	0.260–0.500	24.0	..	21.0	..	..	..
3003-H18	0.010–0.024	27.0	..	24.0	..	2	..
	0.025–0.049	27.0	..	24.0	..	3	2
	0.050–0.259	27.0	..	24.0	..	5	3
	0.260–0.500	27.0	..	24.0	..	..	..
3003–H113⑪	0.010–0.500	14.0	..	5.0	..	..	..

Mechanical Property Limits—Drawn Tube (*continued*)

ALLOY AND TEMPER	SPECIFIED WALL THICKNESS in.	TENSILE STRENGTH—ksi				ELONGATION percent min in 2 in. or 4D②	
		ULTIMATE		YIELD		FULL-SECTION SPECIMEN③	CUT-OUT SPECIMEN④
		min	max	min	max		
ALCLAD 3003⑤							
Alclad 3003-0	0.010–0.024	13.0	19.0	4.5	..	..	..
	0.025–0.049	13.0	19.0	4.5	..	30	20
	0.050–0.259	13.0	19.0	4.5	..	35	25
	0.260–0.500	13.0	19.0	4.5	..	..	30
Alclad 3003-H14	0.010–0.024	19.0	..	16.0	..	..	..
	0.025–0.049	19.0	..	16.0	..	5	3
	0.050–0.259	19.0	..	16.0	..	8	4
	0.260–0.500	19.0	..	16.0	..	..	..
Alclad 3003-H18	0.010–0.500	26.0	..	23.0	..	..	..
Alclad 3003-H113⑪	0.010–0.500	13.0	..	4.5	..	..	..
3004⑤							
3004-0	0.018–0.450	23.0	29.0	8.5	..	..	..
3004-H34	0.018–0.450	32.0	..	25.0	..	..	..
3004-H36	0.018–0.450	35.0	..	28.0	..	..	..
3004-H38	0.018–0.450	38.0	..	30.0	..	..	..
5050⑤							
5050-0	0.010–0.500	18.0	24.0	6.0	..	..	..
5050-H32	0.010–0.500	22.0	..	16.0	..	..	..
5050-H34	0.010–0.500	25.0	..	20.0	..	..	..
5050-H36	0.010–0.500	27.0	..	22.0	..	..	..
5050-H38	0.010–0.500	29.0	..	24.0	..	..	..
5052⑤							
5052-0	0.010–0.450	25.0	35.0	10.0	..	..	..
5052-H32	0.010–0.450	31.0	..	23.0	..	..	..
5052-H34	0.010–0.450	34.0	..	26.0	..	..	..
5052-H36	0.010–0.450	37.0	..	29.0	..	..	..
5052-H38	0.010–0.450	39.0	..	31.0	..	..	..
5086⑤							
5086-0	0.010–0.450	35.0	46.0	14.0	..	..	..
5086-H32	0.010–0.450	40.0	..	28.0	..	..	..
5086-H34	0.010–0.450	44.0	..	34.0	..	..	..
5086-H36	0.010–0.450	47.0	..	38.0	..	..	..
5154⑤							
5154-0	0.010–0.500	30.0	41.0	11.0	..	10	10
5154-H34	0.010–0.500	39.0	..	29.0	..	5	5
5154-H38	0.010–0.250	45.0	..	34.0	..	..	..
6061							
6061-0	0.018–0.500	..	22.0	..	14.0	15	15
6061-T4	0.025–0.049	30.0	..	16.0	..	16	14
	0.050–0.259	30.0	..	16.0	..	18	16
	0.260–0.500	30.0	..	16.0	..	20	18
6061-T42⑥⑧	0.025–0.049	30.0	..	14.0	..	16	14
	0.050–0.259	30.0	..	14.0	..	18	16
	0.260–0.500	30.0	..	14.0	..	20	18
6061-T6 and T62⑥⑧	0.025–0.049	42.0	..	35.0	..	10	8
	0.050–0.259	42.0	..	35.0	..	12	10
	0.260–0.500	42.0	..	35.0	..	14	12
6063							
6063-0	0.018–0.500	..	19.0	..	..	..	..
6063-T4 and T42⑥⑧	0.025–0.049	22.0	..	10.0	..	16	14
	0.050–0.259	22.0	..	10.0	..	18	16
	0.260–0.500	22.0	..	10.0	..	20	18

[1] The data base and criteria upon which these mechanical property limits are established are outlined on page 93 of the Standard under "Mechanical Properties."

[2] D represents diameter of cut-out specimen.

[3] Round tube 2 in. or less in outside diameter and square tube $1\frac{1}{2}$ in. or less on a side are tested in full section unless the limitations of the testing machine preclude the use of such a specimen.

Mechanical Property Limits—Drawn Tube (*concluded*)

ALLOY AND TEMPER	SPECIFIED WALL THICKNESS in.	TENSILE STRENGTH—ksi				ELONGATION percent min in 2 in. or 4D②	
		ULTIMATE		YIELD		FULL-SECTION SPECIMEN③	CUT-OUT SPECIMEN④
		min	max	min	max		
6063 (continued)							
6063-T6 and T62⑥⑧	0.025–0.049 0.050–0.259 0.260–0.500	33.0 33.0 33.0	. . . . . .	28.0 28.0 28.0	. . . . . .	12 14 16	8 10 12
6063–T83	0.025–0.259	33.0	. .	30.0	. .	5	. .
6063-T831	0.025–0.259	28.0	. .	25.0	. .	5	. .
6063-T832	0.025–0.049 0.050–0.259	41.0 40.0	. . . .	36.0 35.0	. . . .	8 8	5 5
6066							
6066-0	0.018–0.500	. .	28.0	. .	18.0	16	16
6066-T4 and T42⑥⑧	0.025–0.500	40.0	. .	25.0	. .	14	12
6066-T6 and T62⑥⑧	0.025–0.050 0.051–0.500	50.0 50.0	. . . .	45.0 45.0	. . . .	8 10	8 8
6262							
6262-T6 and T62⑥⑧	0.025–0.049 0.050–0.259 0.260–0.500	42.0 42.0 42.0	. . . . . .	35.0 35.0 35.0	. . . . . .	10 12 14	8 10 12
6262-T9	0.025–0.375	48.0	. .	44.0	. .	5	4
7075							
7075-0	0.025–0.049 0.050–0.500	. . . .	40.0 40.0	. . . .	21.0⑨ 21.0⑨	10 12	8 10
7075-T6 and T62⑥⑧	0.025–0.259 0.260–0.500	77.0 77.0	. . . .	66.0 66.0	. . . .	8 9	7 8
7075-T73⑩	0.025–0.259 0.260–0.500	66.0 66.0	. . . .	56.0 56.0	. . . .	10 12	8 10

[4] For round tube over 2 in. in diameter, for square tube over $1\frac{1}{2}$ in. on a side, for all sizes of tube other than round or square, or in those cases when a full-section specimen cannot be used, a cut-out specimen is used.

[5] In this alloy, tube other than round is produced only in the O, F, and H113 tempers. Properties for the F temper are not specified or guaranteed.

[6] These properties can usually be obtained by the user when the material is properly solution heat treated or solution and precipitation heated treated from the O (annealed) or F (as fabricated) temper. These properties also apply to samples of material in the O or F tempers, which are solution heat treated or solution and precipitation treated by the producer to determine that the material will respond to proper heat treatment. Properties attained by the user, however, may be lower than those listed if the material has been formed or otherwise cold or hot worked, particularly in the annealed temper, prior to solution heat treatment.

[7] Processes such as flattening, levelling, or straightening coiled products subsequent to shipment by the producer may alter the mechanical properties of the metal (refer to Certification, Section 4 of the Standard).

[8] This temper is not available from the material producer.

[9] Applicable only to round tube. The maximum yield strength for other-than-round tube shall be negotiated.

[10] Material in this temper, 0.750 in. and thicker, when tested in accordance with ASTM G47 in the short transverse direction at a stress level of 75% of the specified minimum yield strength will exhibit no evidence of stress corrosion cracking. Capability of individual lots to resist stress corrosion is determined by testing the previously selected tensile test sample in accordance with the applicable lot acceptance criteria outlined on page 100 of the Standard.

[11] This temper applies to other than round tube which is fabricated from annealed round tube.

A.6 CHANNELS AND I-BEAMS

Aluminum Association Standard Channels—Dimensions, Areas, Weights, and Section Properties[4]

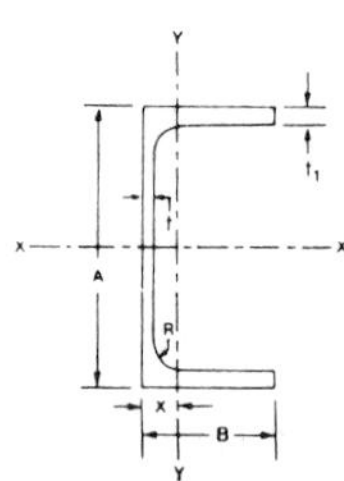

Size		Area①	Weight②	Flange Thickness	Web Thickness	Fillet Radius	Section Properties③						
							Axis X-X			Axis Y-Y			
Depth A	Width B			t_1	t	R	I	S	r	I	S	r	x
in.	in.	in.^{2}	lb/ft	in.	in.	in.	in.^{4}	in.^{3}	in.	in.^{4}	in.^{3}	in.	in.
2.00	1.00	0.491	0.557	0.13	0.13	0.10	0.288	0.288	0.766	0.045	0.064	0.303	0.298
2.00	1.25	0.911	1.071	0.26	0.17	0.15	0.546	0.546	0.774	0.139	0.178	0.391	0.471
3.00	1.50	0.965	1.135	0.20	0.13	0.25	1.41	0.94	1.21	0.22	0.22	0.47	0.49
3.00	1.75	1.358	1.597	0.26	0.17	0.25	1.97	1.31	1.20	0.42	0.37	0.55	0.62
4.00	2.00	1.478	1.738	0.23	0.15	0.25	3.91	1.95	1.63	0.60	0.45	0.64	0.65
4.00	2.25	1.982	2.331	0.29	0.19	0.25	5.21	2.60	1.62	1.02	0.69	0.72	0.78
5.00	2.25	1.881	2.212	0.26	0.15	0.30	7.88	3.15	2.05	0.98	0.64	0.72	0.73
5.00	2.75	2.627	3.089	0.32	0.19	0.30	11.14	4.45	2.06	2.05	1.14	0.88	0.95
6.00	2.50	2.410	2.834	0.29	0.17	0.30	14.35	4.78	2.44	1.53	0.90	0.80	0.79
6.00	3.25	3.427	4.030	0.35	0.21	0.30	21.04	7.01	2.48	3.76	1.76	1.05	1.12
7.00	2.75	2.725	3.205	0.29	0.17	0.30	22.09	6.31	2.85	2.10	1.10	0.88	0.84
7.00	3.50	4.009	4.715	0.38	0.21	0.30	33.79	9.65	2.90	5.13	2.23	1.13	1.20
8.00	3.00	3.526	4.147	0.35	0.19	0.30	37.40	9.35	3.26	3.25	1.57	0.96	0.93
8.00	3.75	4.923	5.789	0.41	0.25	0.35	52.69	13.17	3.27	7.13	2.82	1.20	1.22
9.00	3.25	4.237	4.983	0.35	0.23	0.35	54.41	12.09	3.58	4.40	1.89	1.02	0.93
9.00	4.00	5.927	6.970	0.44	0.29	0.35	78.31	17.40	3.63	9.61	3.49	1.27	1.25
10.00	3.50	5.218	6.136	0.41	0.25	0.35	83.22	16.64	3.99	6.33	2.56	1.10	1.02
10.00	4.25	7.109	8.360	0.50	0.31	0.40	116.15	23.23	4.04	13.02	4.47	1.35	1.34
12.00	4.00	7.036	8.274	0.47	0.29	0.40	159.76	26.63	4.77	11.03	3.86	1.25	1.14
12.00	5.00	10.053	11.822	0.62	0.35	0.45	239.69	39.95	4.88	25.74	7.60	1.60	1.61

[1]Areas listed are based on nominal dimensions.

[2]Weights per foot are based on nominal dimensions and a density of 0.098 pound per cubic inch which is the density of alloy 6061.

[3]I = moment of inertia; S = section modulus; r = radius of gyration.

[4]Users are encouraged to ascertain current availability of particular structural shapes through inquiries to their suppliers.

Aluminum Association Standard I-Beams—Dimensions, Areas, Weights and Section Properties[4]

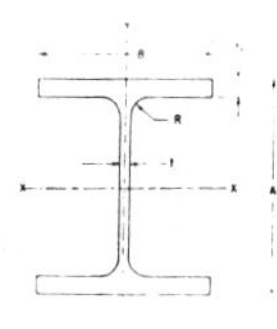

Size		Area①	Weight②	Flange Thickness t_1	Web Thickness t	Fillet Radius R	Section Properties③					
							Axis X-X			Axis Y-Y		
Depth A	Width B						I	S	r	I	S	r
in.	in.	in.^{2}	lb/ft	in.	in.	in.	in.^{4}	in.^{3}	in.	in.^{4}	in.^{3}	in.
3.00	2.50	1.392	1.637	0.20	0.13	0.25	2.24	1.49	1.27	0.52	0.42	0.61
3.00	2.50	1.726	2.030	0.26	0.15	0.25	2.71	1.81	1.25	0.68	0.54	0.63
4.00	3.00	1.965	2.311	0.23	0.15	0.25	5.62	2.81	1.69	1.04	0.69	0.73
4.00	3.00	2.375	2.793	0.29	0.17	0.25	6.71	3.36	1.68	1.31	0.87	0.74
5.00	3.50	3.146	3.700	0.32	0.19	0.30	13.94	5.58	2.11	2.29	1.31	0.85
6.00	4.00	3.427	4.030	0.29	0.19	0.30	21.99	7.33	2.53	3.10	1.55	0.95
6.00	4.00	3.990	4.692	0.35	0.21	0.30	25.50	8.50	2.53	3.74	1.87	0.97
7.00	4.50	4.932	5.800	0.38	0.23	0.30	42.89	12.25	2.95	5.78	2.57	1.08
8.00	5.00	5.256	6.181	0.35	0.23	0.30	59.69	14.92	3.37	7.30	2.92	1.18
8.00	5.00	5.972	7.023	0.41	0.25	0.30	67.78	16.94	3.37	8.55	3.42	1.20
9.00	5.50	7.110	8.361	0.44	0.27	0.30	102.02	22.67	3.79	12.22	4.44	1.31
10.00	6.00	7.352	8.646	0.41	0.25	0.40	132.09	26.42	4.24	14.78	4.93	1.42
10.00	6.00	8.747	10.286	0.50	0.29	0.40	155.79	31.16	4.22	18.03	6.01	1.44
12.00	7.00	9.925	11.672	0.47	0.29	0.40	255.57	42.60	5.07	26.90	7.69	1.65
12.00	7.00	12.153	14.292	0.62	0.31	0.40	317.33	52.89	5.11	35.48	10.14	1.71

[1]Areas listed are based on nominal dimensions.

[2]Weights per foot are based on nominal dimensions and a density of 0.098 pound per cubic inch which is the density of alloy 6061.

[3]I = moment of inertia; S = section modulus; r = radius of gyration.

[4]Users are encouraged to ascertain current availability of particular structural shapes through inquiries to their suppliers.

A.7 FORGINGS

Mechanical Property Limits—Die Forgings[6]

ALLOY AND TEMPER	SPECIFIED THICKNESS② in.	SPECIMEN AXIS PARALLEL TO DIRECTION OF GRAIN FLOW				SPECIMEN AXIS NOT PARALLEL TO DIRECTION OF GRAIN FLOW			BRINELL HARDNESS⑤
		TENSILE STRENGTH ksi min		ELONGATION percent min. in 2 in. or 4D③		TENSILE STRENGTH ksi min		ELONGATION percent min. in 2 in. or 4D③	500Kg Load—10 mm ball min
		ULTIMATE	YIELD	COUPON	FORGING	ULTIMATE	YIELD	FORGING	
1100–H112④	Up thru 4.000	11.0	4.0	25	18	..	..	..	20
2014–T4	Up thru 4.000	55.0	30.0	16	11	..	..	..	100
2014–T6	Up thru 1.000	65.0	56.0	8	6	64.0	55.0	3	125
	1.001–2.000	65.0	56.0	①	6	64.0	55.0	2	125
	2.001–3.000	65.0	55.0	①	6	63.0	54.0	2	125
	3.001–4.000	63.0	55.0	①	6	63.0	54.0	2	125
2018–T61	Up thru 4.000	55.0	40.0	10	7	..	..	..	100
2025–T6	Up thru 4.000	52.0	33.0	16	11	..	..	..	100
2218–T61	Up thru 4.000	55.0	40.0	10	7	..	..	..	100
2218–T72	Up thru 4.000	38.0	29.0	8	5	..	..	..	85
2219–T6	Up thru 4.000	58.0	38.0	10	8	56.0	36.0	4	100
2618–T61	Up thru 4.000	58.0	45.0	6	4	55.0	42.0	4	115
3003–H112④	Up thru 4.000	14.0	5.0	25	18	..	..	..	25
4032–T6	Up thru 4.000	52.0	42.0	5	3	..	..	..	115
5083–H111④	Up thru 4.000	42.0	22.0	..	14	39.0	20.0	12	..
5083–H112④	Up thru 4.000	40.0	18.0	..	16	39.0	16.0	14	..
5456–H112④	Up thru 4.000	44.0	20.0	..	16	..	..	..	..
6053–T6	Up thru 4.000	36.0	30.0	16	11	..	..	..	75
6061–T6	Up thru 4.000	38.0	35.0	10	7	38.0	35.0	5	80
6066–T6	Up thru 4.000	50.0	45.0	12	8	..	..	..	100
6151–T6	Up thru 4.000	44.0	37.0	14	10	44.0	37.0	6	90
7049–T73⑦	Up thru 1.000	72.0	62.0	10	7	71.0	61.0	3	135
	1.001–2.000	72.0	62.0	10	7	70.0	60.0	3	135
	2.001–3.000	71.0	61.0	10	7	70.0	60.0	3	135
	3.001–4.000	71.0	61.0	10	7	70.0	60.0	2	135
	4.001–5.000	70.0	60.0	10	7	68.0	58.0	2	135
7050–T74⑧⑨	Up thru 2.000	72.0	62.0	..	7	68.0	56.0	5	..
	2.001–4.000	71.0	61.0	..	7	67.0	55.0	4	..
	4.001–5.000	70.0	60.0	..	7	66.0	54.0	3	..
	5.001–6.000	70.0	59.0	..	7	66.0	54.0	3	..
7075–T6	Up thru 1.000	75.0	64.0	10	7	71.0	61.0	3	135
	1.001–2.000	74.0	63.0	①	7	71.0	61.0	3	135
	2.001–3.000	74.0	63.0	①	7	70.0	60.0	3	135
	3.001–4.000	73.0	62.0	①	7	70.0	60.0	2	135
7075–T73⑦	Up thru 3.000	66.0	56.0	..	7	62.0	53.0	3	125
	3.001–4.000	64.0	55.0	..	7	61.0	52.0	2	125
7075–T7352⑦	Up thru 3.000	66.0	56.0	..	7	62.0	51.0	3	125
	3.001–4.000	64.0	53.0	..	7	61.0	49.0	2	125

[1]When separately forged coupons are used to verify acceptability of forgings in the indicated thicknesses, the properties shown for thicknesses "Up thru 1 inch," including the test coupon elongation, apply.

[2]As-forged thickness. When forgings are machined prior to heat treatment the properties will also apply to the machined heat treat thickness provided the machined thickness is not less than one-half the original (as-forged) thickness.

[3]D equals specimen diameter.

[4]Properties of H111 and H112 temper forgings are dependent on the equivalent cold work in the forgings. The properties listed should be attainable in any forging within the prescribed thickness range and may be considerably exceeded in some cases.

[5]For information only: The Brinell Hardness is usually measured on the surface of a heat-treated forging using a 500-kg load and a 10-mm penetrator ball.

[6]The data base and criteria upon which these mechanical property limits are established are outlined on page 93 of the Standard under "Mechanical Properties."

[7]Material in this temper, 0.750 in. and thicker, when tested in accordance with ASTM G47 in the short transverse direction at a stress level of 75% of the specified minimum yield strength will exhibit no evidence of stress corrosion cracking. Capability of individual lots to resist stress corrosion is determined by testing the previously selected tensile test sample in accordance with the applicable lot acceptance criteria outlined on page 100 of the Standard.

[8]T74 type tempers although not previously registered have appeared in the literature and in some specifications as T736 type tempers.

[9]Material in this temper when tested at any plane in accordance with ASTM G34–72 will exhibit exfoliation less than that shown in Category B, Figure 2 of ASTM G34–72. Also, material, 0.750 in. and thicker, when tested in accordance with ASTM G47 in the short transverse direction at a stress level of 35 ksi will exhibit no evidence of stress corrosion cracking. Capability of individual lots to resist exfoliation corrosion and stress corrosion cracking is determined by testing the previously selected tensile test sample in accordance with the applicable lot acceptance criteria outlined on page 100 of the Standard.

Mechanical Property Limits—Hand Forgings[1,2,5]

ALLOY AND TEMPER	SPECIFIED THICKNESS③ in.	AXIS OF TEXT SPECIMEN	TENSILE STRENGTH ksi min		ELONGATION percent min in 2 in. or 4D④
			ULTIMATE	YIELD	
2014					
2014-T6	Up thru 2.000	Longitudinal*	65.0	56.0	8
		Long transverse	65.0	56.0	3
	2.001–3.000	Longitudinal*	64.0	56.0	8
		Long transverse	64.0	55.0	3
		Short transverse	62.0	55.0	2
	3.001–4.000	Longitudinal*	63.0	55.0	8
		Long transverse	63.0	55.0	3
		Short transverse	61.0	54.0	2
	4.001–5.000	Longitudinal*	62.0	54.0	7
		Long transverse	62.0	54.0	2
		Short transverse	60.0	53.0	1
	5.001–6.000	Longitudinal*	61.0	53.0	7
		Long transverse	61.0	53.0	2
		Short transverse	59.0	53.0	1
	6.001–7.000	Longitudinal*	60.0	52.0	6
		Long transverse	60.0	52.0	2
		Short transverse	58.0	52.0	1
	7.001–8.000	Longitudinal*	59.0	51.0	6
		Long transverse	59.0	51.0	2
		Short transverse	57.0	51.0	1
2014-T652	Up thru 2.000	Longitudinal*	65.0	56.0	8
		Long transverse	65.0	56.0	3
	2.001–3.000	Longitudinal*	64.0	56.0	8
		Long transverse	64.0	55.0	3
		Short transverse	62.0	52.0	2
	3.001–4.000	Longitudinal*	63.0	55.0	8
		Long transverse	63.0	55.0	3
		Short transverse	61.0	51.0	2
	4.001–5.000	Longitudinal*	62.0	54.0	7
		Long transverse	62.0	54.0	2
		Short transverse	60.0	50.0	1
	5.001–6.000	Longitudinal*	61.0	53.0	7
		Long transverse	61.0	53.0	2
		Short transverse	59.0	50.0	1
	6.001–7.000	Longitudinal*	60.0	52.0	6
		Long transverse	60.0	52.0	2
		Short transverse	58.0	49.0	1
	7.001–8.000	Longitudinal*	59.0	51.0	6
		Long transverse	59.0	51.0	2
		Short transverse	57.0	48.0	1
2219					
2219-T6	Up thru 4.000	Longitudinal*	58.0	40.0	6
		Long transverse	55.0	37.0	4
		Short transverse⑦	53.0	35.0	2
2219-T852	Up thru 4.000	Longitudinal*	62.0	50.0	6
		Long transverse	62.0	49.0	4
		Short transverse⑦	60.0	46.0	3
2618					
2618-T61	Up thru 2.000	Longitudinal*	58.0	47.0	7
		Long transverse	55.0	42.0	5
	2.001–3.000	Longitudinal*	57.0	46.0	7
		Long transverse	55.0	42.0	5
		Short transverse	52.0	42.0	4
	3.001–4.000	Longitudinal*	56.0	45.0	7
		Long transverse	53.0	40.0	5
		Short transverse	51.0	39.0	4
5083					
5083-H111	Up thru 4.000	Longitudinal*	42.0	22.0	14
		Long transverse	39.0	20.0	12
5083-H112	Up thru 4.000	Longitudinal*	40.0	18.0	16
		Long transverse	39.0	16.0	14
5456					
5456-H112	Up thru 3.000	Longitudinal*	44.0	20.0	16
		Long transverse	42.0	18.0	14
6061					
6061-T6	Up thru 4.000	Longitudinal*	38.0	35.0	10
		Long transverse	38.0	35.0	8
		Short transverse⑦	37.0	33.0	5
	4.001–8.000	Longitudinal*	37.0	34.0	8
		Long transverse	37.0	34.0	6
		Short transverse	35.0	32.0	4

Mechanical Property Limits—Hand Forgings (*continued*)

ALLOY AND TEMPER	SPECIFIED THICKNESS③ in.	AXIS OF TEXT SPECIMEN	TENSILE STRENGTH ksi min		ELONGATION percent min in 2 in. or 4D④
			ULTIMATE	YIELD	
		7049			
7049-T73⑥	2.001–3.000	Longitudinal*	71.0	61.0	9
		Long transverse	71.0	59.0	4
		Short transverse	69.0	58.0	3
	3.001–4.000	Longitudinal*	69.0	59.0	8
		Long transverse	69.0	57.0	3
		Short transverse	67.0	56.0	2
	4.001–5.000	Longitudinal*	67.0	56.0	7
		Long transverse	67.0	56.0	3
		Short transverse	66.0	55.0	2
7049-T7352⑥	Up thru 2.000	Longitudinal*	71.0	59.0	9
		Long transverse	71.0	57.0	4
	2.001–3.000	Longitudinal*	71.0	59.0	9
		Long transverse	71.0	57.0	4
		Short transverse	69.0	56.0	3
	3001–4.000	Longitudinal*	69.0	57.0	8
		Long transverse	69.0	54.0	3
		Short transverse	67.0	53.0	2
	4.001–5.000	Longitudinal*	67.0	54.0	7
		Long transverse	67.0	53.0	3
		Short transverse	66.0	51.0	2
		7050			
7050-T7452⑧	Up thru 2.000	Longitudinal*	72.0	63.0	9
		Long transverse	71.0	61.0	5
	2.001–3.000	Longitudinal*	72.0	62.0	9
		Long transverse	70.0	60.0	5
		Short transverse	67.0	55.0	4
	3.001–4.000	Longitudinal*	71.0	61.0	9
		Long transverse	70.0	59.0	5
		Short transverse	67.0	55.0	4
	4.001–5.000	Longitudinal*	70.0	60.0	9
		Long transverse	69.0	58.0	4
		Short transverse	66.0	54.0	3
	5.001–6.000	Longitudinal*	69.0	59.0	9
		Long transverse	68.0	56.0	4
		Short transverse	66.0	53.0	3
	6.001–7.000	Longitudinal	68.0	58.0	9
		Long transverse	67.0	54.0	4
		Short transverse	65.0	51.0	3
	7.001–8.000	Longitudinal*	67.0	57.0	9
		Long transverse	66.0	52.0	4
		Short transverse	64.0	50.0	3
		7075			
7075-T6	Up thru 2.000	Longitudinal*	74.0	63.0	9
		Long transverse	73.0	61.0	4
	2.001–3.000	Longitudinal*	73.0	61.0	9
		Long transverse	71.0	59.0	4
		Short transverse	69.0	58.0	3
	3.001–4.000	Longitudinal*	71.0	60.0	8
		Long transverse	70.0	58.0	3
		Short transverse	68.0	57.0	2
	4.001–5.000	Longitudinal*	69.0	58.0	7
		Long transverse	68.0	56.0	3
		Short transverse	66.0	56.0	2
	5.001–6.000	Longitudinal*	68.0	56.0	6
		Long transverse	66.0	55.0	3
		Short transverse	65.0	55.0	2
7075-T652	Up thru 2.000	Longitudinal*	74.0	63.0	9
		Long transverse	73.0	61.0	4
	2.001–3.000	Longitudinal*	73.0	61.0	9
		Long transverse	71.0	59.0	4
		Short transverse	69.0	57.0	2
	3.001–4.000	Longitudinal*	71.0	60.0	8
		Long transverse	70.0	58.0	3
		Short transverse	68.0	56.0	1
	4.001–5.000	Longitudinal*	69.0	58.0	7
		Long transverse	68.0	56.0	3
		Short transverse	66.0	55.0	1
	5.001–6.000	Longitudinal*	68.0	56.0	6
		Long transverse	66.0	55.0	3
		Short transverse	65.0	54.0	1

Mechanical Property Limits—Hand Forgings (*concluded*)

ALLOY AND TEMPER	SPECIFIED THICKNESS③ in.	AXIS OF TEXT SPECIMEN	TENSILE STRENGTH ksi min		ELONGATION percent min in 2 in. or 4D④
			ULTIMATE	YIELD	
7075 (Continued)					
7075-T73⑥	Up thru 3.000	Longitudinal*	66.0	56.0	7
		Long transverse	64.0	54.0	4
		Short transverse	61.0	52.0	3
	3.001–4.000	Longitudinal*	64.0	55.0	7
		Long transverse	63.0	53.0	3
		Short transverse	60.0	51.0	2
	4.001–5.000	Longitudinal*	62.0	53.0	7
		Long transverse	61.0	51.0	3
		Short transverse	58.0	50.0	2
	5.001–6.000	Longitudinal*	61.0	51.0	6
		Long transverse	59.0	50.0	3
		Short transverse	57.0	49.0	2
7075-T7352⑥	Up thru 3.000	Longitudinal*	66.0	54.0	7
		Long transverse	64.0	52.0	4
		Short transverse⑦	610	50.0	3
	3.001–4.000	Longitudial*	64.0	53.0	7
		Long transverse	63.0	50.0	3
		Short transverse	60.0	48.0	2
	4.001–5.000	Longitudinal*	62.0	51.0	7
		Long transverse	61.0	48.0	3
		Short transverse	58.0	46.0	2
	5.001–6.000	Longitudinal*	61.0	49.0	6
		Long transverse	59.0	46.0	3
		Short transverse	57.0	44.0	2

*Tensile tests are performed and properties are guaranteed only when specifically required by purchase order or contract.

[1]Maximum cross-sectional area 256 in.^{2}, except for 2618-T61, is 144 in.^{2}.

[2]These properties are not applicable to upset biscuit forgings or to rolled forged rings.

[3]As-forged thickness. When forgings are machined prior to heat treatment, the properties will also apply to the machined heat treat thickness provided the original (as-forged) thickness does not exceed the maximum thickness for the alloy as listed.

[4]D represents specimen diameter.

[5]The data base and criteria upon which these mechanical property limits are established are outlined on page 89 of the Standard under "Mechanical Properties."

[6]Material in this temper, 0.750 in. and thicker, when tested in accordance with ASTM G47 in the short transverse direction at a stress level of 75% of the specified minimum yield strength will exhibit no evidence of stress corrosion cracking. Capability of individual lots to resist stress corrosion is determined by testing the previously selected tensile test sample in accordance with the applicable lot acceptance criteria outlined on page 96 of the Standard.

[7]Short transverse properties not applicable to thicknesses 2 in. or less.

[8]T74 type tempers although not previously registered have appeared in the literature and in some specifications as T36 type tempers.

[9]Material in this temper when tested at any plane in accordance with ASTM G34-72 will exhibit exfoliation less than that shown in Category B, Figure 2 of ASTM G34-72. Also, material, 0.750 in. and thicker, when tested in accordance with ASTM G47 in the short transverse direction at a stress level of 35 ksi will exhibit no evidence of stress corrosion cracking. Capability of individual lots to resist exfoliation corrosion and stress corrosion cracking is determined by testing the previously selected tensile test sample in accordance with the applicable lot acceptance criteria outlined on page 96 of the Standard.

INDEX